R. J. Mayer, A. Ciechanover,
M. Rechsteiner (Eds.)
Protein Degradation

Further Titles of Interest

Mayer, R. J., Ciechanover, A., Rechsteiner, M. (Eds.)

Protein Degradation, Vol. 1

2005
ISBN 3-527-30837-7

Buchner, J., Kiefhaber, T. (Eds.)

Protein Folding Handbook

2004
ISBN 3-527-30784-2

Sanchez, J.-C., Corthals, G. L., Hochstrasser, D. F. (Eds.)

Biomedical Application of Proteomics

2004
ISBN 3-527-30807-5

Nierhaus, K. H., Wilson, D. N. (Eds.)

Protein Synthesis and Ribosome Structure

2004
ISBN 3-527-30638-2

Cesareni, G., Gimona, M., Sudol, M., Yaffe, M. (Eds.)

Modular Protein Domains

2004
ISBN 3-527-30813-X

R. John Mayer, Aaron Ciechanover, and Martin Rechsteiner (Eds.)

Protein Degradation

The Ubiquitin-Proteasome System

Volume 2

WILEY-VCH Verlag GmbH & Co. KGaA

Editors

Prof. Dr. R. John Mayer
University of Nottingham
School of Biomedical Sciences
Queen's Medical Centre
Nottingham, NG7 2UH
UK

Prof. Dr. Aaron Ciechanover
Technion-Israel Institute
of Technology
Department of Biochemistry
Afron Street, Bat Galim
Haifa 31096
Israel

Prof. Dr. Martin Rechsteiner
University of Utah Mecial School
Department of Biochemistry
50 N. Medical Drive
Salt Lake City, UT 84132
USA

Library of Congress Card No.: applied for
British Library Cataloging-in-Publication Data: A catalogue record for this book is available from the British Library.

Bibliographic information published by Die Deutsche Bibliothek
Die Deutsche Bibliothek lists this publication in the Deutsche Nationalbibliografie; detailed bibliographic data is available in the Internet at ⟨http://dnb.ddb.de⟩.

Printed in the Federal Republic of Germany.
Printed on acid-free paper.

Typesetting Asco Typesetters, Hong Kong
Printing betz-druck GmbH, Darmstadt
Binding Litges & Dopf Buchbinderei GmbH, Heppenheim

ISBN-13 978-3-527-31130-9
ISBN-10 3-527-31130-0

Contents

Protein Degradation, Vol. 2: The Ubiquitin-Proteasome System.
Edited by R. J. Mayer, A. Ciechanover, M. Rechsteiner

ISBN: 3-527-31130-0

Preface

There is an incredible amount of current global research activity devoted to understanding the chemistry of life. The genomic revolution means that we now have the basic genetic information in order to understand in full the molecular basis of the life process. However, we are still in the early stages of trying to understand the specific mechanisms and pathways that regulate cellular activities. Occasionally discoveries are made that radically change the way in which we view cellular activities. One of the best examples would be the finding that reversible phosphorylation of proteins is a key regulatory mechanism with a plethora of downstream consequences. Now the seminal discovery of another post-translational modification, protein ubiquitylation, is leading to a radical revision of our understanding of cell physiology. It is becoming ever more clear that protein ubiquitylation is as important as protein phosphorylation in regulating cellular activities. One consequence of protein ubiquitylation is protein degradation by the 26S proteasome. However, we are just beginning to understand the full physiological consequences of covalent modification of proteins, not only by ubiquitin, but also by ubiquitin-related proteins.

Because the Ubiquitin Proteasome System (UPS) is a relatively young field of study, there is ample room to speculate on possible future developments. Today a handful of diseases, particularly neurodegenerative ones, are known to be caused by malfunction of the UPS. With perhaps as many as 1000 human genes encoding components of ubiquitin and ubiquitin-related modification pathways, it is almost certain that many more diseases will be found to arise from genetic errors in the UPS or by pathogen subversion of the system. This opens several avenues for the development of new therapies. Already the proteasome inhibitor Velcade is producing clinical success in the fight against multiple myeloma. Other therapies based on the inhibition or activation of specific ubiquitin ligases, the substrate recognition components of the UPS, are likely to be forthcoming. At the fundamental research level there are a number of possible discoveries especially given the surprising range of biochemical reactions involving ubiquitin and its cousins. Who would have guessed that the small highly conserved protein would be involved in endocytosis or that its relative Atg8 would form covalent bonds to a phospholipid during autophagy? We suspect that few students of ubiquitin will be surprised if it or a

Protein Degradation, Vol. 2: The Ubiquitin-Proteasome System.
Edited by R. J. Mayer, A. Ciechanover, M. Rechsteiner

ISBN: 3-527-31130-0

ubiquitin-like protein is one day found to be covalently attached to a nucleic acid for some biological purpose.

We are regularly informed by the ubiquitin community that the initiation of this series of books on the UPS is extremely timely. Even though the field is young, it has now reached the point at which the biomedical scientific community at large needs reference works in which contributing authors indicate the fundamental roles of the ubiquitin proteasome system in all cellular processes. We have attempted to draw together contributions from experts in the field to illustrate the comprehensive manner in which the ubiquitin proteasome system regulates cell physiology. There is no doubt then when the full implications of protein modification by ubiquitin and ubiquitin-like molecules are fully understood we will have gained fundamental new insights into the life process. We will also have come to understand those pathological processes resulting from UPS malfunction. The medical implications should have considerable impact on the pharmaceutical industry and should open new avenues for therapeutic intervention in human and animal diseases. The extensive physiological ramifications of the ubiquitin proteasome system warrant a series of books of which this is the first one.

Aaron Ciechanover
Marty Rechsteiner
John Mayer

List of Contributors

Monika Bajorek
Department of Chemistry
Technion – Israel Institute of Technology
Efron Street, Bat Galim
Haifa 31096
Israel

Wolfgang Baumeister
Max-Planck-Institut für Biochemie
Am Klopferspitz 18a
82152 Martinsried
Germany

Nadia Benaroudj
Pasteur Institute
Unit of Protein Folding and Modeling
25–28 rue du Dr Roux
75724 Paris
Cedex 15
France

Jürgen Bosch
University of Washington
Department of Biochemistry
Structural Genomics of Pathogenic Protozoa
1705 NE Pacific Street
Seattle, WA 98195-7742
USA

Jean E. O'Donoghue
MRC Human Genetics Unit
Western General Hospital
Crewe Road
Edinburgh, EH4 2XU
Scotland
UK

Andreas Förster
University of Utah School of Medicine
Department of Biochemistry
1900E, 20N, Room 211
Salt Lake City, UT 84132-3201
USA

Michael H. Glickman
Department of Chemistry
Technion – Israel Institute of Technology
Efron Street, Bat Galim
Haifa 31096
Israel

Alfred Goldberg
Harvard Medical School
Department of Cell Biology
240 Longwood Avenue
Boston, MA 02115
USA

Colin Gordon
MRC Human Genetics Unit
Western General Hospital
Crewe Road
Edinburgh, EH4 2XU
Scotland
UK

Christopher P. Hill
University of Utah School of Medicine
Department of Biochemistry
1900E, 20N, Room 211
Salt Lake City, UT 84132-3201
USA

Mark Hochstrasser
Department of Molecular Biophysics
and biochemistry
Yale University
266 Whitney Avenue
P.O. Box 208114
Bass 224
New Haven, CT 06520-8114
USA

Jörg Höhfeld
Institut für Zellbiologie
Rheinische Friedrich-Wilhelms-Universität
Bonn
Ulrich-Haberland-Str. 61a
53121 Bonn
Germany

Tony Hunter
Molecular and Cell Biology Laboratory
The Salk Institute for Biological Studies
10010 N. Torrey Pines Road
La Jolla, CA 92037
USA

Benedikt M. Kessler
Department of Pathology
Harvard Medical School
New Research Building
77 Avenue Louis Pasteur
Boston, MA 02115
USA

Zhimin Lu
University of Texas
M.D. Anderson Cancer Center
Department of Neuro-Oncology
1515 Holcombe Blvd., Unit 316
Houston, TX 77030
USA

Shigeo Murata
Department of Molecular Oncology
Tokyo Metropolitan Institute of
Medical Science
Honkomagome 3-18-22, Bunkyo-ku
Tokyo 113-8613
Japan

Yoshinori Ohsumi
National Institute for Basic Biology
Department of Cell Biology
Myodaiji-cho
Okazaki 444-8585
Japan

Huib Ovaa
Department of Cellular Biochemistry
Netherlands Cancer Institute
Plesmanlaan 121
1066 CX Amsterdam
The Netherlands

Herman S. Overkleeft
Leiden Institute of Chemistry
Leiden University
Einsteinweg 55
2300 RA Leiden
The Netherlands

Cam Patterson
University of North Carolina
Chapel Hill, NC 27599
USA

Hidde L. Ploegh
77 Avenue Louis Pasteur
New Research Building, Room 836F
Boston, MA 02115
USA

Beate Rockel
Max-Planck-Institute of Biochemistry
Department of Molecular Structural Biology
Am Klopferspitz 18
82152 Martinsried
Germany

David Smith
Harvard Medical School
Department of Cell Biology
240 Longwood Avenue
Boston, MA 02115
USA

Keiji Tanaka
Department of Molecular Oncology
Tokyo Metropolitan Institute of
Medical Science
Honkomagome 3-18-22, Bunkyo-ku
Tokyo 113-8613
Japan

Hideki Yashiroda
Department of Molecular Oncology
Tokyo Metropolitan Institute of
Medical Science
Honkomagome 3-18-22, Bunkyo-ku
Tokyo 113-8613
Japan

1
Molecular Chaperones and the Ubiquitin–Proteasome System

Cam Patterson and Jörg Höhfeld

Abstract

A role for the ubiquitin–proteasome system in the removal of misfolded and abnormal proteins is well established. Nevertheless, very little is known about how abnormal proteins are recognized for degradation by the proteasome. Recent advances suggest that substrate recognition and processing require a close cooperation of the ubiquitin–proteasome system with molecular chaperones. Chaperones are defined by their ability to recognize nonnative conformations of other proteins and are therefore ideally suited to distinguish between native and abnormal proteins during substrate selection. Here we discuss molecular mechanisms that underlie the cooperation of molecular chaperones with the ubiquitin–proteasome system. Advancing our knowledge about such mechanisms may open up opportunities to modulate chaperone–proteasome cooperation in human diseases.

1.1
Introduction

The biological activity of a protein is defined by its unique three-dimensional structure. Attaining this structure, however, is a delicate process. A recent study suggests that up to 30% of all newly synthesized proteins never reach their native state [1]. As protein misfolding poses a major threat to cell function and viability, molecular mechanisms must have evolved to prevent the accumulation of misfolded proteins and thus aggregate formation. Two protective strategies appear to be followed. Molecular chaperones are employed to stabilize nonnative protein conformations and to promote folding to the native state whenever possible. Alternatively, misfolded proteins are removed by degradation, involving, for example, the ubiquitin–proteasome system. For a long time molecular chaperones and cellular degradation systems were therefore viewed as opposing forces. However, recent evidence suggests that certain chaperones (in particular members of the 70- and 90-kDa heat shock protein families) are able to cooperate with the ubiquitin–

Protein Degradation, Vol. 2: The Ubiquitin-Proteasome System.
Edited by R. J. Mayer, A. Ciechanover, M. Rechsteiner

ISBN: 3-527-31130-0

proteasome system. Protein fate thus appears to be determined by a tight interplay of cellular protein-folding and protein-degradation systems.

1.2
A Biomedical Perspective

The aggregation and accumulation of misfolded proteins is now recognized as a common characteristic of a number of degenerative disorders, many of which have neurological manifestations [2, 3]. These diseases include prionopathies, Alzheimer's and Parkinson's diseases, and polyglutamine expansion diseases such as Huntington's disease and spinocerebellar ataxia. At the cellular level, these diseases are characterized by the accumulation of aberrant proteins either intracellularly or extracellularly in specific groups of cells that subsequently undergo death. The precise association between protein accumulation and cell death remains incompletely understood and may vary from disease to disease. In some cases, misfolded protein accumulations may themselves be toxic or exert spatial constraints on cells that affect their ability to function normally. In other cases, the sequestering of proteins in aggregates may itself be a protective mechanism, and it is the overwhelming of pathways that consolidate aberrant proteins that is the toxic event. In either case, lessons learned from genetically determined neurodegenerative diseases have helped us to understand the inciting events of protein aggregation that ultimately lead to degenerative diseases.

Mutations resulting in neurodegenerative diseases fall into two broad classes. The first class comprises mutations that affect proteins, irrespective of their native function, and cause them to misfold. The classic example of this is Huntington's disease [4, 5]. The protein encoded by the huntingtin gene contains a stretch of glutamine residues (or polyglutamine repeat), and the genomic DNA sequence that codes for this polyglutamine repeat is subject to misreading and expansion. When the length of the polyglutamine repeat in huntingtin reaches a critical threshold of approximately 35 residues, the protein becomes prone to misfolding and aggregation [6]. This appears to be the proximate cause of neurotoxicity in this invariably fatal disease [7, 8]. A number of other neurodegenerative diseases are caused by polyglutamine expansions [9, 10]. For example, spinocerebellar ataxia is caused by polyglutamine expansions in the protein ataxin-1 [11]. In other diseases, protein misfolding occurs due to other mutations that induce misfolding and aggregation; for example, mutations in superoxide dismutase-1 lead to aggregation and neurotoxicity in amyotrophic lateral sclerosis [12, 13].

Other mutations that result in neurodegenerative diseases are instructive in that they directly implicate the ubiquitin–proteasome system in the pathogenesis of these diseases [14]. For example, mutations in the gene encoding the protein parkin are associated with juvenile-onset Parkinson's disease [15, 16]. Parkin is a RING finger–containing ubiquitin ligase, and mutations in this ubiquitin ligase cause accumulation of target proteins that ultimately result in the neurotoxicity and motor dysfunction associated with Parkinson's disease [17–20].

Repressor screens of neurodegeneration phenotypes in animal models have also linked the molecular chaperone machinery to neurodegeneration [21–24]. Taken together, the pathophysiology of neurodegenerative diseases provides a compelling demonstration of the importance of the regulated metabolism of misfolded proteins and provides direct evidence of the role of both molecular chaperones and the ubiquitin–proteasome system in guarding against protein misfolding and its consequent toxicity.

1.3 Molecular Chaperones: Mode of Action and Cellular Functions

Molecular chaperones are defined by their ability to bind and stabilize nonnative conformations of other proteins [25, 26]. Although they are an amazingly diverse group of conserved and ubiquitous proteins, they are also among the most abundant intracellular proteins. The classical function of chaperones is to facilitate protein folding, inhibit misfolding, and prevent aggregation. These folding events are regulated by interactions between chaperones and ancillary proteins, the co-chaperones, which in general assist in cycling unfolded substrate proteins on and off the active chaperone complex [25, 27, 28]. In agreement with their essential function under normal growth conditions, chaperones are ubiquitously expressed and are found in all cellular compartments of the eukaryotic cell (except for peroxisomes). In addition, cells greatly increase chaperone concentration as a response to diverse stresses, when proteins become unfolded and require protection and stabilization [29]. Accordingly, many chaperones are heat shock proteins (Hsps). Four main families of cytoplasmic chaperones can be distinguished: the Hsp70 family, the Hsp90 family, the small heat shock proteins, and the chaperonins.

1.3.1 The Hsp70 Family

The Hsp70 proteins bind to misfolded proteins promiscuously during translation or after stress-mediated protein damage [26, 30]. Members of this family are highly conserved throughout evolution and are found throughout the prokaryotic and eukaryotic phylogeny. It is common for a single cell to contain multiple homologues, even within a single cellular compartment; for example, mammalian cells express two inducible homologues (Hsp70.1 and Hsp70.3) and a constitutive homologue (Hsc70) in the cytoplasm. These homologues have overlapping but not totally redundant cellular functions. Members of this family are typically in the range of 70 kDa in size and contain three functional domains: an amino-terminal ATPase domain, a central peptide-binding cleft, and a carboxyl terminus that seems to form a lid over the peptide-binding cleft [28] (Figure 1.1). The chaperones recognize short segments of the protein substrate, which are composed of clusters of hydrophobic amino acids flanked by basic residues [31]. Such binding motifs occur frequently within protein sequences and are found exposed on nonnative proteins. In fact,

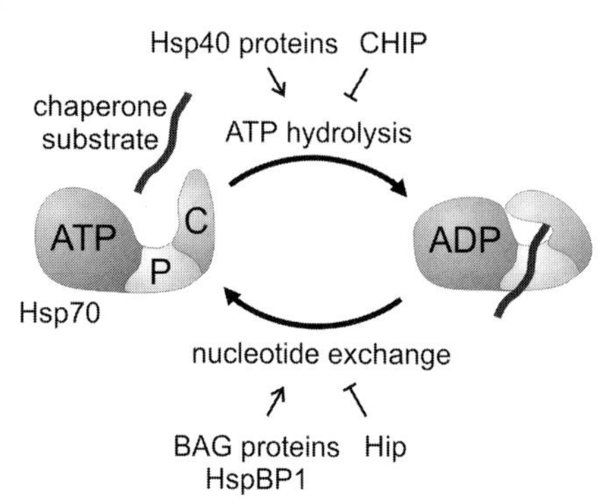

Fig. 1.1. Schematic presentation of the domain architecture and chaperone cycle of Hsp70. Hsp70 proteins display a characteristic domain structure comprising an amino-terminal ATPase domain (ATP), a peptide-binding domain (P), and a carboxyl-terminal domain (C) that is supposed to form a lid over the peptide-binding domain. In the ATP-bound conformation, the binding pocket is open, resulting in a low affinity for the binding of a chaperone substrate. ATP hydrolysis induces stable substrate binding through a closure of the peptide-binding pocket. Substrate release is induced upon nucleotide exchange. ATP hydrolysis and nucleotide exchange are regulated by diverse co-chaperones.

mammalian Hsp70 binds to a wide range of nascent and newly synthesized proteins, comprising about 15–20% of total protein [32]. This percentage is most likely further increased under stress conditions. Hsp70 proteins apparently prevent protein aggregation and promote proper folding by shielding hydrophobic segments of the protein substrate. The hydrophobic segments are recognized by the central peptide-binding domain of Hsp70 proteins (Figure 1.1). The domain is composed of two sheets of β strands that together with connecting loops form a cleft to accommodate extended peptides of about seven amino acids in length, as revealed in crystallographic studies of bacterial Hsp70 [33]. In the obtained crystal structure, the adjacent carboxyl-terminal domain of Hsp70 folds back over the β sandwich, suggesting that the domain may function as a lid in permitting entry and release of protein substrates (Figure 1.1). According to this model, ATP binding and hydrolysis by the amino-terminal ATPase domain of Hsp70 induce conformational changes of the carboxyl terminus, which lead to lid opening and closure [28]. In the ATP-bound conformation of Hsp70, the peptide-binding pocket is open, resulting in rapid binding and release of the substrate and consequently in a low binding affinity (Figure 1.1). Stable holding of the protein substrate requires closing of the binding pocket, which is induced upon ATP hydrolysis and conversion of Hsp70 to the ADP-bound conformation. The dynamic association of Hsp70 with nonnative polypeptide substrates thus depends on ongoing cycles of ATP binding, hydrolysis, and nucleotide exchange. Importantly, ancillary co-chaperones are employed to regulate the ATPase cycle [27, 30]. Co-chaperones of the Hsp40 family (also termed J proteins due to their founding member bacterial DnaJ) stimulate the ATP hydrolysis step within the Hsp70 reaction cycle and in this way promote substrate binding [34] (Figure 1.1). In contrast, the carboxyl terminus of Hsp70-interacting protein CHIP attenuates ATP hydrolysis [35]. Similarly, nucleo-

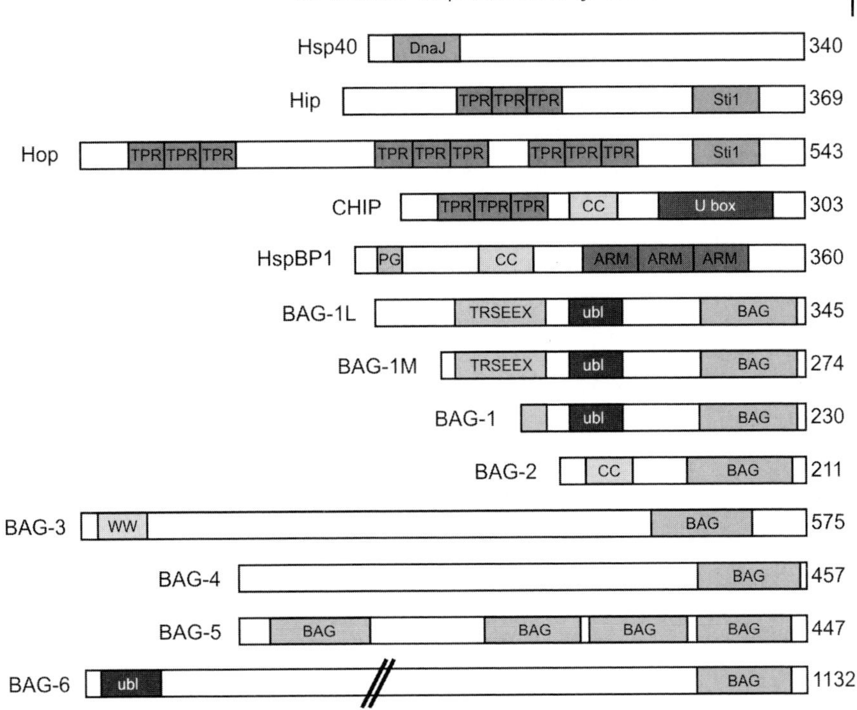

Fig. 1.2. Domain architecture of diverse co-chaperones of Hsp70. DnaJ: domain related to the bacterial co-chaperone DnaJ; TPR: tetratricopeptide repeat; Sti1: domain related to the yeast co-chaperone Sti1; CC: coiled-coil domain; U box: E2-interacting domain present in certain ubiquitin ligases; PG: polyglycine region; ARM: armadillo repeat; TRSEEX: repeat motif found at the amino terminus of BAG-1 isoforms; ubl: ubiquitin-like domain; BAG: Hsp70-binding domain present in BAG proteins; WW: protein interaction domain.

tide exchange on Hsp70 is under the control of stimulating and inhibiting co-chaperones. The Hsp70-interacting protein Hip slows down nucleotide exchange by stabilizing the ADP-bound conformation of the chaperone [36], whereas nucleotide exchange is stimulated by the co-chaperone BAG-1 (Bcl-2-associated athanogene 1), which assists substrate unloading from Hsp70 [37–39]. By altering the ATPase cycle, the co-chaperones directly modulate the folding activity of Hsp70. In addition to chaperone-recognition motifs, co-chaperones often possess other functional domains and therefore link chaperone activity to distinct cellular processes [27, 40] (Figure 1.2). Indeed, as discussed below, the co-chaperones BAG-1 and CHIP apparently modulate Hsp70 function during protein degradation.

1.3.2
The Hsp90 Family

The 90-kDa cytoplasmic chaperones are members of the Hsp90 family, and in mammals two isoforms exist: Hsp90α and Hsp90β. The Hsp70 and Hsp90 families exhibit several common features: both possess ATPase activity and are regulated by ATP binding and hydrolysis, and both are further regulated by ancillary co-chaperones [41–48]. Unlike Hsp70, however, cytoplasmic Hsp90 is not generally involved in the folding of newly synthesized polypeptide chains. Instead it plays a key role in the regulation of signal transduction networks, as most of the known substrates of Hsp90 are signaling proteins, the classical examples being steroid hormone receptors and signaling kinases. On a molecular level, Hsp90 binds to substrates at a late stage of the folding pathway, when the substrate is poised for activation by ligand binding or associations with other factors. Consequently, Hsp90 accepts partially folded conformations from Hsp70 for further processing. In the case of the chaperone-assisted activation of the glucocorticoid hormone receptor and also of the progesterone receptor, the sequence of events leading to attaining an active conformation is fairly well understood [49–53]. It appears that the receptors are initially recognized by Hsp40 and are then delivered to Hsp70 [54] (Figure 1.3). Subsequent transfer onto Hsp90 requires the Hsp70/Hsp90-organizing protein Hop, which possesses non-overlapping binding sites for Hsp70 and Hsp90 and therefore acts as a coupling factor between the two chaperones [55]. In conjunction with p23 and different cyclophilins, Hsp90 eventually medi-

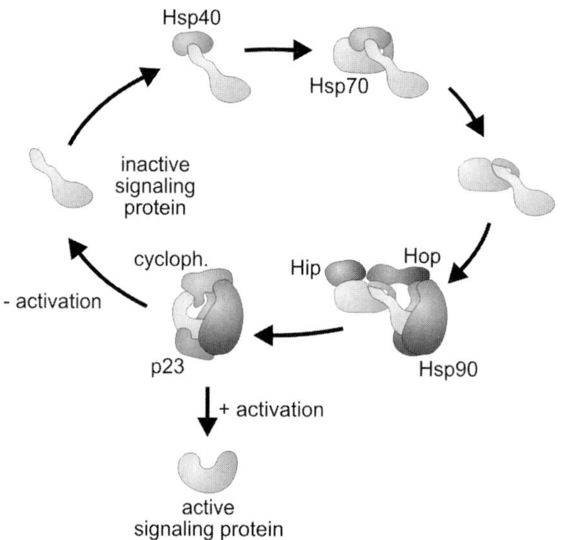

Fig. 1.3. Cooperation of Hsp70 and Hsp90 during the regulation of signal transduction pathways. The inactive signaling protein, e.g., a steroid hormone receptor, is initially recognized by Hsp40 and delivered to Hsp70. Subsequently, a multi-chaperone complex assembles that contains the Hsp70 co-chaperone Hip and the Hsp70/Hsp90-organizing protein Hop. Hop stimulates recruitment of an Hsp90 dimer that accepts the substrate from Hsp70. At the final stage of the chaperone pathway, Hsp90 associates with p23 and diverse cyclophilins (cycloph.) to mediate conformational changes of the signaling protein necessary to reach an activatable state. Upon activation, i.e., hormone binding in the case of the steroid receptor, the signaling protein is released from Hsp90. In the absence of an activating stimulus, the signaling protein folds back to the inactive state when released and enters a new cycle of chaperone binding.

ates conformational changes that enable the receptor to reach a high-affinity state for ligand binding. On other signaling pathways Hsp90 serves as a scaffolding factor to permit interactions between kinases and their substrates, as is the case for Akt kinase and endothelial nitric oxide synthase [56]. Since many of the Hsp90 substrate proteins are involved in regulating cell proliferation and cell death, it is not surprising that the chaperone recently emerged as a drug target in tumor therapy [57–59]. The antibiotics geldanamycin and radicicol specifically bind to Hsp90 in mammalian cells and inhibit the function of the chaperone by occupying its ATP-binding pocket [60–63]. Drugs based on these compounds are now being developed as anticancer agents, as they potentially inactivate multiple signaling pathways that drive carcinogenesis. Remarkably, drug-induced inhibition of Hsp90 blocks the chaperone-assisted activation of signaling proteins and leads to their rapid degradation via the ubiquitin–proteasome pathway [64–69] (Figure 1.4). Hsp90 inhibitors therefore have emerged as helpful tools to study chaperone–proteasome cooperation.

1.3.3
The Small Heat Shock Proteins

The precise functions of small heat shock proteins (sHsps) including Hsp27 and the eye-lens protein αB-crystallin are incompletely understood. However, they

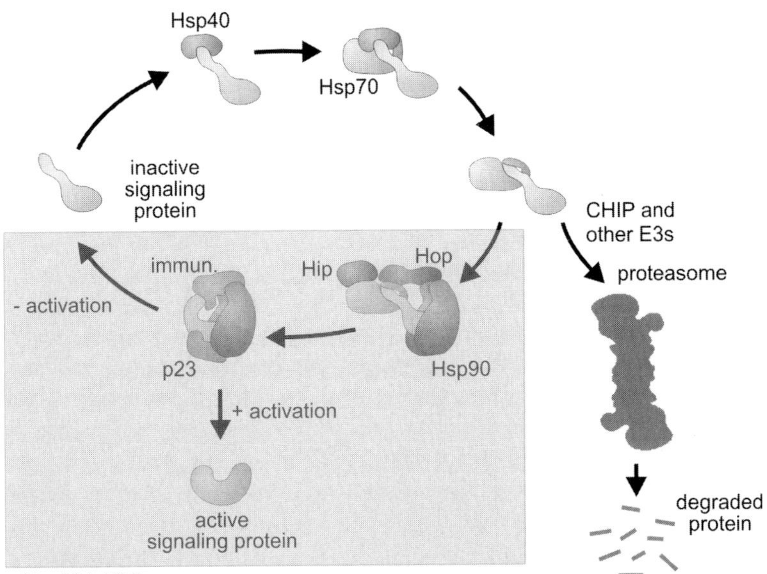

Fig. 1.4. Alteration of chaperone action during signal transduction induced by Hsp90 inhibitors such as geldanamycin and radicicol. In the presence of the inhibitors the activation pathway is blocked, and signaling proteins are targeted to the proteasome for degradation in a process that involves the co-chaperone CHIP and other E3 ubiquitin ligases that remain to be identified.

seem to play a major role in preventing protein aggregation under conditions of cellular stress [70–73]. All members investigated so far form large oligomeric complexes of spherical or cylindrical appearance [74, 75]. Complex formation is independent of ATP binding and hydrolysis, but appears to be regulated by temperature and phosphorylation. The structural analysis of wheat Hsp16.9 suggested that the oligomeric complex acts as a storage form rather than an enclosure for substrates, as the active chaperone appears to be a dimer [75]. In agreement with this notion, dissociation of the oligomeric complex formed by yeast Hsp26 was found to be a prerequisite for efficient chaperone activity [76]. Subsequent refolding may occur spontaneously or may involve cooperation with other chaperones such as Hsp70 [77].

1.3.4 Chaperonins

The chaperone proteins best understood with regard to their mode of action are certainly the so-called chaperonins, which are defined by a barrel-shaped, double-ring structure [25, 28]. Members include bacterial GroEL, Hsp60 of mitochondria and chloroplasts, and the TriC–CCT complex localized in the eukaryotic cytoplasm. Based on their characteristic ring structure, a central cavity is formed, which accommodates nonnative proteins via hydrophobic interactions. Conformational changes of the chaperonin subunits induced through ATP hydrolysis change the inner lining of the cavity from a hydrophobic to a hydrophilic character [78–80]. As a consequence the unfolded polypeptide is released into the central chamber and can proceed on its folding pathway in a protected environment [81]. The chaperonins are therefore capable of folding proteins such as actin that cannot be properly folded via other mechanisms [82].

1.4 Chaperones: Central Players During Protein Quality Control

Due to their ability to recognize nonnative conformations of other proteins, molecular chaperones are of central importance during protein quality control. This was elegantly revealed in studies on the influence of the Hsp70 chaperone system on polyglutamine diseases using the fruit fly *Drosophila melanogaster* as a model organism (reviewed in Refs. [23] and [83]). Hallmarks of the polyglutamine disease spinocerebellar ataxia type 3 (SCA3), for example, were recapitulated in transgenic flies that expressed a pathological polyQ tract of the ataxin-3 protein in the eye disc [84]. Transgene expression caused formation of abnormal protein inclusions and progressive neuronal degeneration. Intriguingly, co-expression of human cytoplasmic Hsp70 suppressed polyQ-induced neurotoxicity. In a similar experimental approach, Hsp40 family members protected neuronal cells against toxic polyQ expression [22]. Enhancing the activity of the Hsp70/Hsp40 chaperone system apparently mitigates cytotoxicity caused by the accumulation of aggregation-prone proteins. These findings obtained in *Drosophila* were confirmed in a mouse model of spinocerebellar ataxia type 1 (SCA1) [85, 86]. Unexpectedly, however, the Hsp70 chaperone system was unable to prevent the formation of protein aggregates in these models of polyglutamine diseases and upon polyQ expression in yeast and mammalian cells [84, 85, 87–89]. Elevating the cellular levels of Hsp70 and of some Hsp40 family members affected the number of protein aggregates and their biochemical properties, but did not inhibit the formation of polyQ aggregates. Notably, Hsp70 and Hsp40 profoundly modulated the aggregation process of polyQ tracts in biochemical experiments; this led to the formation of amorphous, SDS-soluble aggregates, instead of the ordered, SDS-insoluble amyloid fibrils that form in the absence of the chaperone system [88]. These biochemical data were confirmed in yeast and mammalian cells [88, 90]. Although unable to prevent the formation of protein aggregates, the Hsp70 chaperone system apparently prevents the ordered oligomerization and fibril growth that is characteristic of the disease process. In an alternate but not mutually exclusive model to explain their protective role, the chaperones may cover potentially dangerous surfaces exposed by polyQ-containing proteins during the oligomerization process or by the final oligomers. Intriguingly, elevated expression of Hsp70 also suppresses the toxicity of the non-polyQ-containing protein α-synuclein in a *Drosophila* model of Parkinson's disease without inhibiting aggregate formation [24]. Hsp70 may thus exert a rather general function in protecting cells against toxic protein aggregation. This raises the exciting possibility that treatment of diverse forms of human neurodegenerative diseases may be achieved through upregulation of Hsp70 activity.

The mentioned examples illustrate that one does not have to evoke the refolding of an aberrant protein to the native state in order to explain the protective activity of Hsp70 observed in models of amyloid diseases. In some cases it might be sufficient for Hsp70 to modulate the aggregation process or to shield interaction surfaces of the misfolded protein to decrease cytotoxic effects. Another option may involve presentation of the misfolded protein to the ubiquitin–proteasome system for degradation.

1.5
Chaperones and Protein Degradation

Hsp70 and Hsp90 family members as well as small heat shock proteins have all been implicated to participate in protein degradation. For example, the small heat shock protein Hsp27 was recently shown to stimulate the degradation of phosphorylated IκBα via the ubiquitin–proteasome pathway, which may account for the antiapoptotic function of Hsp27 [91]. Similarly, Hsp27 facilitates the proteasomal degradation of phosphorylated tau, a microtubule-binding protein and component of protein deposits in Alzheimer's disease [92]. Hsp70 participates in the degradation of apolipoprotein B100 (apoB), which is essential for the assembly and secretion of very low-density lipoproteins from the liver [93]. Under conditions of limited availability of core lipids, apoB translocation across the ER membrane is attenuated, resulting in the exposure of some domains of the protein into the cytoplasm and their recognition by Hsp70. This is followed by the degradation of apoB via the ubiquitin–proteasome pathway. Elevating cellular Hsp70 levels stimulated the degradation of the membrane protein, suggesting that the chaperone facilitates sorting to the proteasome. Genetic studies in yeast indicate that cytoplasmic Hsp70 may fulfill a rather general role in the degradation of ER-membrane proteins that display large domains into the cytoplasm [94]. In agreement with this notion, Hsp70 also takes part in the degradation of immaturely glycosylated and aberrantly folded forms of the cystic fibrosis transmembrane conductance regulator (CFTR) [95–98]. CFTR is an ion channel localized at the apical surface of epithelial cells. Its functional absence causes cystic fibrosis, the most common fatal genetic disease in Caucasians [99, 100]. The disease-causing allele, ΔF508, which is expressed in more than 70% of all patients, drastically interferes with the protein's ability to fold, essentially barring it from functional expression in the plasma membrane. However, wild-type CFTR also folds very inefficiently, and less than 30% of the protein reaches the plasma membrane [99]. While trafficking from the endoplasmic reticulum (ER) to the Golgi apparatus, immature forms of CFTR are recognized by quality-control systems and are eventually directed to the proteasome for degradation [101–104]. A critical step during CFTR biogenesis is the inefficient folding of the first of two cytoplasmically exposed nucleotide-binding domains (NBD1) of the membrane protein [105, 106]. The disease-causing ΔF508 mutation localizes to NBD1 and further decreases the folding propensity of this domain. During the co-translational insertion of CFTR into the ER membrane, cytoplasmic Hsp70 and its co-chaperone Hdj-2 bind to NBD1 and facilitate intramolecular interactions between the domain and another cytoplasmic region of CFTR, the regulatory R-domain [96, 107]. However, Hsp70 is also able to present CFTR to the ubiquitin–proteasome system [97], and heterologous expression of CFTR in yeast revealed an essential role of cytoplasmic Hsp70 in CFTR turnover [98]. Hsp70 is thus a key player in the cellular surveillance system that monitors the folded state of CFTR at the ER membrane.

Interestingly, CFTR and the disease form ΔF508 are deposited in distinct pericentriolar structures, termed aggresomes, upon overexpression or proteasome inhibition [108]. Subsequent studies established that aggresomes are induced upon ectopic expression of many different aggregation-prone proteins (reviewed in Refs. [109] and [110]). Aggresomes form near the microtubule-organizing center in a manner dependent on the microtubule-associated motor protein dynein, and are surrounded by a "cage" of filamentous vimentin [108, 111]. Aggresome formation is apparently a specific and active cellular response when production of misfolded proteins exceeds the capacity of the ubiquitin–proteasome system to tag and remove these proteins. They likely serve to protect the cell from toxic "gain-of-function" activities acquired by misfolded proteins. Aggresomes are also of clinical relevance as they share remarkable biochemical and structural features, for example, with Lewy bodies, the cytoplasmic inclusion bodies found in neurons affected by Parkinson's disease [112]. The pathways that regulate aggresome assembly are only now being explicated. Histone deacetylase 6 (HDAC6) appears to be a key regulator of aggresome assembly [113]. HDAC6 is a microtubule-associated deacetylase that has the capacity to bind both multi-ubiquitinated proteins and dynein motors and is believed to recruit misfolded proteins to the pericentriolar region for aggresome assembly. Deletion of HDAC6 prevents aggresome formation and sensitizes cells to the toxic effects of misfolded proteins, which supports the hypothesis that aggresomes sequester misfolded proteins to protect against their toxic activities. Components of the ubiquitin–proteasome system and chaperones such as Hsp70 are abundantly present in and are actively recruited to aggresomes [114–116]. Furthermore, elevating cellular Hsp70 levels can reduce aggresome formation by stimulating proteasomal degradation [117]. It appears that these subcellular structures are major sites of chaperone–proteasome cooperation to mediate the metabolism of misfolded proteins.

The formation of aggresome-like structures is also observed in dendritic cells that present foreign antigens to other immune cells [118]. Immature dendritic cells are located in tissues throughout the body, including skin and gut. When they encounter invading microbes, the pathogens are endocytosed and processed in a manner that involves the generation of antigenic peptides by the ubiquitin–proteasome system. Upon induction of dendritic cell maturation, ubiquitinated proteins transiently accumulate in large cytosolic structures that resemble aggresomes and were therefore termed DALIS (dendritic cell aggresome-like induced structures). It was speculated that DALIS formation may enable dendritic cells to regulate antigen processing and presentation. DALIS contain components of the ubiquitin–proteasome machinery as well as Hsp70 and the co-chaperone CHIP [118, 119]. Again, an interplay of molecular chaperones and the ubiquitin–proteasome system during regulated protein turnover is suggested.

The cellular function of molecular chaperones is apparently not restricted to mediating protein folding; instead, chaperones emerge also as vital components on protein-degradation pathways. Remarkably, the balance between folding and degradation activities of chaperones can be manipulated. In cells treated with Hsp90 inhibitors, for example, with geldanamycin (see above), the chaperone-assisted activation of signaling proteins is abrogated and chaperone substrates such as the protein kinases Raf-1 and ErbB2 are rapidly degraded by the ubiquitin–proteasome system [64–69, 120]. This appears to be due, in part, to transfer of the substrates back to Hsp70 and progression toward the ubiquitin-dependent degradation pathway.

Substrate interactions with chaperones – and consequently their commitment either toward the folding pathway or to their degradation via the ubiquitin–proteasome machinery – apparently serve as an essential post-translational protein quality-control mechanism within eukaryotic cells. The partitioning of proteins to either one of these mutually exclusive pathways is referred to as "protein triage" [121]. Although some misfolded proteins may be directly recognized by the proteasome [122], specific pathways within the ubiquitin–proteasome system are probably relied on for the degradation of most misfolded and damaged proteins. For example, E2 enzymes of the Ubc4/5 family selectively mediate the ubiquitylation of abnormal proteins as revealed in genetic studies in *Saccharomyces cerevisiae* [123].

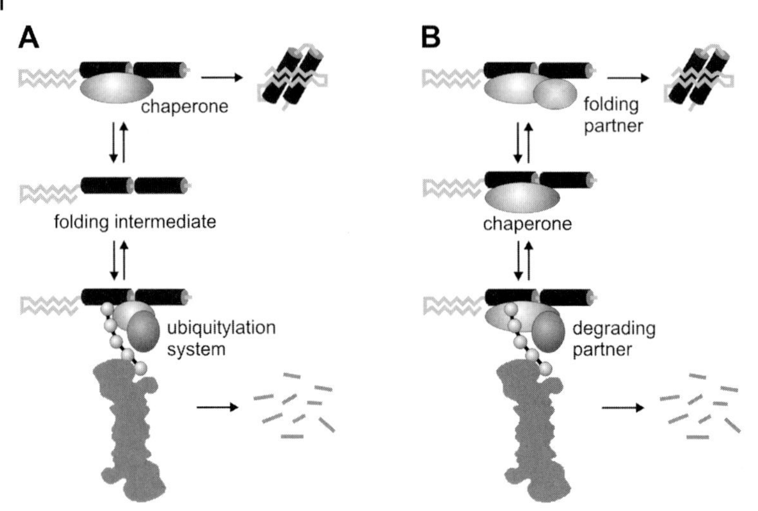

Fig. 1.5. Interplay of molecular chaperones with the ubiquitin–proteasome system. (A) Chaperones and the degradation machinery (i.e., ubiquitylation systems) compete with each other in the recognition of folding intermediates. Interaction with the chaperones directs the substrate towards folding. However, when the protein substrate is unable to attain a folded conformation, the chaperones maintain the folding intermediate in a soluble state that can be recognized by the degradation machinery. (B) The chaperones are actively involved in protein degradation. Through an association with certain components of the ubiquitin conjugation machinery (degrading partner), the chaperones participate in the targeting of protein substrates to the proteasome. A competition between degrading partners and folding partners determines chaperone action and the fate of the protein substrate.

It is well accepted that chaperones play a central role in the triage decision; however, less well understood are the events that lead to the cessation of efforts to fold a substrate, and the diversion of the substrate to the terminal degradative pathway. It is possible that chaperones and components of the ubiquitin–proteasome pathway exist in a state of competition for these substrates and that repeated cycling of a substrate on and off a chaperone maintains the substrate in a soluble state and increases, in a stochastic fashion, its likelihood of interactions with the ubiquitin machinery (Figure 1.5A). However, some data argue for a more direct role of the chaperones in the degradation process. Hsp70 plays an active and necessary role in the ubiquitylation of some substrates [124]; this activity of Hsp70 requires its chaperone function, indicating that conformational changes within substrates may facilitate recognition by the ubiquitylation machinery. Plausible hypotheses to explain these observations include direct associations between the chaperone and ubiquitin–proteasome machinery to facilitate transfer of a substrate from one pathway to the other, or conversion of the chaperone itself to a ubiquitylation complex (Figure 1.5B). It is also entirely possible that several quality-control pathways may exist and that the endogenous triage decision may involve aspects of each of these hypotheses.

1.6 The CHIP Ubiquitin Ligase: A Link Between Folding and Degradation Systems

Major insights into molecular mechanisms that underlie the cooperation of molecular chaperones with the ubiquitin–proteasome system were obtained through the functional characterization of the co-chaperone CHIP (reviewed in Ref. [40]). CHIP was initially identified in a screen for proteins containing tetratricopeptide repeat (TPR) domains, which are found in several co-chaperones – including Hip, Hop, and the cyclophilins – as chaperone-binding domains [27, 55] (Figure 1.2). CHIP contains three TPR domains at its amino terminus, which are used for binding to Hsp70 and Hsp90 [35, 125]. Besides the TPR domains, CHIP possesses a U-box domain at its carboxyl terminus [35] (Figure 1.2). U-box domains are similar to RING finger domains, but they lack the metal-chelating residues and instead are structured by intramolecular interactions [126]. The predicted structural similarity suggests that U boxes, like RING fingers, may also play a role in targeting proteins for ubiquitylation and subsequent proteasome-dependent degradation, and this possibility is borne out in functional analyses of U box–containing proteins [127, 128]. The TPR and U-box domains in CHIP are separated by a central domain rich in charged residues. The charged domain of CHIP is necessary for TPR-dependent interactions with Hsp70 [35] and is also required for homodimerization of CHIP [129].

The tissue distribution of CHIP supports the notion that it participates in protein folding and degradation decisions, as it is most highly expressed in tissues with high metabolic activity and protein turnover: skeletal muscle, heart, and brain. Although it is also present in all other organs, including pancreas, lung, liver, placenta, and kidney, the expression levels are much lower. CHIP is also detectable in most cultured cells, and is particularly abundant in muscle and neuronal cells and in tumor-derived cell lines [35]. Intracellularly, CHIP is primarily localized to the cytoplasm under quiescent conditions [35], although a fraction of CHIP is present in the nucleus [97]. In addition, cytoplasmic CHIP traffics into the nucleus in response to environmental challenge in cultured cells, which may serve as a protective mechanism or to regulate transcriptional responses in the setting of stress [130].

CHIP is distinguished among co-chaperones in that it is a bona fide interaction partner with both of the major cytoplasmic chaperones Hsp90 and Hsp70, based on their interactions with CHIP in the yeast two-hybrid system and *in vivo* binding assays [35, 125]. CHIP interacts with the terminal-terminal EEVD motifs of Hsp70 and Hsp90, similar to other TPR domain–containing co-chaperones such as Hop [55, 131, 132]. When bound to Hsp70, CHIP inhibits ATP hydrolysis and therefore attenuates substrate binding and refolding, resulting in inhibition of the "forward" Hsp70 substrate folding/refolding pathway, at least in *in vitro* assays [35]. The cellular consequences of this "anti-chaperone" function are not yet clear, and in fact CHIP may actually facilitate protein folding under conditions of stress, perhaps by slowing the Hsc70 reaction cycle [130, 133]. CHIP interacts with Hsp90 with approximately equivalent affinity to its interaction with Hsp70 [125]. This interaction results in remodeling of Hsp90 chaperone complexes, such that the co-chaperone p23, which is required for the appropriate activation of many, if not all, Hsp90 client proteins, is excluded. The mechanism for this activity is unclear – p23 and CHIP bind Hsp90 through different sites – yet the consequence of this action is predictable: CHIP should inhibit the function of proteins that require Hsp90 for conformational activation. The glucocorticoid receptor is an Hsp90 client that undergoes activation through a well-described sequence of events that depend on interactions of the glucocorticoid receptor with Hsp90 and various Hsp90 co-chaperones, including p23, making it an excellent model to test this prediction. Indeed, CHIP inhibits glucocorticoid receptor substrate binding and steroid-dependent transactivation ability [125]. Surprisingly, this effect of CHIP is accompanied by decreased steady-state levels of glucocorticoid receptor, and CHIP induces ubiquitylation of the glucocorticoid receptor *in vivo* and *in vitro*, as well as subsequent proteasome-dependent degradation. This effect is both U-box- and TPR-domain-dependent, suggesting that CHIP's effects on GR require direct interaction with Hsp90 and direct ubiquitylation of GR and delivery to the proteasome.

These observations are not limited to the glucocorticoid receptor. ErbB2, another Hsp90 client, is also degraded by CHIP in a proteasome-dependent fashion [120]. Nor are they limited to Hsp90 clients. For example, CHIP cooperates with Hsp70 during the degradation of immature forms of the CFTR protein at the ER membrane and during the ubiquitylation of phosphorylated forms of the microtubule-binding protein tau, which is of clinical importance due to its role in the pathology of Alzheimer's disease [97, 134]. The effects of CHIP are dependent on both the TPR domain, indicating a necessity for interactions with molecular chaperones, and the U box, which suggests that the U box is most likely the "business end" with respect to ubiquitylation. The means by which CHIP-dependent ubiquitylation occurs is not clear. In the case of ErbB2, ubiquitylation depends on a transfer of the client protein from Hsp90 to Hsp70 [120], indicating that the final ubiquitylation complex consists of CHIP, Hsp70 (but not Hsp90), and the client protein. In any event, the studies are consistent in supporting a role for CHIP as a key component of the chaperone-dependent quality-control mechanism. CHIP efficiently targets client proteins, particularly when they are partially unfolded (as is the case for most Hsp90 clients when bound to the chaperone) or frankly misfolded (as is the case for most proteins binding to Hsp70 through exposed hydrophobic residues).

Once the ubiquitylation activity of CHIP was recognized, it was logical to speculate that its U box might function in a manner analogous to that of RING fingers, which have recently been appreciated as key components of the largest family of ubiquitin ligases. If CHIP is a ubiquitin ligase, then its ability to ubiquitylate a substrate should be reconstituted *in vitro* when a substrate is added in the presence of CHIP, E1, an E2, and ubiquitin. Indeed, this is the case [135–137] (Figure 1.6). CHIP is thus the first described chaperone-associated E3 ligase. The ubiquitin ligase activity of CHIP depends on functional and physical interactions with a specific family of E2 enzymes, the Ubc4/Ubc5 family, which in humans comprises the E2s UbcH5a, UbcH5b, and UbcH5c. Of interest is the fact that the Ubc4/Ubc5 E2s

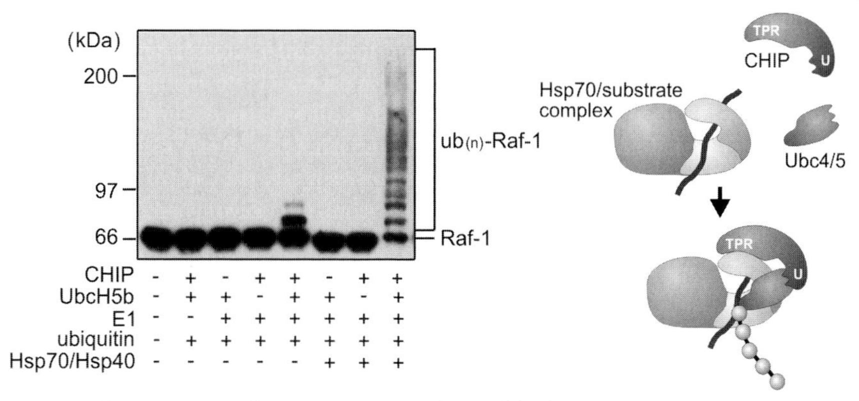

Fig. 1.6. Characterization of CHIP as a chaperone-associated ubiquitin ligase. Purified CHIP, UbcH5b, the ubiquitin-activating enzyme E1, ubiquitin, and the Hsp70–Hsp40 chaperone system were incubated with the bacterially expressed protein kinase Raf-1 (for details, see Ref. [137]). Raf-1 and ubiquitylated forms of the kinase ($ub_{(n)}$-Raf-1) were detected by immunoblotting using a specific anti-Raf-1 antibody. Efficient ubiquitylation of Raf-1 through the CHIP conjugation machinery depends on the recognition of the chaperone substrate by Hsp70, which presents the kinase to the conjugation machinery.

are stress-activated, ubiquitin-conjugating enzymes [123]. CHIP can therefore be seen as a co-chaperone that, in addition to inhibiting traditional chaperone activity, converts chaperone complexes into chaperone-dependent ubiquitin ligases. Indeed, the chaperones themselves seem to act as the main substrate-recognition components of these ubiquitin ligase complexes, as efficient ubiquitylation of chaperone substrates by CHIP depends on the presence of Hsp70 or Hsp90 in reconstituted systems [136, 137] (Figure 1.6). The chaperones apparently function in a manner analogous to F-box proteins, which are required as substrate recognition modules in many RING finger–containing ubiquitin ligase complexes [138–140].

Recently, another surprising function for CHIP has been identified, that of activation of the stress-responsive transcription factor heat shock factor-1 (HSF1) [130]. Through this association, CHIP regulates the expression of chaperones such as Hsp70 independently of its ability to modify their function through direct interactions. The mechanisms through which CHIP activates HSF1 are not entirely clear, but they are dependent in part on the induction of HSF1 trimerization, which is required for nuclear import and DNA binding. In addition, activation of HSF1 by CHIP seems to be independent of CHIP's ubiquitin ligase activity. The consequences of this activation are important for the response to stress, in that cells lacking CHIP are prone to stress-dependent apoptosis and mice deficient in CHIP (through homologous recombination) succumb rapidly to thermal challenge. These data indicate that CHIP plays a heretofore unsuspected role in coordinating the response to stress, not only by serving as a rate-limiting step in the degradation of damaged proteins but also by increasing the buffering capacity of the chaperone system to guard against stress-dependent proteotoxicity.

1.7
Other Proteins That May Influence the Balance Between Chaperone-assisted Folding and Degradation

CHIP is ideally suited to mediate chaperone–proteasome cooperation, as it combines a chaperone-binding motif and a domain that functions in ubiquitin-dependent degradation within its protein structure (Figure 1.2). Some other co-chaperones display a similar structural arrangement [40]. For example, BAG-1 contacts Hsp70 through a BAG-domain located at its carboxyl terminus and, in addition, possesses a central ubiquitin-like domain that is used for binding to the proteasome [141] (Figure 1.2). The co-chaperone thus belongs to a family of ubiquitin domain proteins (UDPs), many of which were shown to be associated with the proteasome [142]. This domain architecture enables BAG-1 to provide a physical link between Hsp70 and the proteasome, and elevating the cellular levels of BAG-1 results in a recruitment of the chaperone to the proteolytic complex. Notably, BAG-1 and CHIP occupy distinct domains on Hsp70 (Figure 1.7). Whereas BAG-1 associates with the amino-terminal ATPase domain, CHIP binds to the carboxyl-terminal EEVD motif of Hsp70 [35, 37]. Ternary complexes that comprise both co-chaperones associated with Hsp70 can be isolated from mammalian cells, suggesting a cooperation of BAG-1 and CHIP in the regulation of Hsp70 activity on certain degradation pathways. In fact, BAG-1 is able to stimulate the CHIP-induced degradation of the glucocorticoid hormone receptor [137]. A cooperation of diverse co-chaperones apparently provides additional levels of regulation to alter chaperone-assisted folding and degradation pathways.

Interestingly, BAG-1 and also Hsp70 and Hsp90 are themselves substrates of the CHIP ubiquitin ligase [135, 143] (J.H. unpublished). Yet, CHIP-mediated ubiquitylation of the chaperones and the co-chaperone does not induce their proteasomal degradation. Instead, it seems to provide additional means to regulate the association of the chaperone systems with the proteasome. In the case of BAG-1, ubiquitylation mediated by CHIP indeed stimulates the binding of the co-chaperone to the proteasome [143]. It remains to be elucidated, however, why Hsp70 and BAG-1 are not degraded when sorted to the proteasome through CHIP-induced ubiqui-

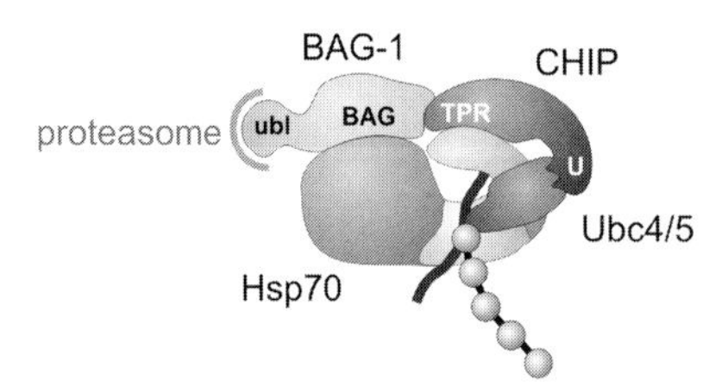

Fig. 1.7. Schematic presentation of the BAG-1–Hsp70–CHIP complex. BAG-1 associates with the ATPase domain of Hsp70, while CHIP is bound to the carboxyl terminus. BAG-1 mediates an association of Hsp70 with the proteasome via its ubiquitin-like domain (ubl), whereas CHIP acts in conjunction with Ubc4/5 as a chaperone-associated ubiquitin ligase to mediate the attachment of a polyubiquitin chain to the chaperone substrate.

tylation, in contrast to chaperone substrates such as the glucocorticoid hormone receptor. Possibly, the folded state of the proteins may serve to distinguish targeting factors and substrates doomed for degradation.

Efficient ubiquitylation of BAG-1 mediated by CHIP is dependent on the formation of the ternary BAG-1–Hsp70–CHIP complex [143]. The formed chaperone complex would thus expose multiple signals for sorting to the proteasome, e.g., the integrated ubiquitin-like domain of BAG-1 and polyubiquitin chains attached to BAG-1, Hsp70, and the bound protein substrate. Such a redundancy of sorting information might be considered unnecessary. Intriguingly, however, several subunits of the regulatory 19S particle of the proteasome are currently thought to act as receptors for polyubiquitin chains and integrated ubiquitin-like domains, including Rpn1, Rpn2, Rpt5, and Rpn10. The Rpn10 subunit was initially identified as a polyubiquitin chain receptor and was later shown to also bind integrated ubiquitin-like domains presented by UDPs [144–146]. Rpn10 possesses two distinct ubiquitin-binding domains, of which only one is used for UDP recognition [145–147]. However, conflicting data exist as to whether the subunit acts as a ubiquitin receptor in the context of the assembled 19S complex [148, 149]. More recently, Rpn1 was identified as a receptor for integrated ubiquitin-like domains [149], and a similar function may be fulfilled by the Rpn1-related subunit Rpn2 [150]. Polyubiquitin chains seem to be recognized by the Rpt5 subunit, one of the AAA ATPases present in the ring-like base of the regulatory 19S complex [151]. Its receptor function was revealed when tetraubiquitin was cross-linked to intact proteasomes [148]. Multiple docking sites for ubiquitin-like domains and polyubiquitin chains are apparently displayed by the regulatory particle of the proteasome. This may provide a structural basis for the recognition of multiple sorting signals exposed by the CHIP–chaperone complex (Figure 1.8). A similar mechanism involving multiple-site binding at the proteasome was recently proposed based on the observation that two unrelated yeast ubiquitin ligases associate with specific subunits of the 19S regulatory complex [152]. In these cases substrate delivery involves interactions of proteasomal subunits with the substrate-bound ubiquitin ligase, with the polyubiquitin chain attached to the substrate, and with the substrate itself. Multiple-site binding may function to slow down dissociation of the substrate from the proteasome and to facilitate transfer into the central proteolytic chamber through ATP-dependent movements of the subunits of the 19S particle.

Human cells contain several BAG-1-related proteins: BAG-2, BAG-3 (CAIR-1; Bis), BAG-4 (SODD), BAG-5, and BAG-6 (Scythe, BAT3) [153] (Figure 1.2). It appears that BAG proteins act as nucleotide-exchange factors to induce substrate unloading from Hsp70 on diverse protein folding, assembly, and degradation pathways. Notably, BAG-6 is another likely candidate for a co-chaperone that regulates protein degradation via the ubiquitin–proteasome pathway. Similar to BAG-1, BAG-6 contains a ubiquitin-like domain that is possibly used for proteasome binding [154]. However, experimental data verifying a role of BAG-6 in protein degradation remain elusive so far.

The cooperation of diverse co-chaperones not only may allow promotion of chaperone-associated degradation but also may provide the means to confine the

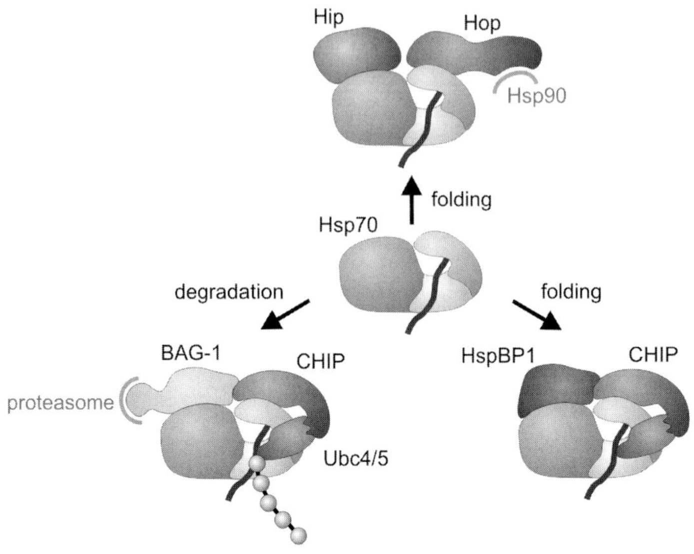

Fig. 1.8. The co-chaperone network that determines folding and degradation activities of Hsp70. BAG-1 and CHIP associate with Hsp70 to induce the proteasomal degradation of a Hsp70-bound protein substrate. When BAG-1 is displaced by binding of HspBP1 to the ATPase domain of Hsp70, the ubiquitin ligase activity of CHIP is attenuated in the formed complex, enabling CHIP to modulate Hsp70 activity without inducing degradation. The ATPase domain can also be occupied by Hip, which stimulates the chaperone activity of Hsp70 and participates in the Hsp70/Hsp90-mediated regulation of signal transduction pathways. At the same time, Hop displaces CHIP from the carboxyl terminus of Hsp70 and recruits Hsp90 to the chaperone complex.

destructive activity of CHIP. The Hsp70-binding protein 1 (HspBP1) seems to fulfill such a regulatory function [155]. HspBP1 was initially identified in a screen for proteins that associate with the ATPase domain of Hsp70 and was shown to stimulate nucleotide release from the chaperone [156, 157]. Notably, association of HspBP1 with the ATPase domain blocks binding of BAG-1 to Hsp70 and at the same time promotes an interaction of CHIP with Hsp70's carboxyl terminus. In the formed ternary HspBP1–Hsp70–CHIP complex, the ubiquitin ligase activity of CHIP is attenuated and Hsp70 as well as a chaperone substrate are no longer efficiently ubiquitylated [155]. By interfering with CHIP-mediated ubiquitylation, HspBP1 stimulates the maturation of CFTR and promotes the sorting of the membrane protein to the cell surface. HspBP1 apparently functions as an antagonist of the CHIP ubiquitin ligase to regulate Hsp70-assisted folding and degradation pathways (Figure 1.8).

The HspBP1-mediated inhibition of the ubiquitin ligase activity may enable CHIP to modulate the Hsp70 ATPase cycle without inducing degradation. In fact, degradation-independent functions of CHIP have recently emerged [130, 133, 158, 159]. CHIP was shown to regulate the chaperone-assisted folding and sorting of the androgen receptor and of endothelial nitric oxide synthase without inducing degradation [158, 159]. Moreover, CHIP fulfills an essential role in the chaperone-mediated regulation of the heat shock transcription factor, independent of its degradation-inducing activity [130]. It remains to be seen, however, whether HspBP1 cooperates with CHIP in these instances, as HspBP1 displayed a certain specificity with regard to chaperone substrates. The co-chaperone interfered with the degradation of CFTR, but did not influence the CHIP-mediated turnover of the glucocorticoid hormone receptor. Such a client specificity may arise in part from the fact that HspBP1 inhibits the ubiquitin ligase activity of CHIP in a complex with Hsc70, but leaves Hsp90-associated ubiquitylation unaffected [155]. In addition, direct interactions between HspBP1 and a subset of chaperone substrates may contribute to substrate selection. In any case, the cooperation of CHIP with other co-chaperones apparently provides a means to regulate chaperone-assisted protein degradation.

It is likely that there are multiple degradation pathways for misfolded proteins in the eukaryotic cytoplasm. Although CHIP participates in the degradation of chaperone substrates induced by applying Hsp90 inhibitors to cell cultures (see above), drug-induced degradation is not abrogated in cells that lack the CHIP ubiquitin ligase [120]. Furthermore, CHIP cooperates with Hsp70 in the ER-associated degradation of CFTR, but the Hsp70-assisted degradation of apoB at the cytoplasmic face of the ER membrane does not involve CHIP [97]. Taken together, these data strongly argue for the existence of other, yet to be identified, ubiquitin ligases that are able to target chaperone substrates to the proteasome. A likely candidate in this regard is Parkin, a RING finger ubiquitin ligase, whose activity is impaired in juvenile forms of Parkinson's disease [17]. Hsp70 and CHIP were found to be associated with Parkin in neuronal cells, suggesting an involvement of Parkin in the proteasomal degradation of chaperone substrates [160]. Interestingly, α-synuclein, the main component of protein deposits observed in dopaminergic neurons of Parkinson patients, and synphilin, a protein that binds α-synuclein and induces deposit formation, both associate with yet other ubiquitin ligases: Siah-1 and Siah-2 [161, 162]. In the case of Siah-1, a link to cytoplasmic chaperone systems is suggested by the finding that the Hsp70 co-chaperone BAG-1 is a binding partner of the ubiquitin ligase and suppresses some of the cellular activities of Siah-1 [163]. Taken together, it is tempting to speculate about a role of Parkin and Siah on chaperone-assisted degradation pathways; yet, this remains to be explored in detail.

1.8
Further Considerations

Although the appreciation of interplay between molecular chaperones and ubiquitin-dependent proteolysis has greatly expanded over the past decade, a number of critical issues remain to be resolved. It is not entirely clear what determines whether a misfolded protein will undergo repeated attempts at misfolding versus diversion to the ubiquitin–proteasome pathway. Recruitment of CHIP into chaperone complexes appears to be a critical component of this reaction, which therefore begs the question as to what regulates this step. Since this step in protein quality control must be both rapidly activated and easily reversible, it is likely that regulation occurs at the post-translational level rather than through changes in steady-state protein levels. The precise sorting mechanisms for ubiquitinated proteins are also unclear. BAG-1 is a player, and it is also likely that overlap exists to some extent for sorting of the cytoplasmic and endoplasmic reticulum quality-control pathways. Nevertheless, much remains to be learned about these steps.

From a broader perspective, it is now also imperative to understand the pathophysiological roles of cytoplasmic quality-control mechanisms regulated by chaperone–proteasome interactions. As mentioned previously, there is a strong association between chaperone dysfunction and accumulations of misfolded proteins that characterizes genetic neurodegenerative diseases. An imbalance between protein folding and degradation may also contribute to some features of senescence and organismal aging. The link between chaperone systems and aging is based on increasing appreciation that modified, misfolded, and aggregated proteins accumulate with age [164]. Dysregulation of chaperone expression has been observed with aging and is therefore implicated in aging-related changes [165]; in general, it is accepted that induction of the major chaperones is impaired with aging, a fact confirmed by recent gene-profiling experiments *in vivo* [166], although given the diversity of chaperones it is probably not surprising that age-related changes in expression are fairly complicated [167]. The mechanism underlying this dysregulation is not entirely clear, but seems to be due in part to impaired activation of the stress-responsive transcription factor HSF1. Overexpression of heat shock proteins in yeast, *C. elegans*, and *Drosophila* leads to increased longevity [168–170]. More recently, conclusive genetic evidence from *C. elegans* indicates that mutation of HSF1 causes a dramatic and significant reduction in lifespan [170, 171], further implicating the accumulation of misfolded proteins in age-related phenotypes.

1.9 Conclusions

The associations between molecular chaperones and the ubiquitin–proteasome system represent a critical step in the response to proteotoxic damage. Whether attempts should be made to refold damaged proteins (thus conserving cellular resources) or degrade them instead (to prevent the possibility of protein aggregation and concomitant toxicity) requires a consideration of cellular economy. Defects in the quality-control mechanisms may have enormous consequences even if only slight imbalances occur between protein folding and degradation, as these imbalances can cause accumulated toxicity over time. The relationship between chaperone–proteasome interactions and pathophysiological events is only now being unraveled. Modulation of this system may provide a unique therapeutic target for degenerative diseases and pathologies associated with aging.

References

1 U. Schubert, L.C. Anton, J. Gibbs, C.C. Norbury, J.W. Yewdell, J.R. Bennink, Rapid degradation of a large fraction of newly synthesized proteins by proteasomes, Nature 404 (2000) 770–774.

2 J.P. Taylor, J. Hardy, K.H. Fischbeck, Toxic proteins in neurodegenerative disease, Science 296 (2002) 1991–1995.

3 A. Horwich, Protein aggregation in disease: a role for folding intermediates forming specific multimeric interactions, J. Clin. Invest. 110 (2002) 1221–1232.

4 E. Scherzinger, R. Lurz, M. Turmaine, L. Mangiarini, B. Hollenbach, R. Hasenbank, G.P. Bates, S.W. Davies, H. Lehrach, E.E. Wanker, Huntingtin-encoded polyglutamine expansions form amyloid-like protein aggregates *in vitro* and *in vivo*, Cell 90 (1997) 549–558.

5 E. Scherzinger, A. Sittler, K. Schweiger, V. Heiser, R. Lurz, R. Hasenbank, G.P. Bates, H. Lehrach, E.E. Wanker, Self-assembly of polyglutamine-containing huntingtin fragments into amyloid-like fibrils: implications for Huntington's disease pathology, Proc. Natl. Acad. Sci. USA 96 (1999) 4604–4609.

6 A. Kazantsev, E. Preisinger, A. Dranovsky, D. Goldgaber, D. Housman, Insoluble detergent-resistant aggregates form between pathological and nonpathological lengths of polyglutamine in mammalian cells, Proc. Natl. Acad. Sci. U S A 96 (1999) 11404–11409.

7 D. Martindale, A. Hackam, A. Wieczorek, L. Ellerby, C. Wellington, K. McCutcheon, R. Singaraja, P. Kazemi-Esfarjani, R. Devon, S.U. Kim, D.E. Bredesen, F. Tufaro, M.R. Hayden, Length of huntingtin and its polyglutamine tract influences localization and frequency of intracellular aggregates. Nat. Genet. 18 (1998) 150–154.

8 I. Sanchez, C. Mahlke, J. Yuan, Pivotal role of oligomerization in expanded polyglutamine neurodegenerative disorders, Nature 421 (2003) 373–379.

9 M.F. Perutz, Glutamine repeats and neurodegenerative diseases: molecular aspects, Trends Biochem. Sci. 24 (1999) 58–63.

10 H.Y. Zoghbi, H.T. Orr, Glutamine repeats and neurodegeneration, Annu. Rev. Neurosci. 23 (2000) 217–247.

11 H.T. Orr, M.Y. Chung, S. Banfi, T.J. Jr. Kwiatkowski, A. Servadio, A.L. Beaudet, A.E. McCall, L.A. Duvick, L.P. Ranum, H.Y. Zoghbi, Expansion of an unstable trinucleotide CAG repeat in spinocerebellar ataxia type 1, Nat. Genet. 4 (1993) 221–226.

12 D.R. Rosen, T. Siddique, D. Patterson, D.A. Figlewicz, P. Sapp, A. Hentati, D. Donaldson, J. Goto, J.P. O'Regan, H.X. Deng, et al. Mutations in Cu/Zn superoxide dismutase gene are associated with familial amyotrophic lateral sclerosis. Nature 362 (1993) 59–62.

13 J.A. Johnston, M.J. Dalton, M.E. Gurney, R.R. Kopito, Formation of high molecular weight complexes of mutant Cu, Zn-superoxide dismutase in a mouse model for familial amyotrophic lateral sclerosis, Proc. Natl. Acad. Sci. USA 97 (2000) 12571–12576.

14 B.I. Giasson, V.M. Lee, Are ubiquitination pathways central to Parkinson's disease? Cell 114 (2003) 1–8.

15 T. Kitada, S. Asakawa, N. Hattori, H. Matsumine, Y. Yamamura, S. Minoshima, M. Yokochi, Y. Mizuno, N. Shimizu, Mutations in the parkin gene cause autosomal recessive juvenile parkinsonism, Nature 392 (1998) 605–608.

16 C.B. Lücking, A. Durr, V. Bonifati, J. Vaughan, G. De Michele, T. Gasser, B.S. Harhangi, G. Meco, P. Denefle, N.W. Wood, Y. Agid, A. Brice, Association between early-onset Parkinson's disease and mutations in the parkin gene, N. Engl. J. Med. 342 (2000) 1560–1567.

17 H. Shimura, N. Hattori, S. Kubo, Y. Mizuno, S. Asakawa, S. Minoshima, N. Shimizu, K. Iwai, T. Chiba, K. Tanaka, T. Suzuki, Familial Parkinson disease gene product, parkin, is a ubiquitin-protein ligase, Nat. Genet. 25 (2000) 302–305.

18 K.K. Chung, Y. Zhang, K.L. Lim, Y. Tanaka, H. Huang, J. Gao, C.A. Ross, V.L. Dawson, T.M. Dawson, Parkin ubiquitinates the α-synuclein-interacting protein, synphilin-1: implications for Lewy-body formation in Parkinson disease, Nat. Med. 7 (2001) 1144–1150.

19 H. Shimura, M.G. Schlossmacher, N. Hattori, M.P. Frosch, A. Trockenbacher, R. Schneider, Y. Mizuno, K.S. Kosik, D.J. Selkoe, Ubiquitination of a new form of α-synuclein by parkin from human brain: implications for Parkinson's disease, Science 293 (2001) 263–269.

20 Y. Imai, M. Soda, H. Inoue, N. Hattori, Y. Mizuno, R. Takahashi, An unfolded putative transmembrane polypeptide, which can lead to endoplasmic reticulum stress, is a substrate of Parkin, Cell 105 (2001) 891–902.

21 P. Fernandez-Funez, M.L. Nino-Rosales, B. de Gouyon, W.C. She, J.M. Luchak, P. Martinez, E. Turiegano, J. Benito, M. Capovilla, P.J. Skinner, A. McCall, I. Canal, H.T. Orr, H.Y. Zoghbi, J. Botas, Identification of genes that modify ataxin-1-induced neurodegeneration, Nature 408 (2000) 101–106.

22 P. Kazemi-Esfarjani, S. Benzer, Genetic suppression of polyglutamine toxicity in Drosophila, Science 287 (2000) 1837–1840.

23 N.M. Bonini, Chaperoning brain degeneration, Proc. Natl. Acad. Sci. U S A 99 Suppl. 4 (2002) 16407–16411.

24 P.K. Auluck, H.Y. Chan, J.Q. Trojanowski, V.M. Lee, N.M. Bonini, Chaperone suppression of α-synuclein toxicity in a Drosophila model for Parkinson's disease, Science 295 (2002) 865–868.

25 F.U. Hartl, M. Hayer-Hartl, Molecular chaperones in the cytosol: from nascent chain to folded protein, Science 295 (2002) 1852–1858.

26 J. Frydman, Folding of newly translated proteins *in vivo*: the role of molecular chaperones, Annu. Rev. Biochem. 70 (2001) 603–647.

27 J. Frydman, J. Höhfeld, Chaperones get in touch: the Hip-Hop connection, Trends Biochem. Sci. 22 (1997) 87–92.

28 B. Bukau, A.L. Horwich, The Hsp70 and Hsp60 chaperone machines, Cell 92 (1998) 351–366.

29 R.I. Morimoto, Regulation of the heat shock transcriptional response: cross talk between a family of heat shock factors, molecular chaperones, and negative regulators. Genes Dev. 12 (1998) 3788–3796.

30 M.P. Mayer, D. Brehmer, C.S. Gässler, B. Bukau, Hsp70 chaperone machines, Adv. Protein Chem. 59 (2001) 1–44.

31 S. Rüdiger, L. Germeroth, J. Schneider-Mergener, B. Bukau, Substrate specificity of the DnaK chaperone determined by screening cellulose-bound peptide libraries, EMBO J. 16 (1997) 1501–1507.

32 V. Thulasiraman, C.F. Yang, J. Frydman, *In vivo* newly translated polypeptides are sequestered in a protected folding environment, EMBO J. 18 (1999) 85–95.

33 X. Zhu, X. Zhao, W.F. Burkholder, A. Gragerov, C.M. Ogata, M.E. Gottesman, W.A. Hendrickson, Structural analysis of substrate binding by the molecular chaperone DnaK, Science 272 (1996) 1606–1614.

34 Y. Minami, J. Höhfeld, K. Ohtsuka, F.U. Hartl, Regulation of the heat-shock protein 70 reaction cycle by the mammalian DnaJ homolog, Hsp40, J. Biol. Chem. 271 (1996) 19617–19624.

35 C.A. Ballinger, P. Connell, Y. Wu, Z. Hu, L.J. Thompson, L.Y. Yin, C. Patterson, Identification of CHIP, a novel tetratricopeptide repeat-containing protein that interacts with heat shock proteins and negatively regulates chaperone functions, Mol. Cell. Biol. 19 (1999) 4535–4545.

36 J. Höhfeld, Y. Minami, F.U. Hartl, Hip, a novel cochaperone involved in the eukaryotic Hsc70/Hsp40 reaction cycle, Cell 83 (1995) 589–598.

37 J. Höhfeld, S. Jentsch, GrpE-like regulation of the hsc70 chaperone by the anti-apoptotic protein BAG-1, Embo J. 16 (1997) 6209–6216.

38 H. Sondermann, C. Scheufler, C. Schneider, J. Höhfeld, F.U. Hartl, I. Moarefi, Structure of a Bag/Hsc70 complex: convergent functional evolution of Hsp70 nucleotide exchange factors, Science 291 (2001) 1553–1557.

39 C.S. Gässler, T. Wiederkehr, D. Brehmer, B. Bukau, M.P. Mayer, Bag-1M accelerates nucleotide release for human Hsc70 and Hsp70 and can act concentration-dependent as positive and negative cofactor, J. Biol. Chem. 276 (2001) 32538–32544.

40 J. Höhfeld, D.M. Cyr, C. Patterson, From the cradle to the grave: molecular chaperones that may choose between folding and degradation, EMBO Rep. 2 (2001) 885–890.

41 W.M. Obermann, H. Sondermann, A.A. Russo, N.P. Pavletich, F.U. Hartl, *In vivo* function of Hsp90 is dependent on ATP binding and ATP hydrolysis, J. Cell Biol. 143 (1998) 901–910.

42 J. Buchner, Hsp90 & Co. – a holding for folding, Trends Biochem. Sci. 24 (1999) 136–141.

43 J.C. Young, I. Moarefi, F.U. Hartl, Hsp90: a specialized but essential protein-folding tool, J. Cell Biol. 154 (2001) 267–273.

44 L.H. Pearl, C. Prodromou, Structur, function and mechanism of the Hsp90 molecular chaperone, Adv. Protein Chem. 59 (2001) 157–186.

45 B. Panaretou, G. Siligardi, P. Meyer, A. Maloney, J.K. Sullivan, S. Singh, S.H. Millson, P.A. Clarke, S. Naaby-Hansen, R. Stein, R. Cramer, M. Mooapour, P. Workman, P.W. Piper, L.H. Peal, C. Prodromou, Activation of the ATPase activity of hsp90 by the stress-regulated cochaperone aha1, Mol. Cell 10 (2002) 1307–1318.

46 J.C. Young, N.J. Hoogenraad, F.U. Hartl, Molecular chaperones Hsp90 and Hsp70 deliver preproteins to the mitochondrial import receptor Tom70, Cell 112 (2003) 41–50.

47 A. Brychzy, T. Rein, K.F. Winklhofer, F.U. Hartl, J.C. Young, W.M. Obermann, Cofactor Tpr2 combines two TPR domains and a J domain to regulate the Hsp70/Hsp90 chaperone system, EMBO J. 22 (2003) 3613–3623.

48 S.M. Roe, M.M. Ali, P. Meyer, C.K. Vaughan, B. Panaretou, P.W. Piper, C. Prodromou, L.H. Pearl, The mechanism of Hsp90 regulation by the protein kinase-specific cochaperone p50 (cdc37), Cell 116 (2004) 87–98.

49 K.D. Dittmar, W.B. Pratt, Folding of the glucocorticoid receptor be the reconstituted Hsp90-based chaperone machinery. The initial hsp90.p60.hsp70 dependent step is sufficient for creating the steroid binding conformation, J. Biol. Chem. 272 (1997) 13047–13054.

50 K.D. Dittmar, D.R. Demady, L.F. Stancato, P. Krishna, W.B. Pratt, Folding of the glucocorticoid receptor by the heat shock protein (hsp) 90-based chaperone machinery. The role of p23 is to stabilize receptor.hsp90 heterocomplexes formed by hsp90.p60.hsp70, J. Biol. Chem. 272 (1997) 21213–21220.

51 S. Chen, V. Prapapanich, R.A. Rimerman, B. Honore, D.F. Smith, Interactions of p60, a mediator of progesterone receptor assembly, with heat shock proteins hsp90 and hsp70. Mol. Endocrinol. 10 (1996) 682–693.

52 H. Kosano, B. Stensgard, M.C. Charlesworth, N. McMahon, D. Toft, The assembly of progesterone receptor-hsp90 complexes using purified proteins, J. Biol. Chem. 273 (1998) 32973–32979.

53 W.B. Pratt, D.O. Toft, Regulation of signaling protein function and trafficking by the hsp90/hsp70-based chaperone machinery, Exp. Biol. Med. 228 (2003) 111–133.

54 M.P. Hernandez, A. Chadli, D.O.

Toft, Hsp40 binding is the first step in the Hsp90 chaperoning pathway for the progesterone receptor, J. Biol. Chem. 277 (2002) 11873–11881.

55 C. Scheufler, A. Brinker, G. Bourenkov, S. Pegoraro, L. Moroder, H. Bartunik, F.U. Hartl, I. Moarefi, Structure of TPR domain-peptide complexes: critical elements in the assembly of the Hsp70-Hsp90 multichaperone machine, Cell 101 (2000) 199–210.

56 J. Fontana, D. Fulton, Y. Chen, T.A. Fairchild, T.J. McCabe, N. Fujita, T. Tsuruo, W.C. Sessa, Domain mapping studies reveal that the M domain of hsp90 serves as a molecular scaffold to regulate Akt-dependent phosphorylation of endothelioal nitric oxide synthase and NO release, Circ. Res. 90 (2002) 866–873.

57 D.B. Solit, H.I. Scher, N. Rosen, Hsp90 as a therapeutic target in prostate cancer, Semin. Oncol. 30 (2003) 709–716.

58 L. Neckers, S.P. Ivy, Heat shock protein 90, Curr. Opin. Oncol. 15 (2003) 419–424.

59 A. Kamal, L. Thao, J. Sensintaffar, L. Zhang, M.F. Boehm, L.C. Fritz, F.J. Burrows, A high-affinity conformation of Hsp90 confers tumour selectivity on Hsp90 inhibitors, Nature 425 (2003) 407–410.

60 C. Prodromou, S.M. Roe, R. O'Brien, J.E. Ladbury, P.W. Piper, L.H. Pearl, Identification and structural characterization of the ATP/ADP-binding site in the Hsp90 molecular chaperone, Cell 90 (1997) 65–75.

61 C.E. Stebbins, A.A. Russo, C. Schneider, N. Rosen, F.U. Hartl, N.P. Pavletich, Crystal structure of an Hsp90-geldanamycin complex: targeting of a protein chaperone by an antitumor agent, Cell 89 (1997) 239–250.

62 S.V. Sharma, T. Agatsuma, H. Nakano, Targeting of the protein chaperone, Hsp90, by the transformation suppressing agent, radicicol, Oncogene 16 (1998) 2639–2645.

63 T.W. Schulte, S. Akinaga, T. Murakata, T. Agatsuma, S. Sugimoto, H. Nakano, Y.S. Lee, B.B. Simen, Y. Argon, S. Felts, D.O. Toft, L.M. Neckers, S.V. Sharma, Interaction of radicicol with members of the heat shock protein 90 family of molecular chaperones, Mol. Endocrinol. 13 (1999) 1435–1448.

64 C. Schneider, L. Sepp-Lorenzino, E. Nimmesgern, O. Ouerfelli, S. Danishefsky, N. Rosen, F.U. Hartl, Pharmacologic shifting of a balance between protein refolding and degradation mediated by Hsp90, Proc. Natl. Acad. Sci. U S A 93 (1996) 14536–14541.

65 E.G. Mimnaugh, C. Chavany, L. Neckers, Polyubiquitination and proteasomal degradation of the p185c-erbB-2 receptor protein-tyrosine kinase induced by geldanamycin, J. Biol. Chem. 271 (1996) 22796–22801.

66 M.G. Marcu, T.W. Schulte, L. Neckers, Novobiocin and related coumarins and depletion of heat shock protein 90-dependent signaling proteins, J. Natl. Cancer Inst. 92 (2000) 242–248.

67 W. Xu, E. Mimnaugh, M.F. Rosser, C. Nicchitta, M. Marcu, Y. Yarden, L. Neckers, Sensitivity of mature Erbb2 to geldanamycin is conferred by its kinase domain and is mediated by the chaperone protein Hsp90, J. Biol. Chem. 276 (2001) 3702–3708.

68 D.B. Solit, F.F. Zheng, M. Drobnjak, P.N. Munster, B. Higgins, D. Verbel, G. Heller, W. Tong, C. Cordon-Cardo, D.B. Agus, H.I. Scher, N. Rosen, 17-Allylamino-17-demethoxygeldanamycin induces the degradation of androgen receptor and HER-2/neu and inhibits the growth of prostate cancer xenografts, Clin. Cancer Res. 8 (2002) 986–993.

69 A. Citri, I. Alroy, S. Lavi, C. Rubin, W. Xu, N. Grammatikakis, C. Patterson, L. Neckers, D.W. Fry, Y. Yarden, Drug-induced ubiquitylation and degradation of ErbB receptor tyrosine kinases: implications for cancer therapy, Embo J. 21 (2002) 2407–2417.

70 R. van Montfort, C. Slingsby, E. Vierling, Structure and function of the small heat shock protein/alpha-crystallin family of molecular chaperones, Adv. Protein Chem. 59 (2001) 105–156.
71 M. Haslbeck, N. Braun, T. Stromer, B. Richter, N. Model, S. Weinkauf, J. Buchner, Hsp42 is the general small heat shock protein in the cytosol of Saccharomyces cerevisiae, EMBO J. 23 (2004) 638–649.
72 T. Stromer, E. Fischer, K. Richter, M. Haslbeck, J. Buchner, Analysis of the regulation of the molecular chaperone Hsp26 by temperature-induced dissociation: the N-terminal domain is important for oligomer assembly and the binding of unfolding proteins. J. Biol. Chem. 279 (2004) 11222–11228.
73 E. Basha, G.J. Lee, L.A. Breci, A.C. Hausrath, N.R. Buan, K.C. Giese, E. Vierling, The identity of proteins associated with a small heat shock protein during heat stress *in vivo* indicates that these chaperones protect a wide range of cellular functions, J. Biol. Chem. 279 (2004) 7566–7575.
74 K.K. Kim, R. Kim, S.H. Kim, Crystal structure of a small heat-shock protein, Nature 394 (1998) 595–599.
75 R.L. van Montfort, E. Basha, K.L. Friedrich, C. Slingsby, E. Vierling, Crystal structure and assembly of a eukaryotic small heat shock protein. Nat. Struct. Biol. 8 (2001) 1025–1030.
76 M. Haslbeck, S. Walke, T. Stromer, M. Ehrnsperger, H.E. White, S. Chen, H.R. Saibil, J. Buchner, Hsp26: a temperature-regulated chaperone, EMBO J. 18 (1999) 6744–6751.
77 G.J. Lee, A.M. Roseman, H.R. Saibil, E. Vierling, A small heat shock protein stably binds heat-denatured model substrates and can maintain a substrate in a folding-competent state, EMBO J. 16 (1997) 659–671.
78 Z. Xu, A.L. Horwich, P.B. Sigler, The crystal structure of the asymmetric GroEL-GroES-(ADP)7 chaperonin complex, Nature 388 (1997) 741–750.
79 H.R. Saibil, N.A. Ranson, The chaperonin folding machine, Trends Biochem. Sci. 27 (2002) 627–632.
80 A.S. Meyer, J.R. Gillespie, D. Walther, I.S. Millet, S. Doniach, J. Frydman, Closing the folding chamber of the eukaryotic chaperonin requires the transition state of ATP hydrolysis, Cell 113 (2003) 369–381.
81 V. Thulasiraman, C.F. Yang, J. Frydman, *In vivo* newly translated polypeptides are sequestered in a protected folding environment, EMBO J. 18 (1999) 85–95.
82 S.A. Lewis, G. Tian, I.E. Vainberg, N.J. Cowan, Chaperonin-mediated folding of actin and tubulin, J. Cell Biol. 132 (1996) 1–4.
83 H. Sakahira, P. Breuer, M.K. Hayer-Hartl, F.U. Hartl, Molecular chaperones as modulators of polyglutamine protein aggregation and toxicity, Proc. Natl. Acad. Sci. USA 99 Suppl. 4 (2002) 16412–16418.
84 J.M. Warrick, H.Y. Chan, G.L. Gray-Board, Y. Chai, H.L. Paulson, N.M. Bonini, Suppression of polyglutamine-mediated neurodegeneration in Drosophila by the molecular chaperone Hsp70. Nat. Genet. 23 (1999) 425–428.
85 C.J. Cummings, M.A. Mancini, B. Antalffy, D.B. DeFranco, H.T. Orr, H.Y. Zoghbi, Chaperone suppression of aggregation and altered subcellular proteasome localization imply protein misfolding in SCA1, Nat. Genet. 19 (1998) 148–154.
86 C.J. Cummings, Y. Sun, P. Opal, B. Antalffy, R. Mestril, H.T. Orr, W.H. Dillmann, H.Y. Zoghbi, Over-expression of inducible HSP70 chaperone suppresses neuropathology and improves motor function in SCA1 mice, Hum. Mol. Genet. 10 (2001) 1511–1518.
87 S. Krobitsch, S. Lindquist, Aggregation of huntingtin in yeast varies with the length of the polyglutamine expansion and the expression of chaperone proteins, Proc. Natl. Acad. Sci. U S A 97 (2000) 1589–1594.
88 P.J. Muchowski, G. Schaffar, A.

Sittler, E.E. Wanker, M.K. Hayer-Hartl, F.U. Hartl, Hsp70 and hsp40 chaperones can inhibit self-assembly of polyglutamine proteins into amyloid-like fibrils, Proc. Natl. Acad. Sci. U S A 97 (2000) 7841–7846.

89 J.Z. Chuang, H. Zhou, M. Zhu, S.H. Li, X.J. Li, C.H. Sung, Characterization of a brain-enriched chaperone, MRJ, that inhibits Huntingtin aggregation and toxicity independently, J. Biol. Chem. 277 (2002) 19831–19838.

90 A. Sittler, R. Lurz, G. Lueder, J. Priller, H. Lehrach, M.K. Hayer-Hartl, F.U. Hartl, E.E. Wanker, Geldanamycin activates a heat shock response and inhibits huntingtin aggregation in a cell culture model of Huntington's disease, Hum. Mol. Genet. 10 (2001) 1307–1315.

91 A. Parcellier, E. Schmitt, S. Gurbuxani, D. Seigneurin-Berny, A. Pance, A. Chantome, S. Plenchette, S. Khochbin, E. Solary, C. Garrido, Hsp27 is a ubiquitin-binding protein involved in IκBα proteasomal degradation, Mol. Cell Biol. 23 (2003) 5790–5802.

92 H. Shimura, Y. Miura-Shimura, K.S. Kosick, Binding of tau to heat shock protein 27 leads to decreased concentration of hyperphosphorylated tau and enhanced cell survival, J. Biol. Chem. 279 (2004) 17957–17962.

93 V. Gusarova, A.J. Caplan, J.L. Brodsky, E.A. Fisher, Apoprotein B degradation is promoted by the molecular chaperones hsp90 and hsp70, J. Biol. Chem. 276 (2001) 24891–24900.

94 C. Taxis, R. Hitt, S.H. Park, P.M. Deak, Z. Kostova, D.H. Wolf, Use of modular substrates demonstrates mechanistic diversity and reveals differences in chaperone requirement of ERAD, J. Biol. Chem. 278 (2003) 35903–35913.

95 Y. Yang, S. Janich, J.A. Cohn, J.M. Wilson, The common variant of cystic fibrosis transmembrane conductance regulator is recognized by hsp70 and degraded in a pre-Golgi nonlysosomal compartment, Proc. Natl. Acad. Sci. U S A 90 (1993) 9480–9484.

96 G.C. Meacham, Z. Lu, S. King, E. Sorscher, A. Tousson, D.M. Cyr, The Hdj-2/Hsc70 chaperone pair facilitates early steps in CFTR biogenesis, Embo J. 18 (1999) 1492–1505.

97 G.C. Meacham, C. Patterson, W. Zhang, J.M. Younger, D.M. Cyr, The Hsc70 co-chaperone CHIP targets immature CFTR for proteasomal degradation, Nat. Cell Biol. 3 (2001) 100–105.

98 Y. Zhang, G. Nijbroek, M.L. Sullivan, A.A. McCracken, S.C. Watkins, S. Michaelis, J.L. Brodsky, Hsp70 molecular chaperone facilitates endoplasmic reticulum-associated protein degradation of cystic fibrosis transmembrane conductance regulator in yeast, Mol. Biol. Cell 12 (2001) 1303–1314.

99 R.R. Kopito, Biosynthesis and degradation of CFTR, Physiol. Rev. 79 (1999) S167–173.

100 J.R. Riordan, Cystic fibrosis as a disease of misprocessing of the cystic fibrosis transmembrane conductance regulator glycoprotein, Am. J. Hum. Genet. 64 (1999) 1499–1504.

101 C.L. Ward, S. Omura, R.R. Kopito, Degradation of CFTR by the ubiquitin–proteasome pathway, Cell 83 (1995) 121–127.

102 T.J. Jensen, M.A. Loo, S. Pind, D.B. Williams, A.L. Goldberg, J.R. Riordan, Multiple proteolytic systems, including the proteasome, contribute to CFTR processing, Cell 83 (1995) 129–135.

103 M.S. Gelman, E.S. Kannegaard, R.R. Kopito, A principal role for the proteasome in endoplasmic reticulum-associated degradation of misfolded intracellular cystic fibrosis transmembrane conductance regulator, J. Biol. Chem. 277 (2002) 11709–11714.

104 Z. Kostova, D.H. Wolf, For whom the bell tolls: protein quality control of the endoplasmic reticulum and the ubiquitin–proteasome connection, EMBO J. 22 (2003) 2309–2317.

105 S. Sato, C.L. Ward, M.E. Krouse, J.J. Wine, R.R. Kopito, Glycerol reverses the misfolding phenotype of the most common cystic fibrosis mutation, J. Biol. Chem. 271 (1996) 635–638.
106 B.H. Qu, P.J. Thomas, Alteration of the cystic fibrosis transmembrane conductance regulator folding pathway, J. Biol. Chem. 271 (1996) 7261–7264.
107 E. Strickland, B.H. Qu, L. Millen, P.J. Thomas, The molecular chapeorne Hsc70 assists the *in vitro* folding of the N-terminal nucleotide-binding domain of the cystic fibrosis transmembrane conductance regulator, J. Biol. Chem. 272 (1997) 25421–25424.
108 J.A. Johnston, C.L. Ward, R.R. Kopito, Aggresomes: a cellular response to misfolded proteins, J. Cell Biol. 143 (1998) 1883–98.
109 R.R. Kopito, Aggresomes, inclusion bodies and protein aggregation, Trends Cell Biol. 10 (2000) 524–530.
110 S. Waelter, A. Boeddrich, R. Lurz, E. Scherzinger, G. Lueder, H. Lehrach, E.E. Wanker, Accumulation of mutant huntingtin fragments in aggresome-like inclusion bodies as a result of insufficient protein degradation, Mol. Biol. Cell 12 (2001) 1393–1407.
111 R. Garcia-Mata, Z. Bebok, E.J. Sorscher, E.S. Sztul, Characterization and dynamics of aggresome formation by a cytosolic GFP-chimera, J. Cell Biol. 146 (1999) 1239–1254.
112 K.S. McNaught, P. Shashidharan, D.P. Perl, P. Jenner, C.W. Olanow, Aggresome-related biogenesis of Lewy bodies, Eur. J. Neurosci. 16 (2002) 2136–2148.
113 Y. Kawaguchi, J.J. Kovacs, A. McLaurin, J.M. Vance, A. Ito, T.P. Yao, The deacetylase HDAC6 regulates aggresome formation and cell viability in response to misfolded protein stress, Cell 115 (2003) 727–738.
114 R. Garcia-Mata, Z. Bebok, E.J. Sorscher, E.S. Sztul, Characterization and dynamics of aggresome formation by a cytosolic GFP-chimera, J. Cell Biol. 146 (1999) 1239–1254.
115 W.C. Wigley, R.P. Fabunmi, M.G. Lee, C.R. Marino, S. Muallem, G.N. DeMartino, P.J. Thomas, Dynamic association of proteasomal machinery with the centrosome, J. Cell Biol. 145 (1999) 481–490.
116 A. Wyttenbach, J. Carmichael, J. Swartz, R.A. Furlong, Y. Narain, J. Rankin, D.C. Rubinsztein, Effects of heat shock, heat shock protein 40 (HDJ-2), and proteasome inhibition on protein aggregation in cellular models of Huntington's disease, Proc. Natl. Acad. Sci. USA 97 (2000) 2898–2903.
117 J.L. Dul, D.P. Davis, E.K. Williamson, F.J. Stevens, Y. Argon, Hsp70 and antifibrillogenic peptides promote degradation and inhibit intracellular aggregation of amyloidogenic light chains, J. Cell Biol. 152 (2001) 705–716.
118 H. Lelouard, E. Gatti, F. Cappello, O. Gresser, V. Camosseto, P. Pierre, Transient aggregation of ubiquitinated proteins during dendritic cell maturation, Nature 417 (2002) 177–182.
119 H. Lelouard, V. Ferrand, D. Marguet, J. Bania, V. Camosseto, A. David, E. Gatti, P. Pierre, Dendritic cell aggresome-like induced structures are dedicated areas for ubiquitination and storage of newly synthesized defective proteins, J. Cell Biol. 164 (2004) 667–675.
120 W. Xu, M. Marcu, X. Yuan, E. Mimnaugh, C. Patterson, L. Neckers, Chaperone-dependent E3 ubiquitin ligase CHIP mediates a degradative pathway for c-ErbB2/Neu, Proc. Natl. Acad. Sci. USA 99 (2002) 12847–12852.
121 S. Wickner, M.R. Maurizi, S. Gottesman, Posttranslational quality control: folding, refolding, and degrading proteins, Science 286 (1999) 1888–1893.
122 C.-W. Liu, M.J. Corboy, G.N. DeMartino, P.J. Thomas, Endoproteolytic activity of the proteasome, Science 299 (2003) 408–411.
123 W. Seufert, S. Jentsch, Ubiquitin-conjugating enzymes UBC4 and UBC5 mediate selective degradation of

short-lived and abnormal proteins, EMBO J. 9 (1990) 543–550.

124 B. Bercovich, I. Stancovski, A. Mayer, N. Blumenfeld, A. Laszlo, A.L. Schwartz, A. Ciechanover, Ubiquitin-dependent degradation of certain protein substrates *in vitro* requires the molecular chaperone Hsc70, J. Biol. Chem. 272 (1997) 9002–9010.

125 P. Connell, C.A. Ballinger, J. Jiang, Y. Wu, L.J. Thompson, J. Höhfeld, C. Patterson, The co-chaperone CHIP regulates protein triage decisions mediated by heat-shock proteins, Nat. Cell Biol. 3 (2001) 93–96.

126 L. Aravind, E.V. Koonin, The U box is a modified RING finger – a common domain in ubiquitination, Curr. Biol. 10 (2000) R132–R134.

127 S. Hatakeyama, M. Yada, M. Matsumoto, N. Ishida, K.I. Nakayama, U box proteins as a new family of ubiquitin-protein ligases, J. Biol. Chem. 276 (2001) 33111–33120.

128 E. Pringa, G. Martinez-Noel, U. Müller, K. Harbers, Interaction of the ring finger-related U-box motif of a nuclear dot protein with ubiquitin-conjugating enzymes, J. Biol. Chem. 276 (2001) 19617–19623.

129 R. Nikolay, T. Wiederkehr, W. Rist, G. Kramer, M.P. Mayer, B. Bukau, Dimerization of the human E3 ligase CHIP via a coiled-coil domain is essential for its activity, J. Biol. Chem. 279 (2004) 2673–2678.

130 Q. Dai, C. Zhang, Y. Wu, H. McDonough, R.A. Whaley, V. Godfrey, H.H. Li, N. Madamanchi, W. Xu, L. Neckers, D. Cyr, C. Patterson, CHIP activates HSF1 and confers protection against apoptosis and cellular stress, EMBO J. 22 (2003) 5446–5458.

131 J. Demand, J. Lüders, J. Höhfeld, The carboxyl-terminal domain of Hsc70 provides binding sites for a distinct set of chaperone cofactors, Mol. Cell. Biol. 18 (1998) 2023–2028.

132 A. Brinker, C. Scheufler, F. Von Der Mulbe, B. Fleckenstein, C. Herrmann, G. Jung, I. Moarefi, F.U. Hartl, Ligand discrimination by TPR domains. Relevance and selectivity of EEVD-recognition in Hsp70 × Hop × Hsp90 complexes, J. Biol. Chem. 277 (2002) 19265–19275.

133 H.H. Kampinga, B. Kanon, F.A. Salomons, A.E. Kabakov, C. Patterson, Overexpression of the cochaperone CHIP enhances Hsp70-dependent folding activity in mammalian cells, Mol. Cell. Biol. 23 (2003) 4948–4958.

134 H. Shimura, D. Schwartz, S.P. Gygi, K.S. Kosik, CHIP-Hsc70 complex ubiquitinates phosphorylated tau and enhances cell survival, J. Biol. Chem. 279 (2004) 4869–4876.

135 J. Jiang, C.A. Ballinger, Y. Wu, Q. Dai, D.M. Cyr, J. Höhfeld, C. Patterson, CHIP is a U-box-dependent E3 ubiquitin ligase: identification of Hsc70 as a target for ubiquitylation, J. Biol. Chem 276 (2001) 42938–42944.

136 S. Murata, Y. Minami, M. Minami, T. Chiba, K. Tanaka, CHIP is a chaperone-dependent E3 ligase that ubiquitylates unfolded protein, EMBO Rep. 2 (2001) 1133–1138.

137 J. Demand, S. Alberti, C. Patterson, J. Höhfeld, Cooperation of a ubiquitin domain protein and an E3 ubiquitin ligase during chaperone/proteasome coupling, Curr. Biol. 11 (2001) 1569–1577.

138 P.K. Jackson, A.G. Eldridge, E. Freed, L. Furstenthal, J.Y. Hsu, B.K. Kaiser, J.D. Reimann, The lore of the RINGs: substrate recognition and catalysis by ubiquitin ligases, Trends Cell. Biol. 10 (2000) 429–439.

139 P.K. Jackson, A.G. Eldridge, The SCF ubiquitin ligase: an extended look, Mol. Cell 9 (2002) 923–925.

140 N. Zheng, B.A. Schulman, L. Song, J.J. Miller, P.D. Jeffrey, P. Wang, C. Chu, D.M. Koepp, S.J. Elledge, M. Pagano, R.C. Conaway, J.W. Conaway, J.W. Harper, N.P. Pavletich, Structure of the Cul1-Rbx1-Skp1-F boxSkp2 SCF ubiquitin ligase complex, Nature 416 (2002) 703–709.

141 J. Lüders, J. Demand, J. Höhfeld,

The ubiquitin-related BAG-1 provides a link between the molecular chaperones Hsc70/Hsp70 and the proteasome, J. Biol. Chem. 275 (2000) 4613–4617.

142 S. Jentsch, G. Pyrowolakis, Ubiquitin and its kin: how close are the family ties? Trends Cell. Biol. 10 (2000) 335–342.

143 S. Alberti, J. Demand, C. Esser, N. Emmerich, H. Schild, J. Höhfeld, Ubiquitylation of BAG-1 suggests a novel regulatory mechanism during the sorting of chaperone substrates to the proteasome, J. Biol. Chem. 277 (2002) 45920–45927; Erratum in J. Biol. Chem. 278 (2003) 15702–15703.

144 Q. Deveraux, V. Ustrell, C. Pickart, M. Rechsteiner, A 26 S protease subunit that binds ubiquitin conjugates, J. Biol. Chem. 269 (1994) 7059–7061.

145 P. Young, Q. Deveraux, R.E. Beal, C.M. Pickart, M. Rechsteiner, Characterization of two polyubiquitin binding sites in the 26 S protease subunit 5a, J. Biol. Chem. 273 (1998) 5461–5467.

146 H. Hiyama, M. Yokoi, C. Masutani, K. Sugasawa, T. Maekawa, K. Tanaka, J.H. Hoeijmakers, F. Hanaoka, Interaction of hHR23 with S5a. The ubiquitin-like domain of hHR23 mediates interaction with S5a subunit of 26 S proteasome, J. Biol. Chem. 274 (1999) 28019–28025.

147 K.J. Walters, M.F. Kleijnen, A.M. Goh, G. Wagner, P.M. Howley, Structural studies of the interaction between ubiquitin family proteins and proteasome subunit S5a, Biochemistry 41 (2002) 1767–1777.

148 Y.A. Lam, T.G. Lawson, M. Velayutham, J.L. Zweier, C.M. Pickart, A proteasomal ATPase subunit recognizes the polyubiquitin degradation signal, Nature 416 (2002) 763–767.

149 S. Elsasser, R.R. Gali, M. Schwickart, C.N. Larsen, D.S. Leggett, B. Muller, M.T. Feng, F. Tubing, G.A. Dittmar, D. Finley, Proteasome subunit Rpn1 binds ubiquitin-like protein domains, Nat. Cell Biol. 4 (2002) 725–730.

150 Y. Saeki, T. Sone, A. Toh-e, H. Yokosawa, Identification of ubiquitin-like protein-binding subunits of the 26S proteasome, Biochem. Biophys. Res. Commun. 296 (2002) 813–819.

151 M.H. Glickman, D.M. Rubin, O. Coux, I. Wefes, G. Pfeifer, Z. Cjeka, W. Baumeister, V.A. Fried, D. Finley, A subcomplex of the proteasome regulatory particle required for ubiquitin-conjugate degradation and related to the COP9-signalosome and eIF3, Cell 94 (1998) 615–623.

152 Y. Xie, A. Varshavsky, UFD4 lacking the proteasome-binding region catalyses ubiquitination but is impaired in proteolysis, Nat. Cell Biol. 4 (2002) 1003–1007.

153 S. Takayama, J.C. Reed, Molecular chaperone targeting and regulation by BAG family proteins, Nat. Cell Biol. 3 (2001) E237–241.

154 K. Thress, J. Song, R.I. Morimoto, S. Kornbluth, Reversible inhibition of Hsp70 chaperone function by Scythe and Reaper, EMBO J. 20 (2001) 1033–1041.

155 S. Alberti, K. Böhse, V. Arndt, A. Schmitz, J. Höhfeld, The co-chaperone HspBP1 inhibits the CHIP ubiquitin ligase and stimulates the maturation of the cystic fibrosis transmembrane conductance regulator, Mol. Biol. Cell (2004), in press.

156 D.A. Raynes, V. Guerriero, Inhibition of Hsp70 ATPase activity and protein renaturation by a novel Hsp70-binding protein, J. Biol. Chem. 273 (1998) 32883–32888.

157 M. Kabani, C. McLellan, D.A. Raynes, V. Guerriero, J.L. Brodsky, HspBP1, a homologue of the yeast Fes1 and Sls1 proteins, is an Hsc70 nucleotide exchange factor, FEBS Lett. 531 (2002) 339–342.

158 C.P. Cardozo, C. Michaud, M.C. Ost, A.E. Fliss, E. Yang, C. Patterson, S.J. Hall, A.J. Caplan, C-terminal Hsp-interacting protein slows androgen receptor synthesis and

reduces its rate of degradation, Arch. Biochem. Biophys. 410 (2003) 134–140.
159 J. Jiang, D. Cyr, R.W. Babbitt, W.C. Sessa, C. Patterson, Chaperone-dependent regulation of endothelial nitric-oxide synthase intracellular trafficking by the co-chaperone/ubiquitin ligase CHIP, J. Biol. Chem. 278 (2003) 49332–49341.
160 Y. Imai, M. Soda, S. Hatakeyama, T. Akagi, T. Hashikawa, K.I. Nakayama, R. Takahashi, CHIP is associated with Parkin, a gene responsible for familial Parkinson's disease, and enhances its ubiquitin ligase activity, Mol. Cell 10 (2002) 55–67.
161 Y. Nagano, H. Yamashita, T. Takahashi, S. Kishida, T. Nakamura, E. Iseki, N. Hattori, Y. Mizuno, A. Kikuchi, M. Matsumoto, Siah-1 facilitates ubiquitination and degradation of synphilin-1, J. Biol. Chem. 278 (2003) 51504–51514.
162 E. Liani, A. Eyal, E. Avraham, R. Shermer, R. Szargel, D. Berg, A. Bornemann, O. Riess, C.A. Ross, R. Rott, S. Engelender, Ubiquitylation of synphilin-1 and alpha-synuclein by SIAH and its presence in cellular inclusions and Lewy bodies imply a role in Parkinson's disease, Proc. Natl. Acad. Sci. USA 101 (2004) 5500–5505.
163 S. Matsuzawa, S. Takayama, B.A. Froesch, J.M. Zapata, J.C. Reed, p53-inducible homologue of Drosophila seven in absentia (Siah) inhibits cell growth: suppression by BAG-1, EMBO J. 17 (1998) 2736–2747.
164 C. Soti, P. Csermely, Aging and molecular chaperones, Exp. Gerontol. 38 (2003) 1037–1040.
165 J. Fargnoli, T. Kunisada, A.J. Jr. Fornace, E.L. Schneider, N.J. Holbrook, Decreased expression of heat shock protein 70 mRNA and protein after heat tretament in cells of aged rats, Proc. Natl. Acad. Sci. USA 87 (1990) 846–850.
166 C.K. Lee, R.G. Klopp, R. Weindruch, T.A. Prolla, Gene expression profile of aging and its retardation by caloric restriction, Science 285 (1999) 1390–1393.
167 S.X. Cao, J.M. Dhahbi, P.L. Mote, S.R. Spindler, Genomic profiling of short- and long-term caloric restriction effects in the liver of aging mice, Proc. Natl. Acad. Sci. USA 98 (2001) 10630–10635.
168 M. Tatar, A.A. Khazaeli, J.W. Curtsinger, Chaperoning extended life, Nature 390 (1997) 30.
169 G. Morrow, R.M. Tanguay, Heat shock proteins and aging in Drosophila melanogaster, Semin. Cell. Dev. Biol. 14 (2003) 291–299.
170 J.F. Morley, R.I. Morimoto, Regulation of longevity in Caenorhabditis elegans by heat shock factor and molecular chaperones, Mol. Biol. Cell 15 (2004) 657–664.
171 A.L. Hsu, C.T. Murphy, C. Kenyon, Regulation of aging and age-related disease by DAF-16 and heat-shock factor, Science 300 (2003) 1142–1145.

2 Molecular Dissection of Autophagy in the Yeast *Saccharomyces cerevisiae*

Yoshinori Ohsumi

2.1 Introduction

More than half a century has passed since C. de Duve discovered lysosomes using cell fractionation procedures [1]. At that time, intracellular bulk protein degradation was believed to occur mostly within this organelle. Eukaryotic cells must elaborate a strategy to segregate dangerous lytic enzymes from biosynthetic sites and cytosol and to restrict the degradative process to a membrane-bound compartment. The process of degradation of cytoplasmic components in lysosomes is called autophagy, in contrast to heterophagy, which is the degradation of extracellular materials through endocytosis. Electron microscopic studies on lysosomes revealed macroautophagy (hereafter referred to as autophagy) as a major route to deliver the cytoplasmic components to the lytic compartment. The first step of autophagy is sequestration of a portion of the cytoplasm or organelle by a membrane sac, the so-called isolation membrane, resulting a double-membrane structure called the autophagosome. Then the autophagosome fuses with the lysosome, gains lytic enzymes, and turns into an autophagolysosome. Lysosomal enzymes disintegrate the inner membrane of the autophagosome and digest its contents. Digestion products are transported back to the cytosol and reutilized for a new round of protein synthesis.

Autophagy is involved in nonselective and bulk degradation of cellular proteins, while the ubiquitin–proteasome system is responsible for the highly selective degradation of short-lived proteins. Since more than 90% of cellular proteins have long lifetimes, the turnover of long-lived proteins is important to the understanding of cell physiology.

Until recently, autophagy in mammals had been studied mostly using electron microscopy by detecting autophagosomes and autophagolysosomes. Since the lysosomal system consists of very dynamic and complicated membrane structures, it was not easy to analyze lysosomes and their related membrane structures biochemically. Many efforts to detect specific proteins on the autophagosome failed, and genes required for autophagy had not been identified.

Protein Degradation, Vol. 2: The Ubiquitin-Proteasome System.
Edited by R. J. Mayer, A. Ciechanover, M. Rechsteiner

ISBN: 3-527-31130-0

In this chapter, I will focus on the recent progress in the molecular dissection of autophagy in the yeast *Saccharomyces cerevisiae* and its relevance in understanding autophagic protein degradation in higher eukaryotes.

2.2 Vacuoles as a Lytic Compartment in Yeast

The vacuole is the most prominent organelle and is easily visible under light microscopy in the budding yeast *S. cerevisiae*. The inside of the vacuole is kept acidic by a V-type proton-translocating ATPase on the vacuolar membrane. The vacuole plays crucial roles in homeostasis of cellular ions and functions as a reservoir of various metabolites such as amino acids. It contains hydrolytic enzymes, proteinases, peptidases, nucleases, phosphatases, mannosidases, and so on. The vacuolar enzymes and their biogenesis have been intensively studied genetically and biochemically. From these facts the vacuole was postulated to function as a lytic compartment like lysosomes in mammalian cells. Actually, it was reported that bulk protein turnover is induced upon nitrogen starvation, which is dependent upon vacuolar enzyme activities [2, 3], suggesting that the vacuole is responsible for bulk protein degradation. Obvious questions were what kind of intracellular proteins are degraded and how they become accessible to the vacuolar enzymes.

2.3 Discovery of Autophagy in Yeast

In 1988, I started studies on the lytic function of yeast vacuoles and found by light microscopy that the yeast cell induces autophagy under nitrogen-starvation conditions. When vacuolar proteinase-deficient mutants grown in a rich medium were shifted to a nitrogen-deprived medium, spherical structures appeared in the vacuole after a short lag, gradually increased in number, and finally filled the vacuole after 10 hours [4]. These structures, called autophagic bodies, were mostly single membrane-bound, occasionally multilamellar structures containing a portion of cytoplasm [4]. Autophagic bodies contained ribosomes and occasionally various other cellular structures including mitochondria and rER [4]. Subsequently, double-membrane structures of a size equivalent to autophagic bodies, autophagosomes, were found in the cytoplasm of the starved cells. Fusion images between the outer membrane of the autophagosome and the vacuolar membrane were obtained by rapid freezing and freeze substitution, as well as by freeze-fracture electron microscopy [5, 6]. The autophagic body is the final membrane structure of autophagy in yeast, which is derived from the inner membrane of the autophagosome. Autophagic bodies are about 300–900 nm in diameter, about 500 nm on average, and deliver about 0.2% of the cytoplasm via one autophagosome. The rate of autophagic protein degradation was estimated at 2–3% total cellular protein per hour. Autophagy in a haploid strain linearly proceeds for up to eight hours, gradually slows down, and reaches a plateau at around 20–30% of degradation [7]. Therefore, there must be negative regulation, but its details are not known yet.

Later, we realized that exactly similar membrane phenomena were induced under carbon, sulfate, phosphate, and single auxotrophic amino acid starvation. These observations strongly indicate that yeast cells take up a portion of cytoplasm to the lytic compartment via autophagosomes in conditions adverse for growth. The membrane dynamics of yeast autophagy is topologically the same as macroautophagy in mammals, though the vacuole is much larger than the lysosome. A schematic drawing of autophagy in yeast is shown in Figure 2.1.

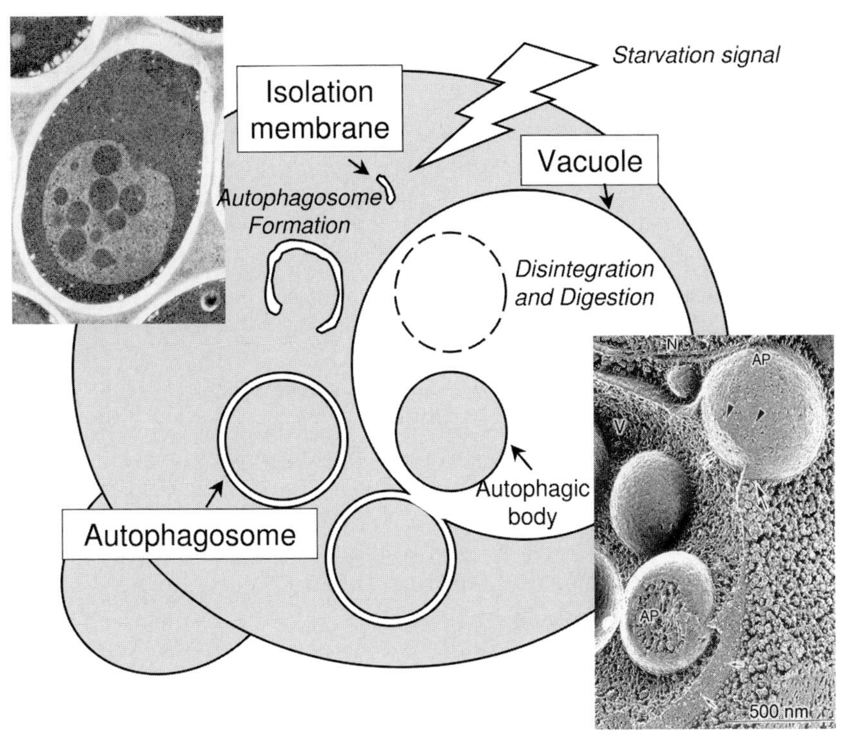

Fig. 2.1. Scheme of autophagy in the yeast *Sac-charomyces cerevisiae*. When yeast cells face various nutrient deficiencies, the isolation membrane encloses a portion of the cytosol and forms a double-membrane structure: the autophagosome (AP). Autophagosomes immediately fuse with the vacuole, and an inner-membrane structure, the autophagic body (AB), is released into the vacuolar lumen. In wild-type cells autophagic bodies are degraded by vacuolar enzymes, but as shown in EM (top left) autophagic bodies are accumulated in vacuolar proteinase-deficient cells. The freeze-fracture image (bottom right) clearly shows fusion of the autophagosome and unique characteristics of the autophagic body.

2.4
Genetic Dissection of Autophagy

To elucidate the molecular mechanism of autophagy, we applied a genetic approach. The most characteristic feature of yeast autophagy is that we are able to monitor the progress of autophagy under the light microscope as the accumulation of autophagic bodies. Taking advantage of this simple technique, we attempted to obtain autophagy-defective mutants. As the first approach, only the morphological changes of the vacuole under starvation were used to screen for mutants. Cells that failed to accumulate autophagic bodies during nitrogen starvation in *pep4* background, deficient of vacuolar enzymes, were selected under light microscopy, and only one mutant, *apg1*, was selected [8]. The *apg1* mutant did not induce bulk protein degradation under starvation, and homozygous *apg1/apg1* diploid cells did not sporulate. This mutant grew normally in a rich medium but could not maintain viability under long nitrogen starvation. To obtain further mutants due to defects in autophagy, the loss of viability under starvation was used for the first screen followed by the morphological examination of vacuoles. In this way about 100 autophagy-defective mutants were isolated and divided into 14 groups (*apg2–apg15*) by complementation analysis. Another approach taken by Thumm and coworkers was the immunoscreening of cells that retain a cytosolic enzyme, fatty acid synthase, after starvation [9]. By these methods six *aut* mutants were originally obtained. Later, two hybrid screens using Apg proteins as bait identified two more *APG* genes [10, 11]. Klionsky's group isolated mutants defective in maturation of aminopeptidase I (API), one of the vacuolar enzymes. API is first synthesized as a proform in the cytosol and then transported to the vacuole and processed to an active form. Most other vacuolar enzymes are incorporated into the ER lumen and transported to the vacuole through the secretory pathway. Transport of API to the vacuole is mediated by the Cvt (cytoplasm-to-vacuole targeting) pathway [12]. The defective mutants in the Cvt pathway, *cvt*, significantly overlapped with autophagy-defective *apg* and *aut* mutants [7, 12], though the two pathways are apparently different; one is degradative and starvation-induced, and the other is biosynthetic and constitutively active. EM analyses of the Cvt pathway clearly showed that the Cvt pathway is mediated by membrane dynamics that is quite similar to that of autophagy [13, 14]. Small double-membrane structures (the Cvt vesicles) specifically enclose an aggregate of API (Cvt complex) and fuse with the vacuolar membrane, releasing small vesicles into the vacuolar lumen.

Later, many groups isolated autophagy-related genes in *S. cerevisiae* and other yeast species and named them differently. To avoid confusion, recently all groups involved agreed to use a novel nomenclature for the autophagy-related gene: *ATG*. The original *APGx* is now renamed as *ATGx* [15]. The genes presently known to be involved in autophagy are shown in Table 2.1.

Table 2.1. Nomenclature of autophagy-related genes and functions of the Apg, Aut, Cvt, and Gsa proteins and mammalian homologues.

Atg	*Apg*	*Aut*	*Cvt*	*Gsa*	*Mammalian*	*Function/localization*
Atg1	Apg1	Aut3	Cvt10	Gsa10	ULK1	Protein kinase, localizes to the PAS
Atg2	Apg2	Aut8		Gsa11	Apg2	Localizes to the PAS
Atg3	Apg3	Aut1		Gsa20	Apg3	Apg8-conjugating enzyme (E2)
Atg4	Apg4	Aut2			Apg4A, Apg4B	Cysteine protease for processing the C-terminus of Apg8
Atg5	Apg5				Apg5	Substrate of Apg12-conjugating reaction, localizes to the PAS
Atg6	Apg6				Beclin-1	Subunit of the PI3-kinase complex, involved in protein sorting to the vacuole as Vps30
Atg7	Apg7		Cvt2	Gsa7	Apg7	Activating enzyme (E1) of Apg8 and Apg12
Atg8	Apg8	Aut7	Cvt5		LC3, GATE16, GABARAP	Ubiquitin-like protein, conjugates with PE, localizes to the PAS and autophagosomes
Atg9	Apg9	Aut9	Cvt7	Gsa14	Apg9?	Transmembrane protein, required for PAS formation
Atg10	Apg10				Apg10	Apg12-conjugating enzyme (E2)
Atg12	Apg12				Apg12	Ubiquitin-like protein, conjugates with Apg5
Atg13	Apg13				?	Subunit of Apg1 kinase, phosphorylated under growing conditions
Atg14	Apg14		Cvt12		?	Subunit of the autophagy-specific PI3-kinase complex
Atg16	Apg16		Cvt11		Apg16L	Binds with Apg12–Apg5 and forms tetramer, required for Apg12–Apg5 recruitment to the PAS, localizes to the PAS
Atg17	Apg17				?	Member of the Apg1 complex, not required for the Cvt pathway
Atg18		Aut10	Cvt18	Gsa12		WD-repeat protein
Atg22		Aut4				Disintegration of autophagic bodies in the vacuole
Atg15		Aut5	Cvt17			Disintegration of autophagic bodies in the vacuole, putative lipase
Atg11			Cvt9	Gsa9		Required only for the Cvt pathway, localizes to the PAS
Atg19			Cvt19			Receptor of aminopeptidase I for the Cvt vesicle, localizes to the PAS
Atg20			Cvt20			Binds to PI3P, required for the Cvt pathway
Atg21			Mai1			Required for the Cvt pathway but not for macroautophagy

PAS: pre-autophagosomal structure

2.5 Characterization of Autophagy-defective Mutants

Autophagy-defective mutants had been isolated as non-conditional mutants, and we now know that almost all of the original *apg* mutants are null-type mutants. They failed to induce bulk protein degradation under various nutrient-depleted conditions, indicating that autophagy is the major pathway of bulk protein degradation. All *apg* and most *aut* mutants grow normally in a nutrient-rich medium, indicating that autophagy is not essential for vegetative growth in yeast. They showed no significant differences in stress responses against heat, osmotic, and salt stress. Several vacuolar functions tested in these autophagy-defective mutants, including secretion and endocytosis, were almost the same as in wild-type cells. One of characteristic features of autophagy-defective mutants is the loss of viability during nitrogen starvation, which was used as screening marker. Autophagy-defective mutants start to die after two days of starvation and almost completely lose viability after one week [8]. Under starvation conditions the cell needs to synthesize essential proteins to adapt to the conditions; consequently, the supply of amino acids by degradation is essential. In nature, yeast cells must face various forms of nutrient starvation; therefore, autophagy-defective mutants may not survive.

Homozygous diploids with any *apg* mutation have been shown not to sporulate [8]. This cell-differentiation process triggered by nitrogen starvation must require bulk protein degradation via autophagy in order to remodel the intracellular structures. Degradation of preexisting proteins by autophagy must be critical for cell survival.

2.6 Cloning of *ATG* Genes

Recently, we finished cloning all of the original *APG* genes. Most genes were cloned from a chromosomal DNA library by complementation of the loss-of-viability phenotype of *apg* mutants by replica plating on agar medium containing phloxine B, which stains dead cells red, then confirmed by the accumulation of autophagic bodies by light microscopy. Some genes were obtained by complementation of sporulation-negative phenotypes as the first step of the screen. The first *ATG* gene cloned, *ATG1*, turned out to encode a Ser/Thr protein kinase [16]. However, since then almost all *ATG* genes have been unidentified genes with unpredictable functions from their sequence data. Autophagy genes had been neglected because they exhibit specific phenotypes only under starvation conditions. Recent systematic analyses of protein interactions by yeast two-hybrid screens or binding assays also clearly indicate that Atg proteins interact with each other but compose an isolated group of proteins.

Autophagy genes turn out to be mostly novel genes, except for *ATG6*, which is required for the vacuolar protein-sorting (Vps) pathway [17]. In yeast, autophagy is almost completely shut off under growing conditions and is strictly induced by starvation, but every *ATG* gene is rather constitutively expressed in the growing conditions. Systematic gene expression analyses suggested that several *ATG* genes are transcriptionally upregulated. However, the protein level of most Atg proteins is not dramatically changed by nutrient conditions. It is unclear whether transcriptional regulation plays an important role in the regulation of autophagy in yeast or not.

2.7 Further Genes Required for Autophagy

Screens for autophagy-defective mutants, like the original *apg* mutants, seem to be nearly finished. However, because of the strategies of screens, mutants with aberrant vacuole morphology, partially defective mutants, and mutants of genes shared with other essential functions were eliminated. It has now become obvious that normal levels of autophagy require more genes of known function as well as unknown genes. Most Gcn proteins appear to be necessary for normal autophagy. Several early *SEC* genes such as *SEC12* an *SEC24* are known to be necessary for autophagy [18, 19]. Several mutants such as *vps35/vam5* and *ypt7* show accumulation of autophagosomes in the cytoplasm under starvation conditions. These mutant cells contain fragmented vacuoles, suggesting that the fusion machinery of the autophagosome with the vacuole shares SNARE molecules similar to other vacuolar homotypic fusion events [18]. In wild-type cells autophagic bodies effectively disappear within one minute. Atg15/Cut5/Cvt17 and Aut4 are involved in this process [20, 21]. Atg15 contains a putative lipase domain, but lipase activity has not been proved yet. Acidification of the vacuole is a requisite for effective digestion of autophagic bodies, since defects in every subunit of the type-ATPase (Vma) cause an accumulation of autophagic bodies in the vacuole [22]. It is still a mystery why the autophagic body membrane disintegrates so quickly in the vacuoles.

2.8 Selectivity of Proteins Degraded

One of the unresolved problems of autophagy is substrate selectivity for sequestration into autophagosomes. Generally, autophagic protein degradation is believed to be nonselective. We showed that isolated vacuoles containing autophagic bodies exhibit similar rates of sequestration of the cytosolic enzymes ADH, PGK, PK, and GluDH. The density of ribosomes is almost the same in autophagic bodies as in the cytoplasm. Immunoelectron microscopy showed the same signal intensities of ADH or PGK among cytosol, autophagosomes, and autophagic bodies [5]. This suggests that sequestration by autophagosomes is a nonselective process at least for these cytosolic energy-metabolism enzymes.

One biosynthetic pathway for vacuolar proteins, the Cvt pathway, has been studied intensively. Aminopeptidase I (Lap4) and α-mannosidase (Ams1) are delivered to the vacuoles via the Cvt pathway, which utilizes all the original *APG* genes. Furthermore, under starvation conditions API and α-mannosidase are selectively sequestered into autophagosomes and delivered to the vacuoles. This selective uptake of API to the Cvt vesicle or autophagosome requires the specific factors Atg11/Cvt9 and Atg19/Cvt19, which may specify the recruitment of the cargo to the vesicles. These facts evoke the possibility that autophagy may involve selective transport of certain proteins to be degraded. Recently, we realized that the cytosolic acetaldehyde dehydrogenase Ald6 is preferentially degraded in the vacuoles via autophagy [23]. The mechanism governing this preferential sequestration into the autophagosome is an interesting problem to be unveiled in the near future.

Glycogen granules, the synthesis of which is induced by nitrogen starvation in the presence of glucose, are mostly excluded from the autophagosomes. It is possible that there is a mechanism for excluding glycogen granules from autophagosomes, but an alternative explanation is that the site of autophagosome formation results in this unevenness, since glycogen granules mostly locate to peripheral regions of the cytosol while autophagosomes form next to the vacuole.

We have occasionally detected mitochondria in autophagic bodies, and we can easily estimate the number of mitochondria taken up in the vacuole by counting mitochondrial DNA in the autophagic bodies with fluorescence microscopy after DAPI staining [4]. Under starvation, a significant proportion of the mitochondria are transported to the vacuoles. Rough ER is also frequently detected in the vacuole [4]. Autophagy may provide the most effective system to degrade whole organelles. Therefore, it has been proposed that autophagy regulates the quantity of organelles and is even involved in their quality control. In *S. cerevisiae* it is hard to conclude whether or not organelle degradation has some selectivity.

In *Pichia pastoris*, cells grown in methanol medium develop large numbers of peroxisomes. When the medium is switched to ethanol or glucose, peroxisomes are selectively degraded via microautophagy and macroautophagy, respectively [24]. Microautophagy is the process of direct wrapping of peroxisomes by the vacuolar membrane. These two pathways seem to be quite different membrane phenomena; however, both require many Atg proteins [25, 26]. Recently, it was shown that micropexophagy (degradation of peroxisomes) is not simply the process of invagination of vacuolar membranes but requires formation of a novel membrane, called MIPA [27]. It is known that several genes are required for specific peroxisome degradation by autophagy. These may confer the molecular mechanism of selective sequestration of this organelle.

2.9
Induction of Autophagy

Under growing conditions, the extent of autophagy is negligibly small. Cells growing in a rich medium are adapted to rapid cell proliferation. High cAMP levels block autophagy, and activated A-kinase mutants do not induce autophagy [28], indicating that autophagy is regulated in an opposite manner to cell growth. Autophagy is induced not only by nitrogen starvation but also by other nutrient starvation, including carbon, sulfate, phosphate, and amino acids [4]. So far, there is no mutant identifying a specific starvation signal. Autophagy is a rather general physiological response to nutrient limitation and may be under the control of several general factors.

However, when rapamycin, a specific inhibitor of Tor kinase, is added to the growing cells in a nutrient-rich medium, cells behave as if they were in a starvation medium and autophagy is induced [28]. Thus, Tor kinase negatively regulates autophagy during growing conditions and may be a master regulator. At present, regulation of Tor and downstream events toward autophagy is not fully understood.

2.10 Membrane Dynamics During Autophagy

The most critical step of autophagy is formation of a new compartment in order to sequester a portion of the cytoplasm to be degraded. For a long time the origin of the autophagosome membrane was proposed to be the ER. We also showed that membrane flow from the ER is necessary for autophagy in yeast [29]. Another group reported that post-Golgi transport is involved in its formation. The autophagosomal membrane has a distinct morphology: it is thinner than any other organelle membranes, and the outer and inner membranes stick together with almost no lumenal space [4, 6]. In freeze-fracture images the inner membrane – the autophagic body membrane – completely lacks intramembrane particles, while the outer membrane contains sparse but significant particles [6] that may participate in targeting and fusion of the autophagosome to the vacuole. This indicates that inner and outer membranes, which should be derived from the same isolation membrane, are somehow differentiated and specialized for delivery of cytoplasmic constituents to the lytic compartment. So far, nobody has shown that membrane vesicles are involved in the membrane elongation step of the isolation membrane. Membrane dynamics during autophagy may be quite different from classical vesicular trafficking events. We proposed that autophagosome formation is not a simple enwrapping process by a preexisting large membrane structure such as the ER, but rather assembly of a new membrane from its constituents.

By electron microscopy we could occasionally detect a cup-shaped intermediate membrane structure [5]. The most important outstanding questions are how the isolation membrane is organized, what influences the morphogenesis of this isolation membrane, and how the isolation membrane seals to form a closed autophagosome.

2.11 Monitoring Methods of Autophagy in the Yeast *S. cerevisiae*

It is not easy to precisely estimate protein degradation through autophagy, since measuring the decrease of bulk protein is technically difficult. Isotope-labeling methods are often used to estimate bulk protein degradation in yeast and mammals. Radioactive valine, leucine, or methionine and cysteine released from prelabeled long-lived proteins are measured. Reutilization of amino acids during starvation may cause underestimation. 3-Methyladenine sensitivity is used for autophagy-dependent degradation. However, in a strict sense it is not clear that 3-methyadenine is a specific inhibitor of autophagy in mammals.

In yeast the vacuole is easily visible under light microscopy. Accumulation of autophagic bodies in the presence of PMSF, a serine protease inhibitor, or in *prb1* or *pep4* strains is the simplest indication of autophagy [4]. In some strains, the accumulation of autophagic bodies is not homogenous among cells, and the number of autophagic bodies is uncountable because of their vigorous Brownian motion: quantitative measurement of autophagy is not straightforward.

Under starvation, aminopeptidase I (Lap4p, API) is mainly sequestered into the autophagosome, delivered to the vacuoles, and processed by the vacuolar enzymes to a mature form [14]. Therefore, the processed form of API during starvation reflects autophagic transport. However, the Cvt complex is transported to the vacuole at once by a single autophagosome, and API maturation is indicative of but not proportional to the extent of autophagy.

The Atg8 protein is entrapped in autophagosomes and delivered to the vacuole via autophagic bodies [30]. When cells expressing GFP-Atg8 cells are shifted to a starvation medium, GFP-Atg8 stains the vacuolar lumen in wild-type cells but not in autophagy-defective mutants, since GFP is fairly resistant to vacuolar enzymes [31]. Therefore, the intensity of fluorescence of GFP in the vacuolar lumen is a good indicator of autophagy. Nobody has analyzed the fluorescence intensity quantitatively.

We have developed a monitoring system for autophagy by genetic manipulation. Alkaline phosphatase (Pho8) is a vacuolar membrane protein with a small N-terminal cytoplasmic tail. We constructed a truncated form of Pho8 lacking a membrane-spanning region at the N-terminal end (Pho8Δ60) that is expressed under the control of a strong constitutive promoter. This proform of Pho8Δ60 is distributed in the cytosol, but under starvation a portion of it is delivered to the vacuoles via autophagy and becomes active in the vacuoles. Since Pho8 is a vacuolar resident protein, it stays stable without further degradation. This assay provides the most reliable estimation of autophagic degradation [32].

2.12
Function of Atg Proteins

As mentioned earlier, the Atg proteins turned out to be mostly novel proteins, but further analyses have revealed that the Atg proteins may be classified into four functional groups: the Atg1 protein kinase complex, the autophagy-specific PI3 kinase complex, the Atg12 protein conjugation system, and the Atg8 lipidation system (Figure 2.2). One of the most remarkable findings is the discovery of two

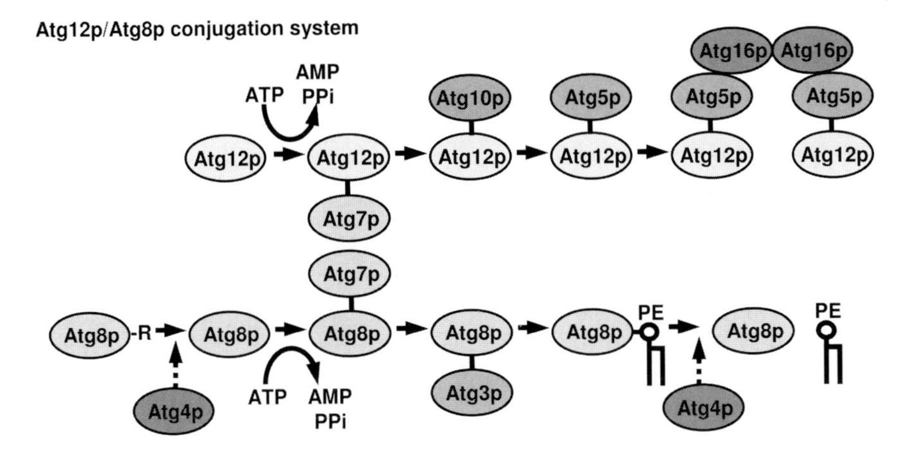

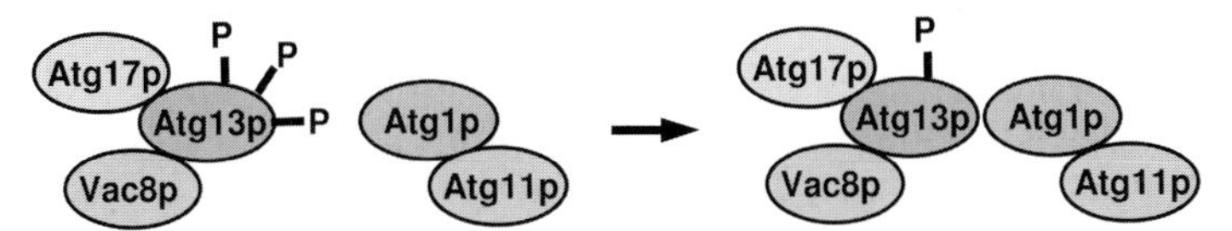

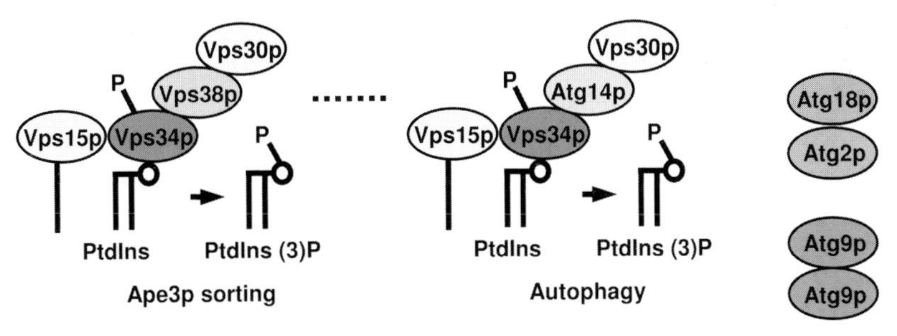

Fig. 2.2. Atg proteins necessary for autophagosome formation. A total of 16 Atg proteins are required for autophagosome formation. The proteins are found in four complexes.

ubiquitin-like conjugation systems for Atg proteins [33]. Half of the original *APG* genes are involved in these novel conjugation systems.

2.12.1
The Atg12 Protein Conjugation System

Atg12 is a small hydrophilic protein of 186 amino acids with no apparent homology to ubiquitin. Western blot analysis of N-terminally HA-tagged Atg12 showed one extra band of about 65 kDa in addition to a band of expected mass of the fusion protein. This high-molecular-mass band did not appear in *atg5*, *atg7*, and *atg10* mutants. HA-tagged Atg5 also showed two bands, and the high-molecular-mass band corresponded exactly to the upper band of Atg12, suggesting a covalent link between Atg12 and Atg5. The C-terminal glycine residue of Atg12 was essential for the Atg12–Atg5 conjugate and also autophagy. Changing the 19 lysine residues of Atg5 to arginine revealed that a lysine residue is the acceptor site of Atg12 [34]. Thus, we concluded that the C-terminal glycine at 149 residue of Atg12 forms an isopeptide bond with the ε-amino group of a lysine residue of Atg5. This conjugate formation was essential for autophagy and was mediated by consecutive reactions such as ubiquitination. The C-terminal glycine residue is first activated by an activating enzyme, Atg7 (E1), and then transferred to a conjugating enzyme, Atg10 (E2), through forming thioester conjugates [35, 36]. Atg7 exhibits a weak but significant homology with E1 enzymes of the ubiquitin system, but Atg10 is a unique E2 enzyme without any homology to known E2 enzymes. Finally, Atg12 is transferred to Atg5 through an isopeptide bond.

The Atg12 conjugation reaction is similar to ubiquitination but has distinct features. Atg12 is synthesized as an active form with a single glycine at the C-terminus, unlike other ubiquitin-like proteins (UBLs), which have C-terminal extensions after double glycine. Atg12 in yeast is much larger than ubiquitin and has no apparent sequence similarity with ubiquitin. However, its secondary structure is predicted to contain a ubiquitin-like domain at the C-terminal region. N-terminal truncation of Atg12 demonstrated that a C-terminal 80-amino-acid fragment is necessary and sufficient for conjugation and also for autophagy, indicating that Atg12 is really a UBL. Atg5 is the only target molecule of the Atg12 modification. Atg12 and Atg5 form a conjugate immediately after synthesis, and free forms of the proteins are hardly detectable in cell lysates. So far, no protease activity to cleave the Atg12–Atg5 linkage has been found, suggesting that this conjugation reaction is irreversible. Conjugate formation is not affected by autophagy-inducing starvation conditions. The Atg12–Atg5 conjugate behaves just like a single polypeptide and functions as part of the machinery of autophagosome formation.

The Atg12–Atg5 conjugate further forms a protein complex with Atg16. Although Atg16 was originally isolated by a two-hybrid screen using Atg12 as bait, it does bind to Atg5 at the N-terminal region and does not bind directly to Atg12 [10]. Atg16 has a coiled-coil region in its C-terminal half and forms an oligomer through this region. Atg12–Atg15 · Atg16 likely forms a tetrameric complex of 350 kDa, which is the functional form essential for autophagy [37]. This large protein complex is stable and is not affected by nutrient conditions. The minimum essential amount of the Atg12–Atg5 · Atg16 complex in the cell may be small, since several mutations in Atg12 or Atg16 that severely reduce the amount of the complex are still nearly normal in autophagy.

2.12.2
The Atg8 System

The *ATG8* gene encodes a small basic protein of 117 amino acids. Atg8 has many homologues in eukaryotes and is part of a large protein family. By immunostaining, Atg8 was shown to be localized to the autophagosomal and the autophagic body. Immunoelectron microscopy shows that Atg8 is localized not only on the isolation membrane and pre-autophagosome structure but also in the lumen of autophagic bodies, providing a good marker for the intermediate membranes during autophagosome formation [30].

C-terminal myc tagging revealed processing of Atg8 at the very C-terminal end. Another Atg factor, Atg4, was responsible for the process. Atg4 is a member of a novel cysteine proteinase family conserved in all eukaryotes. Mutational analyses of Atg8 indicated that Atg4 cleaves a single arginine residue from the nascent Atg8 and exposes a glycine residue at the C-terminus. Atg12 and Atg8 show significant homology in the C-terminal region. Further analyses indicated that Atg8 is also a ubiquitin-like protein that is activated by Atg7. It is then transferred to Atg3, an E2 enzyme. Thus, Atg7 is a unique E1 enzyme that activates two different UBLs, Atg12 and Atg8, and transfers them to an E2 enzyme, Atg10 and Atg3, respectively [38].

The next apparent question was, what is the target of Atg8? Atg8 shows a single band in SDS-PAGE but was realized to be in two forms; one form is loosely membrane-bound, and the other is tightly membrane-bound. The formation of the tightly membrane-bound form of Atg8 requires Atg7, Atg3, the C-terminal glycine of Atg8, and Atg4, suggesting that it is generated by a conjugation reaction. By SDS-PAGE in the presence of 6 M urea, two forms of Atg8 were found to be separable. The modified form of Atg8 showed faster mobility than Atg8 itself. Mass spectrometry of the modified form of Atg8 clearly showed that it is a covalent conjugate of Atg8 with a membrane phospholipid, phosphatidylethanolamine (PE) [39]. The fatty acids of PE were mostly palmitoyl and oleic acids, quite abundant in yeast membrane.

This indicates that ubiquitin-like modification is not restricted to protein–protein linkages but also occurs in protein–lipid linkages. Importantly, Atg8–PE formation was reversible and the processing enzyme, Atg4, played a role on this process. Thus, Atg4 is a processing enzyme and also a deconjugating enzyme [38]. The cycle of Atg8 lipidation and delipidation is necessary for normal autophagic activity.

To further understand the role of this interesting phospholipid modification, it is necessary to identify the site of Atg8–PE formation and to characterize the structures containing Atg8–PE.

The Atg12 and Atg8 conjugation systems work concertedly; not only do they share the same E1 enzyme, Atg7, but the proteins also function together because the Atg8–PE level is severely reduced in mutants of the Atg12 system, *atg5*, *atg10*, and *atg12*. Transcription of *ATG8* is known to be highly upregulated during nitrogen starvation. Certainly, Atg8 levels increase under starvation, but not so dramatically [30]. So far it is still not known whether upregulation of Atg8 is necessary for autophagy.

We do not know the precise role of lipidation reactions. Recently we succeeded in reconstituting the *in vitro* lipidation reaction using purified Atg8ΔR, Atg7, Atg3, and PE-containing liposome [40]. Further work will elucidate the molecular details of this interesting reaction system. It is still not clear whether only the modified molecule is essential for its function or whether the unmodified form still has some function. In *Pichia pastoris*, disruption of the Atg8 homologue (Paz2) shows a defect in the early stages of micropexophagy, indicating that the unprocessed form also has a physiological role [26]. In higher eukaryotes there are many homologues of Atg8, but their functions are not clear. Lipidated Atg8 homologues probably have a role during biogenesis of new membranes.

2.12.3
The Atg1 Kinase Complex

The third protein complex required for autophagy is the Atg1 protein kinase complex. Atg1 is a serine/threonine protein kinase [16, 41]. Its N-terminus region contains a protein kinase domain, and kinase activity has been detected *in vitro*. A kinase-negative Atg1 mutant could not induce autophagy, implying that kinase activity is essential for the function of the enzyme [11, 16]. The C-terminal region of Atg1 has no apparent sequence homology to other known proteins and is necessary for autophagy. Atg1 kinase activity is enhanced during induction of autophagy, and thus the level of kinase activity seems to be important for the regulation of autophagosome formation [11].

Atg1 physically interacts with Atg13, Atg17, and Cvt9. The Atg13 protein is a highly phosphorylated protein under nutrient-rich condition. Upon starvation or addition of rapamycin, a specific inhibitor of Tor kinase, it is dephosphorylated by a still-unknown phosphatase [11]. Oppositely, upon addition of nutrients to starved cells, Atg13 is rapidly hyperphosphorylated. The phosphorylation state of Atg13 is controlled by the nutrient conditions through the Tor signaling pathway. A genetic interaction exists between Atg1 and Atg13, since overproduction of Atg1 partially suppresses the autophagy defect of the *apg13* null mutant [42]. In its central region, Atg13 physically binds with Atg1 [11]. Under starvation, Atg13 is tightly associated with Atg1, while under nutrient-rich conditions, the affinity is lowered [11]. In addition, in the *atg13Δ* mutant, the kinase activity of Atg1 becomes low. These results suggest that Atg13 is a positive regulator for the Atg1 protein kinase. Transport of API is completely blocked when the *atg13* null mutant is grown in a nutrient-rich medium, but the block could be partially overcome by incubation in starvation conditions [43]. In an *atg13* mutant that lacks most of its Atg1-binding region, the Cvt pathway was normal but autophagy was completely defective [11].

Thus, Atg13 may regulate autophagy and the Cvt pathway through the Atg1 protein kinase. It is known that Atg13 also associates with Vac8 via its C-terminal region [44].

2.12.4 Autophagy-specific PI3 Kinase Complex

The fourth complex is the autophagy-specific PI3 kinase complex. Cloning and characterization of *ATG6* revealed that it is allelic to *VPS30*. Vps30/Atg6 has dual functions for vacuolar protein sorting and autophagy [45]. Atg14 is a possible coiled-coil protein and is associated with Vps30. Overexpression of Atg14 partially suppressed the autophagic defect of a truncated mutant of Vps30, but it does not suppress the defect of the deletion allele of *VPS30*. This suggests that Atg14 binds to Vps30 to exert its function for autophagy. In contrast with the *vps30* mutant, the *atg14* mutant does not show a defect in vacuolar protein sorting [45].

Later it was found that Vps30 forms two distinct protein complexes [45]. One complex of Vps30, Atg14, Vps34, and Vps15 is necessary for autophagy. The other complex of Vps30, Vps38, Vps34, and Vps15 is required for vacuolar protein sorting. These complexes share three factors. Vps34 is the sole phosphatidylinositol 3-kinase in yeast, and Vps15 is a regulatory protein kinase of Vps30. Vps30 is a possible coiled-coil protein and is peripherally membrane associated. Lack of Vps34 or Vps15 results in solubilization of Vps30. Atg14 is a specific factor in the autophagy-specific PI3-kinase complex; therefore, it may play an important role in determining the specificity of the PI3-kinase complex [46]. Atg14 is peripherally associated to an unknown membrane [45].

2.12.5 Other Atg Proteins

There are three remaining Atg proteins, Atg2, Atg9, and Atg18 that do not form a stable complex with the above factors. Their precise functions are not known yet. However, they may play important roles in linking the four reaction systems. Atg2 is a large, soluble protein [47] and has been shown to interact with Atg18. Atg9 is a putative multi-membrane-spanning protein, but its localization does not fit with known organelle markers [48].

2.13 Site of Atg Protein Functioning: The Pre-autophagosomal Structure

All *atg* mutants do not accumulate autophagosomes in the cytoplasm during starvation, indicating that all genes have functions at or before the formation of autophagosomes. So far, studies on the Atg proteins indicate that all these proteins function at the autophagosome-formation step. There are many fundamental problems to be solved. What is the origin of the autophagosome membrane? How does the membrane assemble to form the spherical structure? What is the fusion machinery for autophagosome formation and fusion with the vacuolar membrane?

As mentioned, all original Apg proteins function closely together in the autophagosome-formation step. Recently, we showed that many Atg proteins are localized to a small area close to the vacuole, called the pre-autophagosomal structure (PAS) [31]. PAS is detected by GFP-Atg8 and colocalizes with (Atg12-)Atg5, the Atg1 kinase complex and Atg2, and presumably Atg14. Lipidation of PE is a requisite for the recruitment of Atg8 to the PAS. In *atg14* or *atg6* mutants, Atg8 and Atg5 do not form a dot structure in the cytosol, indicating that the autophagy-specific PI3-kinase complex plays an important role in the organization of PAS [31]. In contrast, defects in the Atg1 kinase complex show little effect on PAS structure. *Atg1ts* mutant cells expressing GFP-Atg8 completely block autophagosome formation; instead, they show a bright PAS structure next to the vacuole at the restrictive temperature. Upon shift down to the permissive temperature, a less brightly fluorescent structure is generated from PAS and fused to the vacuole; consequently, the vacuolar lumen is stained brightly. PAS seems to be an organizing center for the autophagosome.

2.14
Atg Proteins in Higher Eukaryotes

As shown in Table 2.1, most of the *ATG* genes are conserved from yeast to mammals and plants, indicating that eukaryotic cells acquired autophagic protein degradation at an early stage of evolution. The two conjugation reactions are especially well conserved. Interestingly, both Atg12 and Atg5 are encoded by single genes, but mammals and plants have many Atg8 homologues. In mammals they are called LC3, GATE16, and GABARAP, LC3 is involved in autophagy, but the other properties of the proteins are not clear yet.

We have shown that in mammalian cells, the Atg12–Atg5 protein conjugate is essential for autophagosome formation. In yeast, the Atg12–Atg5 conjugate interacts with a small coiled-coil protein, Atg16, to form a ~350-kDa multimeric complex [37]. We have demonstrated that the mouse Atg12–Atg5 conjugate forms a ~800-kDa protein complex containing a novel WD-repeat protein [49]. As the N-terminal region of this novel WD-repeat protein shows homology with yeast Atg16, we have designated it mouse Atg16-like protein (Atg16L). Atg16L has a large C-terminal domain containing seven WD repeats and is well conserved in all eukaryotes. The N-terminal region of Atg16L interacts with both Atg5 and Atg16L monomers, but the WD-repeat domain does not. In conjunction with Atg12–Atg5, Atg16L associates with the autophagic isolation membrane for the duration of autophagosome formation, indicating that Atg16L is the functional counterpart of the yeast Atg16. We also found that membrane targeting of Atg16L requires Atg5 but not Atg12 [49]. As WD-repeat proteins provide a platform for protein–protein interactions, the ~800-kDa complex is expected to function in autophagosome formation, further interacting with other proteins in mammalian cells.

Still, several Atg proteins are not identified in mammals or in plants. So far, an Atg1 kinase homologue is reported, but its regulators Atg13 and Atg17 are missing. A requirement of PI3-kinase activity for autophagy is also reported in mammals, but the autophagy-specific component, Atg14, has not been found. Possibly, as in the case of Atg16, sequence homology alone may not sufficient to find their counterparts, or they may be yeast-specific factors.

These proteins provide the most powerful tools for analysis of autophagy in higher eukaryotes.

2.15
Atg Proteins as Markers for Autophagy in Mammalian Cells

In yeast, autophagy is required for cell survival during starvation and is necessary for spore formation. In contrast, the role of autophagy in mammals is still poorly understood. Although the possible involvement of autophagy in development, cell death, and pathogenesis has been repeatedly pointed out, systematic analysis has not been performed, mainly due to limitations of monitoring methods. Moreover, in *S. cerevisiae* autophagy is solely a starvation response, but in multicellular organisms it could be regulated in a different manner. Our recent studies have made available several marker proteins for autophagosomes. To understand where and when autophagy occurs *in vivo*, we have generated transgenic mice systemically expressing GFP fused to LC3, which is a mammalian ortholog of yeast Atg8 [50, 51]. Cryosections of various organs were prepared and the occurrence of autophagy was examined by fluorescence microscopy. Active autophagy was observed in various tissues, such as the skeletal muscle, liver, heart, exocrine glands, thymic epithelial cells, lens epithelial cells, and podocytes. Patterns of induction of autophagy in different tissues are clearly distinct. In brain, autophagosomes were hardly detectable; under starvation conditions, brain cells may not suffer from nutrient limitation. In some tissues, autophagy even occurs spontaneously. Our results suggest that the regulation of autophagy is organ-dependent and that the role of autophagy is not restricted to the starvation response. This transgenic mouse is a useful tool for studying mammalian autophagy.

2.16
Physiological Role of Autophagy in Multicellular Organisms

The elucidation of genes essential for autophagy in yeast has facilitated work on autophagy in various organisms including *Dictyostelium discoideum* (slime mold), *Caenorhabditis elegans* (worm), *Drosophila melanogaster* (fly), *Arabidopsis thaliana* (plants), mouse, and humans [52–57]. Knockout of Atg genes showed severe phenotypes, mostly embryonically lethal at certain stages. These results indicate that autophagy probably has important roles in the development or cell differentiation of multicellular organisms.

2.17 Perspectives

Many researchers now pay attention to autophagy, but it is still a developing field of biology. Further studies on the function of Atg proteins not only will unveil the mystery of autophagosome formation but also may provide new insights into membrane dynamics within cell. Studies using different systems will provide a variety of physiological functions of autophagy in the near future. Finally, further work will define the true meaning of protein and organelle turnover more precisely.

References

1 de Duve C, Lysosomes, a new group of cytoplasmic particles, in Subcellular Particles; Hayashi, T. (ed): Ronald, New York, pp. 128–159.
2 G.S. Zubenko, E.W. Jones, *Genetics* **1981**, *97*, 45–64.
3 R. Egner, M. Thumm, M. Straub, A. Simeon, H.J. Schuller, D.H. Wolf, *J. Biol. Chem.* **1993**, *268*, 27269–27276.
4 K. Takeshige, M. Baba, S. Tsuboi, T. Noda, Y. Ohsumi, *J. Cell Biol.* **1992**, *119*, 301–311.
5 M. Baba, K. Takeshige, N. Baba, Y. Ohsumi, *J. Cell Biol.* **1994**, *124*, 903–913.
6 M. Baba, M. Osumi, Y. Ohsumi, *Cell Struct. Funct.* **1995**, *20*, 465–471.
7 S.V. Scott, A. Hefner-Gravink, K.A. Morano, T. Noda, Y. Ohsumi, D.J. Klionsky, *Proc. Natl. Acad. Sci. U. S. A.* **1996**, *93*, 12304–12308.
8 M. Tsukada, Y. Ohsumi, *FEBS Lett.* **1993**, *333*, 169–174.
9 M. Thumm, R. Egner, B. Koch, M. Schlumpberger, M. Straub, M. Veenhuis, D.H. Wolf, *FEBS Lett.* **1994**, *349*, 275–280.
10 N. Mizushima, T. Noda, Y. Ohsumi, *EMBO J.* **1999**, *18*, 3888–3896.
11 Y. Kamada, T. Funakoshi, T. Shintani, K. Nagano, M. Ohsumi, Y. Ohsumi, *J. Cell Biol.* **2000**, *150*, 1507–1513.
12 T.M. Harding, K.A. Morano, S.V. Scott, D.J. Klionsky, *J. Cell Biol.* **1995**, *131*, 591–602.
13 S.V. Scott, M. Baba, Y. Ohsumi, D.J. Klionsky, *J. Cell Biol.* **1997**, *138*, 37–44.
14 M. Baba, M. Osumi, S.V. Scott, D.J. Klionsky, Y. Ohsumi, *J. Cell Biol.* **1997**, *139*, 1687–1695.
15 D.J. Klionsky, J.M. Cregg, W.A. Jr Dunn, S.D. Emr, Y. Sakai, I.V. Sandoval, A. Sibirny, S. Subramani, M. Thumm, M. Veenhuis, Y. Ohsumi, *Dev. Cell* **2003**, *5*, 539–545.
16 A. Matsuura, M. Tsukada, Y. Wada, Y. Ohsumi, *Gene* **1997**, *192*, 245–250.
17 S. Kametaka, A. Matsuura, Y. Wada, Y. Ohsumi, *Gene* **1996**, *178*, 139–143.
18 N. Ishihara, M. Hamasaki, S. Yokota, K. Suzuki, Y. Kamada, A. Kihara, T. Yoshimori, T. Noda, Y. Ohsumi, *Mol. Biol. Cell* **2001**, *12*, 3690–3702.
19 M. Hamasaki, T. Noda, Y. Ohsumi, *Cell Struct. Funct.* **2003**, *28*, 49–54.
20 I. Suriapranata, U.D. Epple, D. Bernreuther, M. Bredschneider, K. Sovarasteanu, M. Thumm, *J. Cell Sci.* **2000**, *113*, 4025–4033.
21 U.D. Epple, I. Suriapranata, E.L. Eskelinen, M. Thumm, *J. Bacteriol.* **2001**, *183*, 5942–5955.
22 N. Nakamura, A. Matsuura, Y. Wada, Y. Ohsumi, *J. Biochem. (Tokyo.)* **1997**, *121*, 338–344.
23 J. Onodera, Y. Ohsumi, *J. Biol. Chem.* **2004**, *279*, 16071–16074.
24 D.J. Klionsky, Y. Ohsumi, *Annu. Rev. Cell Dev. Biol.* **1999**, *15*, 1–32.

25 H. Mukaiyama, M. Oku, M. Baba, T. Samizo, A.T. Hammond, B.S. Glick, N. Kato, Y. Sakai, *Genes. Cells.* **2002**, *7*, 75–90.

26 H. Mukaiyama, M. Baba, M. Osumi, S. Aoyagi, N. Kato, Y. Ohsumi, Y. Sakai, *Mol. Biol. Cell.* **2004**, *15*, 58–70.

27 M. Oku, D. Warnecke, T. Noda, F. Muller, E. Heinz, H. Mukaiyama, N. Kato, Y. Sakai Y., *EMBO J.* **2003**, *22*, 3231–3241.

28 T. Noda, Y. Ohsumi, *J. Biol. Chem.* **1998**, *273*, 3963–3966.

29 M. Hamasaki, T. Noda, M. Baba, Y. Ohsumi, *Traffic* **2005**, in press.

30 T. Kirisako, M. Baba, N. Ishihara, K. Miyazawa, M. Ohsumi, T. Yoshimori, T. Noda, Y. Ohsumi, *J. Cell. Biol.* **1999**, *147*, 435–446.

31 K. Suzuki, T. Kirisako, Y. Kamada, N. Mizushima, T. Noda, Y. Ohsumi, *EMBO. J.* **2001**, *20*, 5971–5981.

32 T. Noda, A. Matsuura, Y. Wada, Y. Ohsumi, *Biochem. Biophys. Res. Commun.* **1995**, *210*, 126–132.

33 Y. Ohsumi, *Nat. Rev. Mol. Cell Biol.* **2001**, *2*, 211–216.

34 N. Mizushima, T. Noda, T. Yoshimori, Y. Tanaka, T. Ishii, M.D. George, D.J. Klionsky, M. Ohsumi, Y. Ohsumi, *Nature.* **1998**, *395*, 395–398.

35 I. Tanida, N. Mizushima, M. Kiyooka, M. Ohsumi, T. Ueno, Y. Ohsumi, E. Kominami, *Mol. Biol. Cell.* **1999**, *10*, 1367–1379.

36 T. Shintani, N. Mizushima, Y. Ogawa, A. Matsuura, T. Noda, Y. Ohsumi, *EMBO. J.* **1999**, *18*, 5234–5241.

37 A. Kuma, N. Mizushima, N. Ishihara, Y. Ohsumi, *J. Biol. Chem.* **2002**, *277*, 18619–18625.

38 T. Kirisako, Y. Ichimura, H. Okada, Y. Kabeya, N. Mizushima, T. Yoshimori, M. Ohsumi, T. Takao, T. Noda, Y. Ohsumi, *J. Cell. Biol.* **2000**, *151*, 263–276.

39 Y. Ichimura, T. Kirisako, T. Takao, Y. Satomi, Y. Shimonishi, N. Ishihara, N. Mizushima, I. Tanida, E. Kominami, M. Ohsumi, T. Noda, Y. Ohsumi, *Nature.* **2000**, *408*, 488–492.

40 Y. Ichimura, Y. Imamura, K. Emoto, M. Umeda, T. Noda, Y. Ohsumi, *J. Biol. Chem.* **2004**, *279*, 40584–40592.

41 M. Straub, M. Bredschneider, M. Thumm, *J. Bacteriol.* **1997**, *179*, 3875–3883.

42 T. Funakoshi, A. Matsuura, T. Noda, Y. Ohsumi, *Gene.* **1997**, *192*, 207–213.

43 H. Abeliovich, W.A. Jr Dunn, J. Kim, D.J. Klionsky, *J. Cell. Biol.* **2000**, *151*, 1025–1034.

44 S.V. Scott, D.C. 3rd Nice, J.J. Nau, L.S. Weisman, Y. Kamada, I. Keizer-Gunnink, T. Funakoshi, M. Veenhuis, Y. Ohsumi, D.J. Klionsky, *J. Biol. Chem.* **2000**, *275*, 25840–25849.

45 S. Kametaka, T. Okano, M. Ohsumi, Y. Ohsumi, *J. Biol. Chem.* **1998**, *273*, 22284–22291.

46 A. Kihara, T. Noda, N. Ishihara, Y. Ohsumi, *J. Cell. Biol.* **2001**, *152*, 519–530.

47 T. Shintani, K. Suzuki, Y. Kamada, T. Noda, Y. Ohsumi, *J. Biol. Chem.* **2001**, *276*, 30452–30460.

48 T. Noda, J. Kim, W.P. Huang, M. Baba, C. Tokunaga, Y. Ohsumi, D.J. Klionsky, *J. Cell. Biol.* **2000**, *148*, 465–480.

49 N. Mizushima, A. Kuma, Y. Kobayashi, A. Yamamoto, M. Matsubae, T. Takao, T. Natsume, Y. Ohsumi, T. Yoshimori, *J. Cell. Sci.* **2003**, *116*, 1679–1688.

50 Y. Kabeya, N. Mizushima, T. Ueno, A. Yamamoto, T. Kirisako, T. Noda, E. Kominami, Y. Ohsumi, T. Yoshimori, *EMBO. J.* **2000**, *19*, 5720–5728.

51 N. Mizushima, A. Yamamoto, M. Matsui, T. Yoshimori, Y. Ohsumi, *Mol. Biol. Cell.* **2004**, *15*, 1101–1111.

52 G.P. Otto, M.Y. Wu, N. Kazgan, O.R. Anderson, R.H. Kessin, *J. Biol. Chem.* **2003**, *278*, 17636–17645.

53 A. Melendez, Z. Talloczy, M. Seaman, E.L. Eskelinen, D.H. Hall, B. Levine, *Science.* **2003**, *301*, 1387–1391.

54 M. Thumm, T. Kadowaki, *Mol. Genet. Genomics.* **2001**, *266*, 657–663.

55 J.H. Doelling, J.M. Walker, E.M.

Friedman, A.R. Thompson, R.D. Vierstra, *J. Biol. Chem.* **2002**, *277*, 33105–33114.

56 X. Qu, J. Yu, G. Bhagat, N. Furuya, H. Hibshoosh, A. Troxel, J. Rosen, E.L. Eskelinen, N. Mizushima, Y. Ohsumi, G. Cattoretti, B. Levine, *J. Clin. Invest.* **2003**, *112*, 1809–1820.

57 K. Yoshimoto, H. Hanaoka, S. Sato, T. Kato, S. Tabata, T. Noda, Y. Ohsumi, *Plant Cell* **2004**, *16*, 2967–2683.

3
Dissecting Intracellular Proteolysis Using Small Molecule Inhibitors and Molecular Probes*

Huib Ovaa, Herman S. Overkleeft, Benedikt M. Kessler, and Hidde L. Ploegh

Abstract

The ubiquitin–proteasome system has emerged as essential sets of reactions involved in many biological processes in addition to the disposal of misfolded and damaged proteins. Studies in different research areas reveal its role in regulating cell growth, differentiation, apoptosis, signaling, and protein targeting. Small molecule inhibitors against the proteasome have been useful in determining the specific role of this enzyme in these processes. Here we review recent progress made in the development and application of molecules that target proteasomal proteolysis. In addition, an increasing number of other enzymes in this pathway, in particular deubiquitinating enzymes (DUBs) and *N*-glycanases, appear to be attractive alternative targets for developing inhibitors that can be used to interfere with biological processes linked to the ubiquitin–proteasome pathway.

3.1
Introduction

Our knowledge of the ubiquitin–proteasome system as a key player in a wide variety of biological processes rests in part on yeast genetics and on our ability to manipulate it pharmacologically with proteasome inhibitors. Some of these inhibitors are cell-permeable and are active *in vivo*, making it possible to interfere with proteasome function in mammalian cells [1–3]. Proteasome inhibitors have now entered the clinic for the treatment of malignancies such as multiple myeloma and are no longer purely investigational tools [4–7].

Although usually considered a springboard for the analysis of mammalian proteasome structure, the prokaryotic proteasome has also come to the fore as a possi-

* A list of abbreviations used can be found at the end of the chapter.

Protein Degradation, Vol. 2: The Ubiquitin-Proteasome System.
Edited by R. J. Mayer, A. Ciechanover, M. Rechsteiner

ISBN: 3-527-31130-0

ble pharmaceutical target for proteasome inhibitors. *Mycobacterium tuberculosis* apparently requires its intact proteasomes to survive the harsh oxidative conditions inside the lung macrophages in which it usually resides. This observation suggests exciting opportunities to treat mycobacterial disease by rendering proteasome inhibitors specific to the mycobacterial proteasome [8, 9].

Originally viewed mostly as an abundant cytoplasmic protease, the proteasome is now considered central to many different aspects of cellular physiology. To maintain steady-state protein levels, polypeptides are continuously synthesized and destroyed. This process is regulated not only at a transcriptional level but also at the level of post-translational modification. Most cellular proteins are continuously synthesized and degraded within the life span of a cell. Protein turnover serves many critical regulatory roles, including quality control, by ensuring the degradation of proteins with abnormal structures that arise from mutation, metabolic damage, or misfolding. A variety of proteases are responsible for cytosolic protein turnover, but degradation of the vast majority of cellular proteins in mammalian cells is carried out by the proteasome, usually after previous tagging of the substrate with a polyubiquitin chain [10]. With few exceptions, proteasomal proteolysis requires substrates to be conjugated with multiple ubiquitin (Ub) molecular [11, 12]. Proteasomes degrade proteins in a processive fashion, generating peptides ranging in length from three to 22 residues [13].

The proteasome itself (Figure 3.1) can be divided into two distinct portions: a catalytic core and accessory subunits that associate with the proteasome at either end of the catalytic core particle [14, 15]. These associated proteins include polypeptides involved in recognition of Ub-conjugated substrates, proteins capable of unfolding protein substrates, enzymes capable of removing Ub from Ub-modified substrates, and at least one enzyme capable of removing *N*-linked glycans from *N*-glycosylated substrates [16–18]. It is likely that the proteasome is at the nexus of other, yet to be discovered, protein interactions. It follows, then, that the concept of proteasome

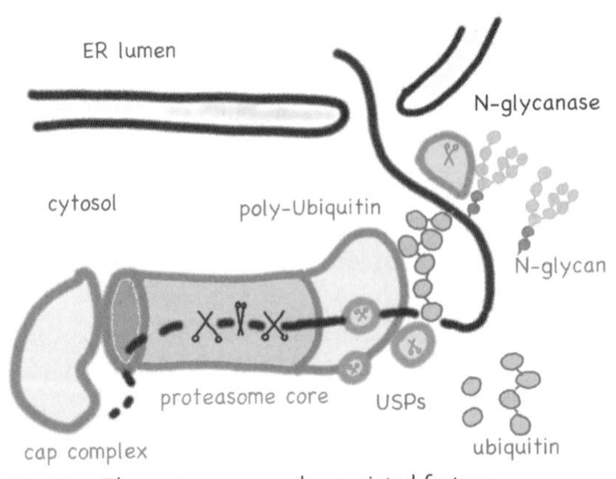

Fig. 3.1. The proteasome and associated factors.

inhibition should be defined to include not only agents that interfere with the catalytic subunits of the core particle but also compounds that target proteasome-associated activities. We consider it likely that many of these activities act in concert and that a pharmacological blockade of any one of them will modulate proteasomal function in a controlled manner. In addition to the protease activities associated with the $\beta1$, $\beta2$, and $\beta5$ subunits of the 20S complex, the full 26S proteasome includes other protease activities, notably the Ub-specific thiol proteases (USP14, UCH37) and the metalloprotease POH1 [19–21].

Modification of proteins with Ub is linked not only to proteolysis but also to targeting of modified proteins to proper intracellular destinations [22]. For instance, internalization of a Ub-modified receptor from the cell surface or the biogenesis of specialized intracellular compartments, such as multivesicular bodies (MVBs) [23–25], is regulated by Ub modification and possibly Ub removal. Even though modification with Ub may be the common theme here, the proteasome is not required for Ub modification to exert its function, nor does it solely degrade ubiquitinated proteins. Whereas internalization of Ub-modified growth hormone receptor requires an intact proteasomal system, this appears not to be the case for MVB formation, even though both processes critically depend on modification of target proteins with Ub [26].

When considering proteasomal proteolysis, it would be a mistake to lump together all aspects of protein degradation. Some proteins are targeted for Ub conjugation and proteasomal proteolysis while still attached to the ribosome: the incorrect incorporation of amino acids, leading to aberrant folding, might be one element that targets the nascent chain for degradation [27, 28]. In other cases, specific signals, such as phosphorylation, are required to polyubiquitinate and destroy the substrate in a carefully timed manner. Proteasomal destruction is usually highly processive and effective, but for some proteins and protein complexes, the proteasome generates the active form from an inactive precursor or protein complex, NF-κB being a case in point [29]. Proteins inserted co-translationally into the lumen of the endoplasmic reticulum (ER) fold in the topological equivalent of extracellular space. In the ER, proteins that fail to fold correctly are recognized and dispatched to the cytoplasm, in a process referred to as dislocation or retrotranslocation [30, 31]. In this process, Ub conjugation of the substrate plays a key role [32]. The Ub-conjugated proteins are then destroyed by the proteasome. The role, if any, of proteasomes in the process of extracting a substrate from the ER membrane is not clear [33]. It is likely that multiple classes of proteasomes can be defined based on their intracellular location and, hence, site of action [34, 35]. Therefore, different classes of proteasomes may differ in their susceptibility to pharmacological inhibition, depending on their interacting partners, their cellular environment, and the pharmacokinetics of the inhibitor used. Furthermore, the function of the proteasomes in the immune system is modulated through the action of cytokines. IFN-γ is a potent inducer of the $\beta1i$, $\beta2i$, and $\beta5i$ subunits, which replace the catalytically active $\beta1$, $\beta2$, and $\beta5$ subunits, respectively, in the mammalian proteasome to generate the immunoproteasome [36, 37]. The immunoproteasome is structurally and functionally distinct from its constitutive counterpart [38, 39] and may be targeted selectively with appropriate inhibitors.

Importantly, both the proteasome and immunoproteasome are critically involved in the generation of a proper immune response in the context of MHC class I–mediated antigen presentation [40]. Inhibition of proteasomal activity strongly affects a variety of cellular processes. The proteasome is now considered a valid target for cancer therapy and treatment of stroke [41]; selective mycobacterial proteasome inhibitors also hold promise for the treatment of tuberculosis. In addition, proteasome inhibitors have been shown to increase the viability of cells treated with anthrax lethal toxin, inhibiting a proteasome-dependant step that is an early event in intoxication with anthrax lethal toxin. Proteasome inhibitors may thus become important in the defense against chemical warfare [42]. On the other hand, the proteasome inhibitor PS341 (VELCADE, Bortezomib) was recently introduced into the clinic for treatment of multiple myeloma [6] and is in clinical trials for a variety of other malignancies [5, 43–45], underscoring the need for research tools that allow determination of the proteasomal mode of action and its activity *in vivo*.

In this chapter we discuss the presently known classes of proteasome inhibitors and some of their applications. Because it would be difficult to view the proteasome in isolation and to disregard some of the proteasome-associated enzymatic activities as key players, we shall also discuss the identity of some of the proteasome-associated factors and the means to manipulate them where appropriate. There are many questions that remain unanswered, not the least of which is how to get a better understanding of the pharmacodynamics and pharmacokinetics of the various inhibitors, especially those presently in use, or considered for use, in the clinic.

3.2 The Proteasome as an Essential Component of Intracellular Proteolysis

To date, manipulation of the proteasome with the aid of small compounds has mainly been achieved through targeting the actual proteolytic activities of the 20S core. Targeting the individual ATPase and USP activities residing within the 19S caps with inhibitors entails an alternative inroad to the manipulation of proteasomal protein degradation. This also holds true for addressing events up- or downstream of the proteasome. These include the action of *N*-glycanase, which is instrumental in the removal of *N*-linked glycans of proteins that have escaped the secretory pathway and that are degraded by the proteasome [18]. In addition, modifications of components of the proteasome complex, such as phosphorylation/dephosphorylation, *O*-GlcNAcylation by *O*-GlcNac transferase (OGT), and *O*-GlcNac removal by *O*-GlcNacase, also modulate proteasome function [46–48]. The latter carbohydrate modification influences substrate entrance to the 26S proteasome by *O*-GlcNac modification of the Rpt2 ATPase subunit that resides in the 19S cap complex. This dynamic glycosyl modification appears to be under direct metabolic control. GlcNacase inhibitors such as streptozotocin (STZ) open opportunities to develop targeting strategies upstream of the proteasome. Examples of this type will be discussed in this chapter.

3.3 Proteasome Structure, Function, and Localization

The 20S proteasome, the inner core of the larger 26S particle that comprises the eukaryotic proteasome complex, is highly conserved in nature [49]. Archaebacterial and eukaryotic 20S proteasomes consist of 28 subunits, arranged in four stacked heptagonal rings, forming a hollow, barrel-shaped protein complex [14]. The proteolytic activity of the 20S particle resides within the two inner rings, containing seven β subunits each. The two outer rings, both assembled from seven α subunits, provide stability to the overall $(\alpha_{(1-7)}\beta_{(1-7)}\beta_{(1-7)}\alpha_{(1-7)})$ complex and serve as docking stations for additional protein complexes. These include the 19S cap (to form the 26S complex) and the interferon-γ-inducible PA28 complex in eukaryotic proteasomes, both of which are regulatory components with different functions that influence the activity and substrate specificity of the core particle. The 19S cap activates proteasomal proteolysis by recognition of proteasome substrates through their polyubiquitin chain and then unfolds them, enabling access to the proteolytic chamber. Ubiquitin molecules are recycled through the action of either of at least two proteasome-associated ubiquitin-specific proteases (USPs) or a zinc-dependent ubiquitin-specific metalloprotease that resides within the 19S complex [20, 21, 50–53].

While retaining its overall shape, the nature of individual α and β subunits within the 20S proteasome has diverged among different species. The *Thermoplasma* 20S proteasome is assembled in a fashion similar to that of the mammalian proteasome. The activity of the proteolytic β subunits resides in the N-terminal threonine residues, with the secondary alcohol of the threonine side chain acting as the nucleophilic species. The free N-terminal amine acts as the base in the catalytic cycle, catalyzing nucleophilic attack on the scissile peptide bonds. Importantly, the catalytically active substrate-binding site is formed only upon specific interactions with adjacent β subunits. Therefore, individual subunits do not show catalytic activity in isolation [54, 55].

Within eukaryotic 20S proteasomes, each of the seven subunits in either α or β rings are unique [39, 56]. Eukaryotic 20S proteasomes contain three distinct proteolytic activities, classified based on the use of synthetic substrates, although other proteolytic specificities were also reported [39, 57]. Individual activities have been analyzed with a variety of tools, including inhibitors, protein substrates, and a panel of specific fluorogenic peptide substrates. The main activities are now commonly referred to as the chymotryptic activity (X, $\beta5$), which is targeted by most proteasome inhibitors and which cleaves preferentially after hydrophobic residues; the tryptic activity (Y, $\beta2$), cleaving after basic residues; and the PGPH or caspase-like activity (Z, $\beta1$), responsible for cleaving after acidic residues. The chymotryptic/tryptic/PGPH classification is somewhat ambiguous, since substrate preference does not accurately reflect catalytic activity, as revealed by studies using longer peptide and protein substrates [58–61]. All three catalytic β subunits show a rather broad substrate tolerance, and we will refer to the individual subunits, responsible for catalytic activity, as $\beta1$, $\beta2$, and $\beta5$ throughout the body of the text.

In higher vertebrates, a distinct 20S proteasome particle, referred to as the immunoproteasome, is expressed in many tissues upon interferon-γ induction [36]. The immunoproteasome contains three unique proteolytically active subunits termed β1i(LMP2), β5i(LMP7), and β2i(MECL1). They are highly homologous to their constitutive counterparts and display similar, yet subtly distinct, substrate specificities [60, 62]. In addition, several hybrid forms of proteasome species that harbor either β5i or β1i subunits, without the other interferon-γ-inducible subunits, have been described in different tissues and cell lines [63, 64]. The role of such proteasome subsets remains to be determined. A recent crystallographic study on eukaryotic proteasomes revealed a possible additional catalytic site associated with the $\beta7$ subunit [39].

The 20S core is found in both the cytoplasm and the nucleus [34, 65]. Associated proteins and complexes may dictate activity, or at least distribution-dependent proteasomal activity. The 19S complexes are involved in ubiquitin recognition and unfolding of the targeted polypeptide, and they facilitate translocation into the proteolytic chamber in an ATP-dependent manner, but additional associations and distribution-dependent tasks may well exist. Other studies suggest that cytoplasmic proteasomes co-localize with intermediate filaments and the endoplasmic reticulum (ER) membrane. This would fit the observation that membrane-bound proteins are degraded by the proteasome, perhaps also involving adaptor molecules that direct proteasomes to the ER or other organelles [66].

20S proteasomes are abundant. It has been estimated that the concentration of free 20S proteasomes is twice as high as that of 26S proteasomes in mammalian cells [67]. Since free 20S particles constitute the vast majority of different forms of proteasomes present in cells, a role for them in proteolysis is suggested based on their abundance alone. Proliferating and transformed cell lines usually have both higher proteasomal content and proteasomal activity than quiescent and non-transformed cells. The 20S core particle is capable of destroying highly oxidized proteins, and this may well be an important mechanism to respond to oxidative stress conditions [68].

The tight assembly of the 20S core particle is reflected by the relative ease with which it can be purified. Although a more demanding task, several groups have accomplished the purification of fully assembled 26S proteasome particles [69–72]. The 20S proteasomes from many different sources as well as the eubacterial HslU/V protease have now been subjected to X-ray structural analysis [39, 56, 73–79]. Crystallized proteasomes retain their enzymatic activity, allowing the structural elucidation of proteasome–inhibitor complexes. In this way the covalent nature of aldehyde inhibitors bound to the catalytic subunits (as a hemiacetal); that of epoxyketones (morpholine adduct), β lactones (ester adduct), and vinyl sulfones (Michael adduct); as well as the noncovalent nature of TCM-95 inhibition have been established unambiguously. Co-crystal structures of inhibitor–proteasome complexes have been used to determine the effects of occupancy of catalytic sites on structural elements more distal to the proteolytic core [77].

3.4 Proteasome Inhibitors as Tools to Study Proteasome Function

Ever since its discovery as a key player in protein turnover, the proteasome has been subjected to studies involving the use of small molecule inhibitors. The aim of such studies is usually twofold. With specific inhibitors, the nature of the individual subunit-associated activities can be charted. Moreover, the ability to partially disable the proteasome allows a study of its role in biological processes.

The ideal proteasome inhibitor would be both cell-permeable and specific, allowing the study of the proteasome in living cells and in live animals. The cleavage preferences of proteasomes can be assessed using fluorescent substrates, but such substrates can be used only in cell extracts. The accuracy of the fluorescent readout cannot be readily extrapolated to proteasomal activity *in vivo*. Standard, commercially available fluorogenic peptides include Z-LLE-βNA (β1-specific), Boc-LRR-AMC (β2-specific), and Suc-LLVY-AMC (β5-specific). By using such substrates, not only subunit specificity but also the kinetics of subunit inhibition can be monitored [80, 81].

Whereas a chemical knockout approach, disabling all activities, is expected to be lethal, disabling of specific catalytically active β subunits, for instance, an immunoproteasome subunit, may be useful to determine its contribution to the generation of MHC class I antigenic repertoires and its ability to modulate immune responses [82]. Inhibition of the chymotrypsin-like site as achieved by most proteasome inhibitors, or its inactivation by mutation alone, causes a large reduction in the rates of protein breakdown *in vitro* [59]. Potent and selective inactivation of trypsin-like or caspase-like sites is more difficult to achieve by small molecule inhibitors [83]. Compounds that selectively target proteasome particles located in specific subcellular compartments, or that show tissue- or species-dependent specificity, will be valuable both for biological research and therapeutic applications [84], but no such compounds have been identified to date. The same holds true for drugs that inhibit targets up- and downstream of proteasomal degradation.

Progress in inhibitor development has been described extensively in recent reviews [2, 85–87]. In the following sections the main classes of existing proteasome inhibitors and some future directions will be described briefly. Approaches that allow interference with targets up- and downstream of the proteasome will be discussed thereafter.

3.4.1 Peptide Aldehydes

Peptide aldehydes are the most popular class of proteasome inhibitors in biomedical and biological research. The most widely used member, Z-Leu-Leu-Leu-Al (**1**,

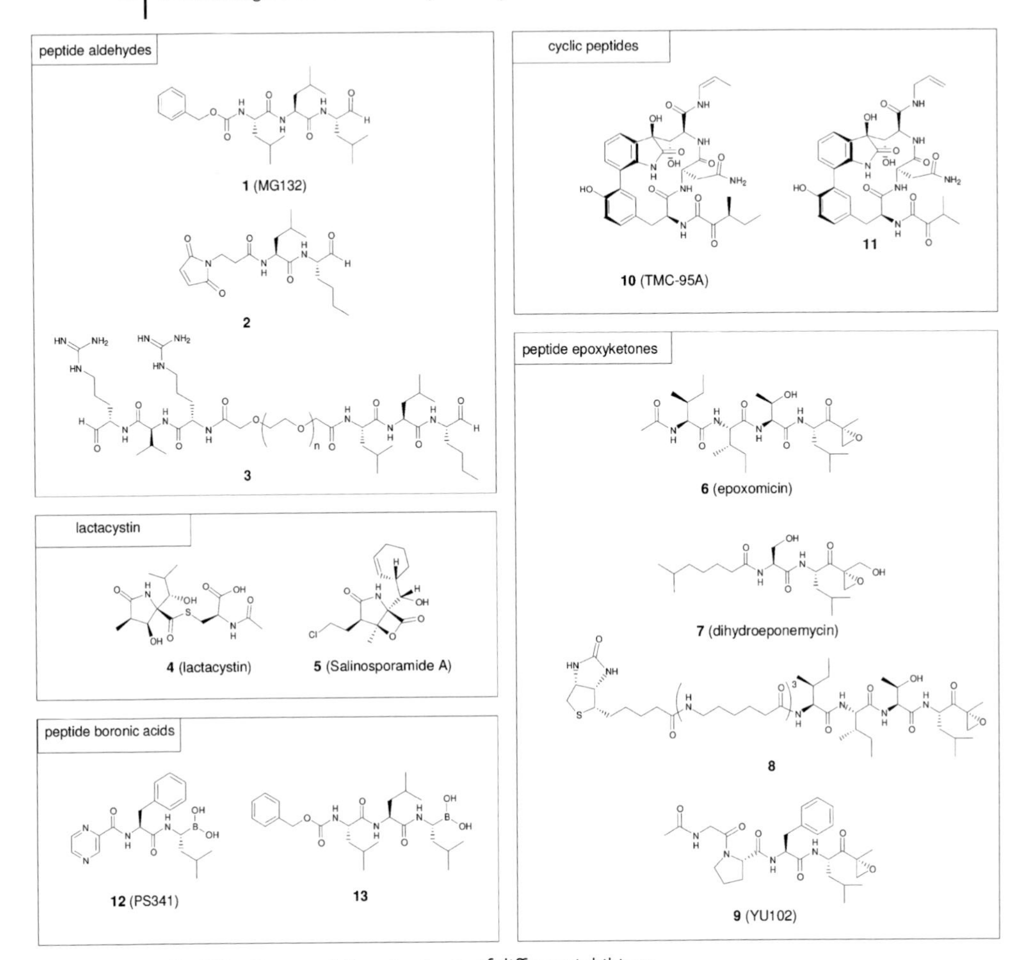

Fig. 3.2. Representative structures of different inhibitors.

MG132, Figure 3.2), is one of the standard tools used to modulate proteasome activity. The catalytic mode of action of peptide aldehydes was first demonstrated by X-ray diffraction of *Thermoplasma* 20S proteasomes in the presence of the peptide aldehyde Ac-Leu-Leu-Nle-Al [55]. The inhibitor's aldehyde moiety presumably forms a hemiacetal linkage with a catalytic threonine residue. Although this linkage is covalent, it can be hydrolyzed under physiological conditions, making peptide aldehydes reversible, competitive inhibitors. Many variations in the peptide backbone have appeared in recent years, including several compounds from natural sources such as tyropeptin A [88, 89]. A major drawback of peptide aldehydes is their propensity to cross-react with other proteolytic activities, primarily cysteine proteases, and their tendency to undergo oxidation to the corresponding acids, rendering the probes inactive under physiological conditions.

Moroder and coworkers developed a set of bifunctional peptide aldehydes, represented by maleimide derivative **2**, with specificity for β_2 over β_1 and β_5 [90]. After initial hemiacetal formation with the N-terminal threonine of β_2, the maleimide moiety undergoes an irreversible Michael reaction with a cysteine residue of the neighboring β_3 subunit, which protrudes into the β_2 active site that then becomes disabled. The reactivity of the maleimido group towards mercaptans in general limits its use, but fine-tuning of the reactivity of the maleimido group may provide a route to β_2-selective inhibitors for broader applications. The development of homo- and heterobifunctional peptide dialdehydes, interspaced with polyethylene glycol (e.g., compound **3**), was reported by the same group, showing a 100-fold increase in potency compared to monovalent counterparts.

3.4.2
Lactacystin

The fungal metabolite lactacystin (**4**), isolated from *Streptomyces*, is a classical proteasome inhibitor and one of the few that does not have a peptoid structure [91, 92]. The β-lactone metabolite of lactacystin, *clasto*-lactacystin-β-lactone (omuralide), is the reactive species. After binding to active sites, the β-lactone reacts with the threonine hydroxyl moiety to result in acylation of the active site. Although this acylation is covalent, hydrolysis of the ester linkage occurs over time, resulting in the loss of inhibition and loss of effective inhibitor. In studies of its mode of action using radiolabeled lactacystin, it was found that at low concentrations it effectively inhibits β_5, whereas β_1 and β_2 active sites are targeted only at higher concentrations [93]. Importantly, hydrolysis of the acylated β_1 and β_2 subunits appears to be faster than that of the corresponding β_5 subunit. Recently, a naturally occurring β-lactone named salinosporamide A (**5**), closely resembling omuralide, was discovered in marine actinomycete bacteria [94]. Based on its resemblance to omuralide, it was tested for its ability to inhibit proteasomes. When tested against purified 20S proteasome, the compound showed an efficient inhibition of the chymotryptic activity, with an IC_{50} value of 1.3 nM. Therefore, this compound is at least 35 times more potent than its structural relatives omuralide and lactacystin. It is likely that the intermediate reaction product, formed upon opening of the β-lactone of salinosporamide A by nucleophilic attack of the threonine hydroxyl moiety, undergoes a second reaction. Recently, Corey and coworkers reported a total synthesis of salinosporamide A [95].

3.4.3
Peptide Epoxyketones

The natural product epoxomicin (**6**) and related structures were isolated and identified based on their anti-tumorigenic properties in pharmacological screens [96]. A very potent proteasome inhibitor, epoxomicin shows strong preference for β5. The analogous natural product eponemycin and its synthetic analogue dihydroeponemycin (**7**) show enhanced affinity for β2. The epoxyketone chemical warhead harbors two, not one, reactive groups. Reaction of the threonine hydroxyl with the β-carbonyl results in a ketal linkage, while subsequent reaction of the free proteasomal N-terminus at the γ position results in a rigid morpholine ring system.

Crews et al. have devoted considerable effort to the generation of synthetic epoxomicin derivatives. With biotinylated epoxomicin derivative **8** they showed the proteasome to be the biological target of epoxomicin [97]. Recent efforts include the synthesis of oligopeptide epoxyketone derivatives with varying amino acid functionalities (including non-natural ones) at positions P1–P4 [83]. From these studies stems YU102 (**9**), to date the only compound that comes close to being a selective β inhibitor. Epoxyketones are relatively selective, metabolically quite inert, and in some cases cell-permeable, and they modify the proteasome irreversibly, thereby enabling affinity tagging and target retrieval [98].

3.4.4 Cyclic Peptides

TMC-95A (**10**) is a synthetically challenging cyclic peptide metabolite of *Apiospora montagnei*. It is a potent competitive proteasome inhibitor [99, 100]. Unlike the aforementioned inhibitors, it appears not to form a covalent link with a threonine moiety of catalytically active β subunits upon fitting into the active site. Rather, it blocks access to the active sites by imposing steric constraints. Therefore, it offers excellent opportunities to modulate its inhibitory profile specific for different subunits, not being hampered by covalent bonds. In addition to TMC-95A, several closely related structures with distinct inhibitory profiles, named TMC-95B, -C, and -D, were isolated from the same source [101–105]. At first glance, the structure of **10** appears rather daunting; one would expect it to be difficult to obtain synthetic analogues. A recent study revealed that the structure can be simplified, as in **11**, by omitting a chiral center and by replacing the difficult to obtain *N*-acyl enamine moiety by an *N*-acyl allylamine moiety [106]. These replacements have no detrimental effects on its potential as an inhibitor. TMC-95A and some other structurally similar compounds are the only proteasome inhibitors that do not covalently bind the threonine active sites [100, 107–109].

3.4.5 Peptide Boronates

Peptide boronates are considered to the most potent inhibitors of the proteasome and is the only class which a member has reached the clinic so far [4, 5, 44, 45, 110]. Boronic acids have a high affinity for hydroxyl groups, displaying an empty p-orbital to threonine oxygen lone-pair electrons. Based on the hard-soft acid-base principle, it is assumed that peptide boronates show a general preference for serine and threonine proteases over cysteine proteases (sulfhydryl moieties). Such assumptions may not always apply; for example, vinyl sulfones (see Section 3.4.6) were described originally as cysteine protease inhibitors [111], but they also proved to be potent and selective proteasome inhibitors [112, 113]. Whatever the exact mechanism, peptide boronic acids fall in the class of covalent, competitive, reversible inhibitors due to the strength of the boron–oxygen bond. The off-rate of the inhibitor, however, is much slower compared to peptide aldehyde inhibitors. Importantly, boronates provide greater metabolic stability. The combination of inhibitory potency, selectivity, and stability makes peptide boronic acids well suited as candidates for clinical use. Indeed, PS341 (**12**) has reached the clinic for the treatment of multiple myeloma and is in clinical trials for treatment of other cancers [114]. Another relevant example of the peptide boronic acid family is derivative **13**. Compound **13**, featuring the ZLLL tripeptidyl core, is far more potent than its analogous peptide aldehyde **1** and peptide vinyl sulfone analogue **15**, demonstrating the potential of the boronic acid moiety as a chemical warhead.

3.4.6
Peptide Vinyl Sulfones

Peptide vinyl sulfones are a prominent class of irreversible proteasome inhibitors [113] (representative structures are given in Figure 3.3). They covalently modify

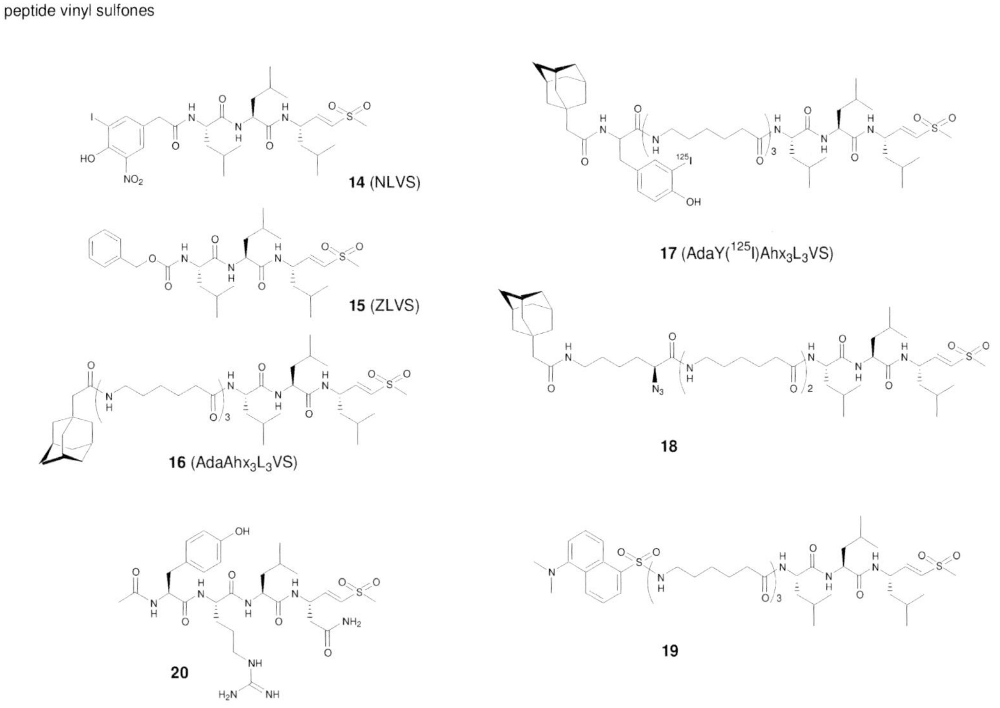

peptide vinyl sulfones

14 (NLVS)

15 (ZLVS)

16 ($AdaAhx_3L_3VS$)

17 ($AdaY(^{125}I)Ahx_3L_3VS$)

18

19

20

Fig. 3.3. Structures of vinyl sulfone–based probes.

the catalytic subunits through Michael reaction of the threonine hydroxyl with the vinyl sulfone moiety, resulting in the formation of a physiologically stable β-sulfonyl ether linkage. The finding that vinyl sulfones have turned out to be such effective proteasome inhibitors is in itself surprising: "hard-soft" acid-base principles dictate that this pharmacophore should have preference for "soft" nucleophilic thiols over "hard" alcohol nucleophiles. Indeed, peptide vinyl sulfones are widely used as cysteine protease inhibitors [115, 116]. The fact that, depending on the nature of the peptide portion attached to the electrophilic trap, the activity of peptide vinyl sulfones can be directed almost exclusively towards the proteasome underscores the importance of the peptide-based recognition elements in attaining protease specificity. Representative peptide vinyl sulfones are NLVS (**14**) and ZL_3VS (**15**), both of which are cell-permeable and show preferential targeting of $\beta5$ [113, 117].

3.4.7
Peptide Vinyl Sulfones as Proteasomal Activity Probes

One limitation of experimental work with proteasome inhibitors is the difficulty in gauging some of the most basic pharmacokinetic and pharmacodynamic parameters. While it is feasible to measure serum half-life, accumulation in various tissues and organs is more difficult to assess. Furthermore, even though the inhibition constant of these inhibitors for isolated proteasomes is well established, the extent to which proteasomal inhibition occurs *in vivo* has been more complicated to estimate.

N-terminal extension of peptide vinyl sulfones, as in $AdaAhx_3L_3VS$ (**16**), has a profound effect on their inhibitory activity. The effective labeling of all proteasomal activities in cultured cells with a single compound remained elusive until recently. Incorporation of three aminohexanoic acid residues and introduction of a large hydrophobic N-terminal cap such as the adamantane acetyl group resulted in a set of compounds capable of inhibiting all catalytically active β subunits of both the constitutive proteasome and the immunoproteasome [61] with comparable efficiency. This is illustrated by the treatment of cell lysates of EL-4 cells (expressing both the constitutive proteasome and the immunoproteasome) with the ^{125}I-labeled $AdaYAhx_3L_3VS$ (**17**, Figure 3.3). Probe **17**, however, is not cell-permeable, due to the presence of the iodotyrosyl residue. To overcome this shortcoming, compound **18** (Figure 3.3), a modification of **16** containing a bio-orthogonal azide moiety, was prepared. The azido group interferes with neither its inhibitory profile nor its cell permeability. Labeling of whole cells with **18** decorates all catalytic activities of the proteasome with an azide as a latent ligation handle. After cell lysis and retrieval and denaturation of the protein content, the azido groups can be addressed by a biotinylated phosphine reagent in a Staudinger ligation reaction, effectively biotinylating active-site subunits [118–120]. Streptavidin-horseradish peroxidase conjugate–mediated Western blot can now reveal proteasomal activity profiles in cultured cells.

Derivatizations such as radioiodination, biotinylation, or introduction of an azide

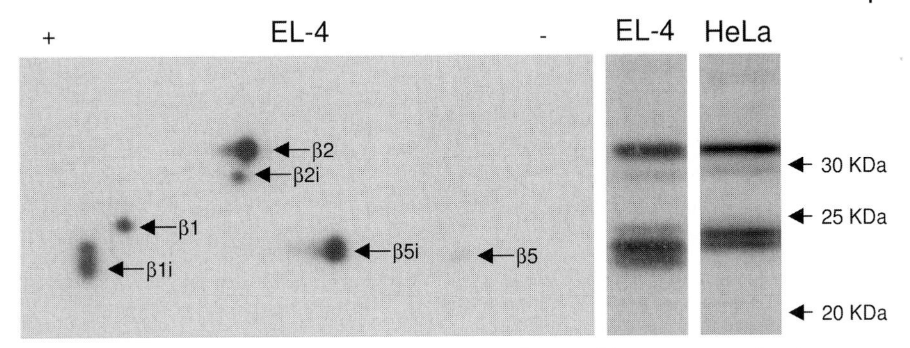

Fig. 3.4. Proteasome labeling in EL4 mouse cell extracts. Incubation of EL4 cell extracts with AdaY(^{125}I)Ahx$_3$L$_3$VS results in the covalent attachment of the radiolabeled probe to the active subunits of the proteasome. Proteins were separated by 2D isoelectric focusing (IEF)-SDS PAGE followed by autoradiography.

moiety, not surprisingly, have both advantages and disadvantages. A biotinyl or tyrosyl moiety, the latter enabling radioiodination, generally interferes with cell permeability. Azide-containing proteasome-specific probes retain cell permeability [121], but the required two-step labeling strategy makes the strategy more demanding for high-throughput applications. Small haptens such as a dansyl moiety may allow retention of cell permeability of probes and may allow detection of labeled enzymes using high-affinity antibodies directed against the incorporated hapten. This approach would allow a high-sensitivity level of detection. For this purpose, cell-permeable dansylated proteasome inhibitor **19** was synthesized. Inhibitor **19** freely reaches cellular targets and modifies covalently and irreversibly all of the proteasome's catalytic subunits (Figure 3.5). The methodology is thus entirely independent of the use of radioisotopes, biotinylation, or secondary chemoselective ligations. This dansylated inhibitor allows accurate assessment of the proteasomal targets hit in living cells by drugs such as **12** (PS341, Velcade, Bortezomib). In principle, this strategy is applicable to other proteases as well.

3.4.8
Future Directions in the Development of Inhibitors of the Proteasome's Proteolytic Activities

Next to broad-spectrum inhibitors, the search for subunit-specific (other than β5) inhibitors remains an important research objective. Several approaches have been made to achieve this goal. Nazif and Bogyo reported an elegant strategy towards peptide vinyl sulfone libraries, based on immobilizing aspartic vinyl sulfone to a matrix through the carboxylic acid side-chain functionality [122]. Positional scanning of amino acids at P2–P4 resulted in the identification of AcYRLNVS (**20**), a selective inhibitor of β2 (Figure 3.3). Reagents like these will help determine the role of individual catalytic subunits in proteasome function, protein degradation, and the generation of antigenic peptides. In this context, it will be of consider-

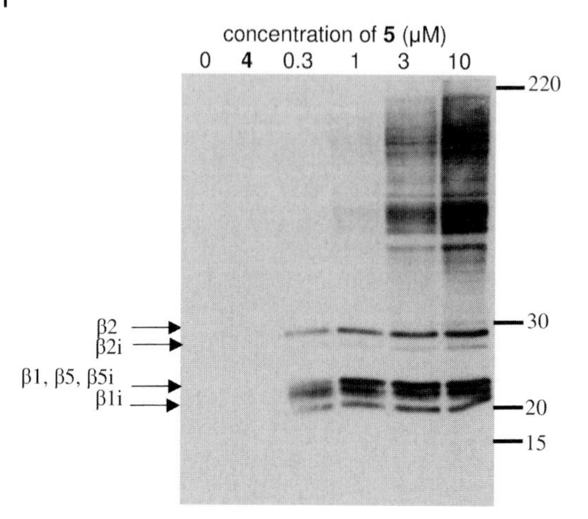

Fig. 3.5. Immunoblot using inhibitor **19**. Labeling pattern observed by immunoblot after incubation of EL4 cell extracts with different concentrations of inhibitor **19** or an inactivated control probe.

able interest to generate inhibitors capable of targeting distinct proteasome species, such as the immunoproteasome or proteasomes from bacterial or fungal origin [81, 83].

3.5 Assessing the Biological Role of the Proteasome With Inhibitors and Probes

Inhibitors of the proteasome have been essential tools in the discovery of many new substrates of the ubiquitin–proteasome pathway and in establishing its role in different biological processes [3].

When treated with otherwise lethal concentrations of NLVS or lactacystin, small cell subpopulations within cultured EL4 cell lines are capable of adapting and proliferating in the presence of this inhibitor. Partial impairment of the proteasome (in the adapted cells, $\beta 5$ proved to be completely disabled, whereas $\beta 1$ and $\beta 2$ remained to a large extent active) can be overcome by a small subpopulation of cells that can outgrow the rest of the culture, resulting in a cell line resistant to inhibition of the $\beta 5$ subunit. As may be expected, adapted EL4 cells are partially compromised in their ability to generate MHC class I antigenic peptides [93, 117, 123, 124]. Tripeptidyl peptidase II (TPPII) appears to compensate in these cells for the loss of proteasomal activity, a finding that may become very important regarding development of resistance in cancer patients treated with proteasome inhibitors [125–127]. Burkitt's lymphoma cells prove to be quite resistant to apoptosis induced by proteasome inhibitors [128]. Although proteasomal peptidase activities are significantly reduced in these cells, the overall rates of protein breakdown barely change. As in NLVS- and lactacystin-adapted cells, in Burkitt's lymphoma cells it was found that the activity of TPPII is increased compared with other cells. This effect appears to be related to expression of the constitutively activated c-*myc* oncogene. Moreover, an inhibitor of TPPII activity, AAFcmk (alanyl-alanyl-phenylalanyl chloromethyl ketone) [129–131], in contrast to proteasome inhibitors, was able to inhibit proliferation of these cells, suggesting that upregulation of TPPII may indeed compensate for the decreased overall activity of proteasomes in these cancer cell lines. The inhibitor butabindide [129] and analogues thereof [132] form superior alternatives to the use of AAFcmk.

3.6 Proteasome-associated Components: The Role of *N*-glycanase

Successful maturation of proteins determines the intracellular fate of secretory and membrane proteins in the endoplasmic reticulum (ER). Failure of adaptor molecules such as calnexin and calreticulin to provide assistance in folding and refolding or assembly of glycosylated proteins can lead to retention in the ER and redirection to the cytosol for degradation by the proteasome of these glycoproteins [131, 133]. Many substrates may be subject to this mode of degradation. For instance, MHC class I molecules are assembled and loaded with antigenic peptide in the ER and subsequently displayed at the cell surface. *N*-linked glycosylation of class I nascent chain that enters the secretory pathway contributes to its proper folding, assembly, and trafficking. The identification of a role for peptide-(*N*-acetyl-*β*-D-glucosaminyl)-asparagine amidase (PNG) activity in the cytosol of mammalian cells emerged from a strategy used by the human cytomegalovirus (HCMV) to evade detection by the immune system of its host by the HCMV gene products US2 and US11 [30, 134–137]. Inhibition of proteasomal proteolysis results in the accumulation of a deglycosylated MHC class I heavy-chain intermediate in the cytosol. This finding is consistent with the action of a peptide:*N*-glycanase (PNGase) on the substrate prior to its destruction by the proteasome [18, 138–144]. Oligosaccharyl transferase [144, 145] and PNG1 play important roles in the degradation of ER proteins. PNG1 is located in the cytosol, where it is thought to assist the proteasome in degradation of ER-derived glycoproteins. PNG1 recognizes glycosylated, preferentially denatured [138] protein substrates and, at least in yeast, may directly associate with the 19S cap subunit mHR23B [143]. It is therefore reasonable to assume that selective *N*-glycanase inhibitors, allosteric *N*-glycanase activators, and oligosaccharyl transferase inhibitors will form useful tools to explore pharmaceutical targets upstream of the proteasome.

A high-throughput screen ($n > 100{,}000$) for small-molecule inhibitors of mammalian PNG revealed the general caspase inhibitor ZVAD(OMe)fmk (benzyloxycarbonyl-valine-alanine-aspartic fluoromethyl ketone) (**21**, Figure 3.6) as an inhibitor of *N*-glycanase. Caspases and PNGases share no obvious structural or functional similarities. At concentrations of ZVAD(OMe)fmk commonly required to block apoptosis, *N*-glycanase is inhibited as well [146]. ZVAD(OMe)fmk inhibits PNG1 with an IC_{50} of about 12 μM in cultured cells. ZVAD(OMe)fmk is *in situ* converted

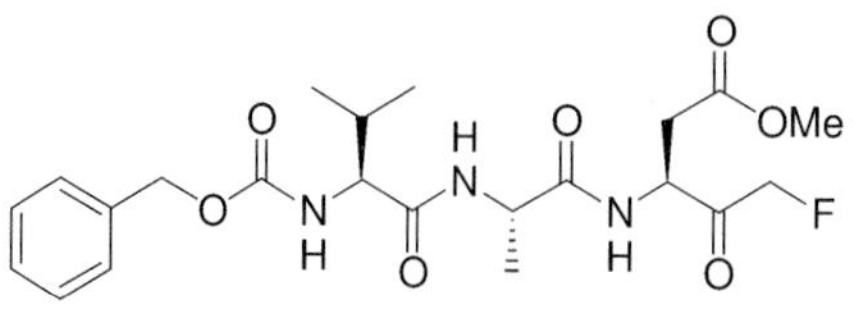

21

Fig. 3.6. The *N*-glycanase and caspase inhibitor ZVAD(OMe)fmk.

into the active inhibitor ZVADfmk by saponification catalyzed by esterases. *In vitro* only the product of saponification, ZVADfmk, exhibits inhibitory activity, but it is not cell-permeable. It is unlikely that this would be a unique example of such unexpected cross-targeting. It will therefore be important to explore cross-reactivities of inhibitors in general.

3.7 A Link Between Proteasomal Proteolysis and Deubiquitination

3.7.1 Reversal of Ub Modification

The biological effect of ubiquitin-specific proteases (USPs) is twofold: they either rescue proteins from destruction or condemn proteins to destruction via proteasomal degradation [147–149]. The steady-state level of Ub conjugates is the result of a subtle balance between the action of ubiquitin ligases and USPs in a manner comparable to the opposing actions of kinases and phosphatases.

3.7.2 Ubiquitin-specific Proteases

Four major subfamilies of ubiquitin-specific proteases have been identified to date [150]. The best-studied subfamilies, characterized by the presence of a catalytically active cysteine residue, are known as ubiquitin-specific processing proteases (UBPs) and ubiquitin carboxy-terminal hydrolases (UCHs). Members of these families possess the signature sequence motifs of a cysteine protease and show characteristic patterns of sequence conservation in their catalytic core domains. USPs can remove Ub from large polypeptides and disassemble poly-Ub chains, whereas UCHs normally target Ub derivatives with C-terminal linear extensions [151]. Ovarian tumor domain (OTU)-containing cysteine proteases form a third large family that shares no obvious homologies with either UBP or UCH families [152–155]. A single JAMM family metalloprotease within the 19S cap of the proteasome, RPN11 (POH-1), was shown to cleave ubiquitin moieties [20, 53].

RPN11 lacks a cysteine protease signature and is insensitive to the classical USP inhibitor Ub aldehyde. Other families of USPs may well exist.

3.7.3
USP Reactive Probes Correlate USP Activity With Proteasomal Proteolysis

Several reports have described the association of USPs with the proteasome. Development of radioiodinated Ub-nitrile led to the discovery of UCH37's association with the mammalian proteasome [50], whereas a radiolabeled ubiquitin probe with a C-terminal vinyl sulfone moiety was crucial for the discovery of the association of USP14, the mammalian homologue of Ubp6, with the 19S proteasome cap [51, 52]. In all cases, labeling of these USPs was abolished by pre-incubation with ubiquitin aldehyde.

USP-reactive probes are mechanism-based and thus provide a convenient tool to examine the enzymatic activity of USPs in response to proteasome inhibition [153, 156–158].

Whereas labeling, and hence activity, of UCH37 does not change upon treatment with proteasome inhibitor, the labeling of proteasome-associated USP14 was increased up to 15-fold in a time-dependent manner [51] in EL4 cell extracts (Figure 3.7). The observed increase in probe modification of USP14 is not unique to pro-

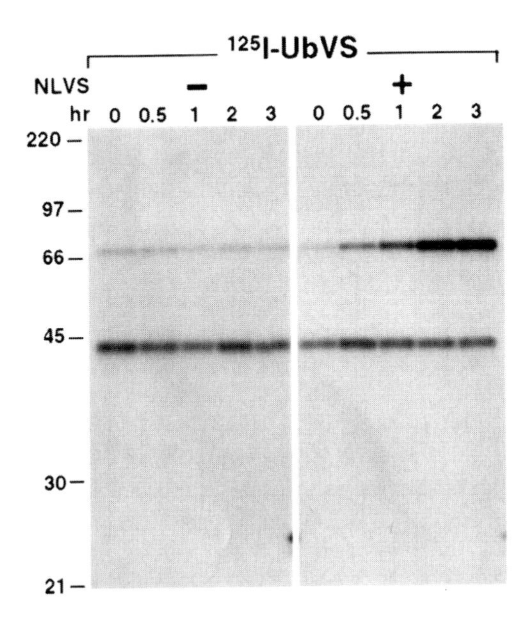

Fig. 3.7. Association of USP14 with the proteasome. Superose-6 fractions were labeled with ^{125}I iodinated ubiquitin vinyl sulfone and 20S complexes were immunoprecipitated with an anti-20S antibody. Fractions containing probe-modified USP14 were detected only in fractions corresponding to 26S proteasome complex and not in fractions with free 20S proteasomes.

teasome inhibition by NLVS, as treatment of cells with other proteasome inhibitors produced similar effects. Complete Ub removal is thought to precede proteasomal proteolysis. When proteasomal proteolysis is blocked, the resultant accumulation of Ub-conjugated substrates may elicit enhanced activity of USPs. In other words, the activities of the proteasome and associated USPs may be interdependent.

The exact reason for this increase in activity upon proteasomal inhibition is not yet fully understood. USP14 was also shown to exist either in a free form or bound to the proteasome. Only the latter is enzymatically active, as demonstrated in labeling experiments with mechanism-based probes. The requirement for USP14 (ubp6 in yeast) to associate with proteasome particles to become active may represent a regulatory mechanism that prevents random deubiquitination of substrates within cells. Removal of ubiquitin from substrates bound to the proteasome prior to their destruction salvages ubiquitin and may be important to maintain a steady-state ubiquitin level [159].

The importance of proper USP14 function is also underscored by the fact that mice deficient in USP14 develop cerebellar ataxia early on in their development [160]. Small molecule inhibitors specific for deubiquitinating enzymes would be an alternative inroad to interfere with targets upstream of proteasome function.

3.8 Future Developments and Final Remarks

Advances in understanding protein degradation and protein-folding pathways have been made possible by inhibitors of distinct activities, directly or indirectly, involved in proteolytic degradation pathways. Development of novel inhibitors may allow a deeper insight into the ubiquitin–proteasome system and will offer new approaches for blockade of up- and downstream events as well as future pharmacological intervention and hence treatment of disease.

Acknowledgments

This work was supported by the National Institutes of Health (H.L.P., and B.M.K.) and the Netherlands Organization for Scientific Research (H.O. and H.S.O.). We apologize to colleagues whose work we could not cite due to restrictions to the length of this review.

Abbreviations

Ub	ubiquitin;
UBL	ubiquitin-like protein;
UCH	ubiquitin carboxy-terminal hydrolase;

UBP ubiquitin-processing protease;
USP ubiquitin-specific protease;
PNG peptide-*N*-glycanase;
ER endoplasmic reticulum;
MVBs multivesicular bodies;
UPS ubiquitin–proteasome system

References

1 Lee, D.H., and A.L. Goldberg. 1998. Proteasome inhibitors: valuable new tools for cell biologists. *Trends Cell Biol* 8:397–403.
2 Goldberg, A.L. 2000. Probing the proteasome pathway. *Nat Biotechnol* 18:494–496.
3 Bogyo, M., and E.W. Wang. 2002. Proteasome inhibitors: complex tools for a complex enzyme. *Curr Top Microbiol Immunol* 268:185–208.
4 Adams, J. 2001. Proteasome inhibition in cancer: development of PS-341. *Semin Oncol* 28:613–619.
5 Gillessen, S., M. Groettup, and T. Cerny. 2002. The proteasome, a new target for cancer therapy. *Onkologie* 25:534–539.
6 Adams, J. 2003. The proteasome: structure, function, and role in the cell. *Cancer Treat Rev* 29 Suppl 1: 3–9.
7 Hideshima, T., P. Richardson, and K.C. Anderson. 2003. Novel therapeutic approaches for multiple myeloma. *Immunol Rev* 194:164–176.
8 Pieters, J., and H. Ploegh. 2003. Microbiology. Chemical warfare and mycobacterial defense. *Science* 302:1900–1902.
9 Darwin, K.H., S. Ehrt, J.C. Gutierrez-Ramos, N. Weich, and C.F. Nathan. 2003. The proteasome of Mycobacterium tuberculosis is required for resistance to nitric oxide. *Science* 302:1963–1966.
10 Hershko, A., and A. Ciechanover. 1998. The ubiquitin system. *Annu Rev Biochem* 67:425–479.
11 Pickart, C.M. 1997. Targeting of substrates to the 26S proteasome. *Faseb J* 11:1055–1066.
12 Orlowski, M., and S. Wilk. 2003. Ubiquitin-independent proteolytic functions of the proteasome. *Arch Biochem Biophys* 415:1–5.
13 Kisselev, A.F., T.N. Akopian, K.M. Woo, and A.L. Goldberg. 1999. The sizes of peptides generated from protein by mammalian 26 and 20 S proteasomes. Implications for understanding the degradative mechanism and antigen presentation. *J Biol Chem* 274:3363–3371.
14 Voges, D., P. Zwickl, and W. Baumeister. 1999. The 26S proteasome: a molecular machine designed for controlled proteolysis. *Annu Rev Biochem* 68:1015–1068.
15 Coux, O. 2002. The 26S proteasome. *Prog Mol Subcell Biol* 29:85–107.
16 Kitzmuller, C., A. Caprini, S.E. Moore, J.P. Frenoy, E. Schwaiger, O. Kellermann, N.E. Ivessa, and M. Ermonval. 2003. Processing of *N*-linked glycans during endoplasmic-reticulum-associated degradation of a short-lived variant of ribophorin I. *Biochem J* 376:687–696.
17 Suzuki, T., M.A. Kwofie, and W.J. Lennarz. 2003. Ngly1, a mouse gene encoding a deglycosylating enzyme implicated in proteasomal degradation: expression, genomic organization, and chromosomal mapping. *Biochem Biophys Res Commun* 304:326–332.
18 Hirsch, C., D. Blom, and H.L. Ploegh. 2003. A role for *N*-glycanase in the cytosolic turnover of glycoproteins. *Embo J* 22:1036–1046.
19 Kim, J.H., K.C. Park, S.S. Chung, O. Bang, and C.H. Chung. 2003. Deubiquitinating enzymes as cellular

regulators. *J Biochem (Tokyo)* 134:9–18.

20 Verma, R., L. Aravind, R. Oania, W.H. McDonald, J.R. Yates, 3rd, E.V. Koonin, and R.J. Deshaies. 2002. Role of Rpn11 metalloprotease in deubiquitination and degradation by the 26S proteasome. *Science* 298:611–615.

21 Yao, T., and R.E. Cohen. 2002. A cryptic protease couples deubiquitination and degradation by the proteasome. *Nature* 419:403–407.

22 Aguilar, R.C., and B. Wendland. 2003. Ubiquitin: not just for proteasomes anymore. *Curr Opin Cell Biol* 15:184–190.

23 Bilodeau, P.S., S.C. Winistorfer, W.R. Kearney, A.D. Robertson, and R.C. Piper. 2003. Vps27-Hse1 and ESCRT-I complexes cooperate to increase efficiency of sorting ubiquitinated proteins at the endosome. *J Cell Biol* 163:237–243.

24 Katzmann, D.J., M. Babst, and S.D. Emr. 2001. Ubiquitin-dependent sorting into the multivesicular body pathway requires the function of a conserved endosomal protein sorting complex, ESCRT-I. *Cell* 106:145–155.

25 Katzmann, D.J., C.J. Stefan, M. Babst, and S.D. Emr. 2003. Vps27 recruits ESCRT machinery to endosomes during MVB sorting. *J Cell Biol* 162:413–423.

26 Hicke, L., and R. Dunn. 2003. Regulation of membrane protein transport by ubiquitin and ubiquitin-binding proteins. *Annu Rev Cell Dev Biol* 19:141–172.

27 Schubert, U., L.C. Anton, J. Gibbs, C.C. Norbury, J.W. Yewdell, and J.R. Bennink. 2000. Rapid degradation of a large fraction of newly synthesized proteins by proteasomes. *Nature* 404:770–774.

28 Yewdell, J.W., L.C. Anton, and J.R. Bennink. 1996. Defective ribosomal products (DRiPs): a major source of antigenic peptides for MHC class I molecules? *J Immunol* 157:1823–1826.

29 Liu, C.W., M.J. Corboy, G.N. DeMartino, and P.J. Thomas. 2003. Endoproteolytic activity of the proteasome. *Science* 299:408–411.

30 Wiertz, E.J., T.R. Jones, L. Sun, M. Bogyo, H.J. Geuze, and H.L. Ploegh. 1996. The human cytomegalovirus US11 gene product dislocates MHC class I heavy chains from the endoplasmic reticulum to the cytosol. *Cell* 84:769–779.

31 Wiertz, E.J., D. Tortorella, M. Bogyo, J. Yu, W. Mothes, T.R. Jones, T.A. Rapoport, and H.L. Ploegh. 1996. Sec61-mediated transfer of a membrane protein from the endoplasmic reticulum to the proteasome for destruction. *Nature* 384:432–438.

32 Kikkert, M., G. Hassink, M. Barel, C. Hirsch, F.J. van der Wal, and E. Wiertz. 2001. Ubiquitination is essential for human cytomegalovirus US11-mediated dislocation of MHC class I molecules from the endoplasmic reticulum to the cytosol. *Biochem J* 358:369–377.

33 Hirsch, C., and H.L. Ploegh. 2000. Intracellular targeting of the proteasome. *Trends Cell Biol* 10:268–272.

34 Enenkel, C., A. Lehmann, and P.M. Kloetzel. 1998. Subcellular distribution of proteasomes implicates a major location of protein degradation in the nuclear envelope-ER network in yeast. *Embo J* 17:6144–6154.

35 Brooks, P., G. Fuertes, R.Z. Murray, S. Bose, E. Knecht, M.C. Rechsteiner, K.B. Hendil, K. Tanaka, J. Dyson, and J. Rivett. 2000. Subcellular localization of proteasomes and their regulatory complexes in mammalian cells. *Biochem J* 346 Pt 1:155–161.

36 Fruh, K., and Y. Yang. 1999. Antigen presentation by MHC class I and its regulation by interferon gamma. *Curr Opin Immunol* 11:76–81.

37 Kloetzel, P.M. 2001. Antigen processing by the proteasome. *Nat Rev Mol Cell Biol* 2:179–187.

38 Sijts, A.J., T. Ruppert, B. Rehermann, M. Schmidt, U. Koszinowski, and P.M. Kloetzel. 2000. Efficient generation of a hepatitis B virus cytotoxic T

lymphocyte epitope requires the structural features of immunoproteasomes. *J Exp Med* 191:503–514.

39 Unno, M., T. Mizushima, Y. Morimoto, Y. Tomisugi, K. Tanaka, N. Yasuoka, and T. Tsukihara. 2002. Structure determination of the constitutive 20S proteasome from bovine liver at 2.75 A resolution. *J Biochem (Tokyo)* 131:171–173.

40 Goldberg, A.L., P. Cascio, T. Saric, and K.L. Rock. 2002. The importance of the proteasome and subsequent proteolytic steps in the generation of antigenic peptides. *Mol Immunol* 39:147–164.

41 Di Napoli, M., and F. Papa. 2003. The proteasome system and proteasome inhibitors in stroke: controlling the inflammatory response. *Curr Opin Investig Drugs* 4:1333–1342.

42 Tang, G., and S.H. Leppla. 1999. Proteasome activity is required for anthrax lethal toxin to kill macrophages. *Infect Immun* 67:3055–3060.

43 Hideshima, T., D. Chauhan, T. Hayashi, M. Akiyama, N. Mitsiades, C. Mitsiades, K. Podar, N.C. Munshi, P.G. Richardson, and K.C. Anderson. 2003. Proteasome inhibitor PS-341 abrogates IL-6 triggered signaling cascades via caspase-dependent downregulation of gp130 in multiple myeloma. *Oncogene* 22:8386–8393.

44 Elliott, P.J., T.M. Zollner, and W.H. Boehncke. 2003. Proteasome inhibition: a new anti-inflammatory strategy. *J Mol Med* 81:235–245.

45 Shah, S.A., M.W. Potter, and M.P. Callery. 2002. Ubiquitin proteasome inhibition and cancer therapy. *Surgery* 131:595–600.

46 Bose, S., G.G. Mason, and A.J. Rivett. 1999. Phosphorylation of proteasomes in mammalian cells. *Mol Biol Rep* 26:11–14.

47 Bose, S., F.L. Stratford, K.I. Broadfoot, G.G. Mason, and A.J. Rivett. 2004. Phosphorylation of 20S proteasome alpha subunit C8 (alpha7) stabilizes the 26S proteasome and plays a role in the regulation of proteasome complexes by gamma-interferon. *Biochem J* 378:177–184.

48 Zhang, F., K. Su, X. Yang, D.B. Bowe, A.J. Paterson, and J.E. Kudlow. 2003. O-GlcNAc modification is an endogenous inhibitor of the proteasome. *Cell* 115:715–725.

49 Baumeister, W., J. Walz, F. Zuhl, and E. Seemuller. 1998. The proteasome: paradigm of a self-compartmentalizing protease. *Cell* 92:367–380.

50 Lam, Y.A., W. Xu, G.N. DeMartino, and R.E. Cohen. 1997. Editing of ubiquitin conjugates by an isopeptidase in the 26S proteasome. *Nature* 385:737–740.

51 Borodovsky, A., B.M. Kessler, R. Casagrande, H.S. Overkleeft, K.D. Wilkinson, and H.L. Ploegh. 2001. A novel active site-directed probe specific for deubiquitylating enzymes reveals proteasome association of USP14. *Embo J* 20:5187–5196.

52 Leggett, D.S., J. Hanna, A. Borodovsky, B. Crosas, M. Schmidt, R.T. Baker, T. Walz, H. Ploegh, and D. Finley. 2002. Multiple associated proteins regulate proteasome structure and function. *Mol Cell* 10:495–507.

53 Berndt, C., D. Bech-Otschir, W. Dubiel, and M. Seeger. 2002. Ubiquitin system: JAMMing in the name of the lid. *Curr Biol* 12:R815–817.

54 Stock, D., L. Ditzel, W. Baumeister, R. Huber, and J. Lowe. 1995. Catalytic mechanism of the 20S proteasome of Thermoplasma acidophilum revealed by X-ray crystallography. *Cold Spring Harb Symp Quant Biol* 60:525–532.

55 Lowe, J., D. Stock, B. Jap, P. Zwickl, W. Baumeister, and R. Huber. 1995. Crystal structure of the 20S proteasome from the archaeon T. acidophilum at 3.4 A resolution. *Science* 268:533–539.

56 Groll, M., L. Ditzel, J. Lowe, D. Stock, M. Bochtler, H.D. Bartunik, and R. Huber. 1997. Structure of 20S proteasome from yeast at 2.4 A resolution. *Nature* 386:463–471.

57 Orlowski, M., C. Cardozo, and C.

Michaud. 1993. Evidence for the presence of five distinct proteolytic components in the pituitary multicatalytic proteinase complex. Properties of two components cleaving bonds on the carboxyl side of branched chain and small neutral amino acids. *Biochemistry* 32:1563–1572.

58 Nussbaum, A.K., T.P. Dick, W. Keilholz, M. Schirle, S. Stevanovic, K. Dietz, W. Heinemeyer, M. Groll, D.H. Wolf, R. Huber, H.G. Rammensee, and H. Schild. 1998. Cleavage motifs of the yeast 20S proteasome beta subunits deduced from digests of enolase 1. *Proc Natl Acad Sci U S A* 95:12504–12509.

59 Dick, T.P., A.K. Nussbaum, M. Deeg, W. Heinemeyer, M. Groll, M. Schirle, W. Keilholz, S. Stevanovic, D.H. Wolf, R. Huber, H.G. Rammensee, and H. Schild. 1998. Contribution of proteasomal beta-subunits to the cleavage of peptide substrates analyzed with yeast mutants. *J Biol Chem* 273:25637–25646.

60 Toes, R.E., A.K. Nussbaum, S. Degermann, M. Schirle, N.P. Emmerich, M. Kraft, C. Laplace, A. Zwinderman, T.P. Dick, J. Muller, B. Schonfisch, C. Schmid, H.J. Fehling, S. Stevanovic, H.G. Rammensee, and H. Schild. 2001. Discrete cleavage motifs of constitutive and immunoproteasomes revealed by quantitative analysis of cleavage products. *J Exp Med* 194:1–12.

61 Kessler, B.M., D. Tortorella, M. Altun, A.F. Kisselev, E. Fiebiger, B.G. Hekking, H.L. Ploegh, and H.S. Overkleeft. 2001. Extended peptide-based inhibitors efficiently target the proteasome and reveal overlapping specificities of the catalytic beta-subunits. *Chem Biol* 8:913–929.

62 Cascio, P., C. Hilton, A.F. Kisselev, K.L. Rock, and A.L. Goldberg. 2001. 26S proteasomes and immunoproteasomes produce mainly N-extended versions of an antigenic peptide. *Embo J* 20:2357–2366.

63 Dahlmann, B., T. Ruppert, P.M. Kloetzel, and L. Kuehn. 2001. Subtypes of 20S proteasomes from skeletal muscle. *Biochimie* 83:295–299.

64 Dahlmann, B., T. Ruppert, L. Kuehn, S. Merforth, and P.M. Kloetzel. 2000. Different proteasome subtypes in a single tissue exhibit different enzymatic properties. *J Mol Biol* 303:643–653.

65 Russell, S.J., K.A. Steger, and S.A. Johnston. 1999. Subcellular localization, stoichiometry, and protein levels of 26 S proteasome subunits in yeast. *J Biol Chem* 274:21943–21952.

66 Cummings, C.J., M.A. Mancini, B. Antalffy, D.B. DeFranco, H.T. Orr, and H.Y. Zoghbi. 1998. Chaperone suppression of aggregation and altered subcellular proteasome localization imply protein misfolding in SCA1. *Nat Genet* 19:148–154.

67 Shringarpure, R., and K.J. Davies. 2002. Protein turnover by the proteasome in aging and disease. *Free Radic Biol Med* 32:1084–1089.

68 Shringarpure, R., T. Grune, J. Mehlhase, and K.J. Davies. 2003. Ubiquitin conjugation is not required for the degradation of oxidized proteins by proteasome. *J Biol Chem* 278:311–318.

69 Kanayama, H.O., T. Tamura, S. Ugai, S. Kagawa, N. Tanahashi, T. Yoshimura, K. Tanaka, and A. Ichihara. 1992. Demonstration that a human 26S proteolytic complex consists of a proteasome and multiple associated protein components and hydrolyzes ATP and ubiquitin-ligated proteins by closely linked mechanisms. *Eur J Biochem* 206:567–578.

70 Peters, J.M., J.R. Harris, and J.A. Kleinschmidt. 1991. Ultrastructure of the approximately 26S complex containing the approximately 20S cylinder particle (multicatalytic proteinase/proteasome). *Eur J Cell Biol* 56:422–432.

71 Dahlmann, B., L. Kuehn, and H. Reinauer. 1995. Studies on the activation by ATP of the 26 S proteasome complex from rat skeletal muscle. *Biochem J* 309 (Pt 1):195–202.

72 Rivett, A.J., G.G. Mason, R.Z. Murray, and J. Reidlinger. 1997. Regulation of proteasome structure and function. *Mol Biol Rep* 24:99–102.

73 Morimoto, Y., T. Mizushima, A. Yagi, N. Tanahashi, K. Tanaka, A. Ichihara, and T. Tsukihara. 1995. Ordered structure of the crystallized bovine 20S proteasome. *J Biochem (Tokyo)* 117:471–474.

74 Bochtler, M., C. Hartmann, H.K. Song, G.P. Bourenkov, H.D. Bartunik, and R. Huber. 2000. The structures of HslU and the ATP-dependent protease HslU-HslV. *Nature* 403:800–805.

75 Tomisugi, Y., M. Unno, T. Mizushima, Y. Morimoto, N. Tanahashi, K. Tanaka, T. Tsukihara, and N. Yasuoka. 2000. New crystal forms and low resolution structure analysis of 20S proteasomes from bovine liver. *J Biochem (Tokyo)* 127:941–943.

76 Jap, B., G. Puhler, H. Lucke, D. Typke, J. Lowe, D. Stock, R. Huber, and W. Baumeister. 1993. Preliminary X-ray crystallographic study of the proteasome from Thermoplasma acidophilum. *J Mol Biol* 234:881–884.

77 Groll, M., T. Nazif, R. Huber, and M. Bogyo. 2002. Probing structural determinants distal to the site of hydrolysis that control substrate specificity of the 20S proteasome. *Chem Biol* 9:655–662.

78 Kwon, A.R., B.M. Kessler, H.S. Overkleeft, and D.B. McKay. 2003. Structure and reactivity of an asymmetric complex between HslV and I-domain deleted HslU, a prokaryotic homolog of the eukaryotic proteasome. *J Mol Biol* 330:185–195.

79 Sousa, M.C., B.M. Kessler, H.S. Overkleeft, and D.B. McKay. 2002. Crystal structure of HslUV complexed with a vinyl sulfone inhibitor: corroboration of a proposed mechanism of allosteric activation of HslV by HslU. *J. Mol. Biol.* 318.

80 Elofsson, M., U. Splittgerber, J. Myung, R. Mohan, and C.M. Crews. 1999. Towards subunit-specific proteasome inhibitors: synthesis and evaluation of peptide alpha',beta'-epoxyketones. *Chem Biol* 6:811–822.

81 Kisselev, A.F., M. Garcia-Calvo, H.S. Overkleeft, E. Peterson, M.W. Pennington, H.L. Ploegh, N.A. Thornberry, and A.L. Goldberg. 2003. The caspase-like sites of proteasomes, their substrate specificity, new inhibitors and substrates, and allosteric interactions with the trypsin-like sites. *J Biol Chem* 278:35869–35877.

82 Stoltze, L., A.K. Nussbaum, A. Sijts, N.P. Emmerich, P.M. Kloetzel, and H. Schild. 2000. The function of the proteasome system in MHC class I antigen processing. *Immunol Today* 21:317–319.

83 Myung, J., K.B. Kim, K. Lindsten, N.P. Dantuma, and C.M. Crews. 2001. Lack of proteasome active site allostery as revealed by subunit-specific inhibitors. *Mol Cell* 7:411–420.

84 Garrett, I.R., D. Chen, G. Gutierrez, M. Zhao, A. Escobedo, G. Rossini, S.E. Harris, W. Gallwitz, K.B. Kim, S. Hu, C.M. Crews, and G.R. Mundy. 2003. Selective inhibitors of the osteoblast proteasome stimulate bone formation *in vivo* and *in vitro*. *J Clin Invest* 111:1771–1782.

85 Rivett, A.J., and R.C. Gardner. 2000. Proteasome inhibitors: from *in vitro* uses to clinical trials. *J Pept Sci* 6:478–488.

86 Kisselev, A.F., and A.L. Goldberg. 2001. Proteasome inhibitors: from research tools to drug candidates. *Chem Biol* 8:739–758.

87 Crews, C.M. 2003. Feeding the machine: mechanisms of proteasome-catalyzed degradation of ubiquitinated proteins. *Curr Opin Chem Biol* 7:534–539.

88 Momose, I., R. Sekizawa, H. Iinuma, and T. Takeuchi. 2002. Inhibition of proteasome activity by tyropeptin A in PC12 cells. *Biosci Biotechnol Biochem* 66:2256–2258.

89 Momose, I., R. Sekizawa, H. Hashizume, N. Kinoshita, Y. Homma, M. Hamada, H. Iinuma, and T. Takeuchi. 2001. Tyropeptins A and

B, new proteasome inhibitors produced by Kitasatospora sp. MK993-dF2. I. Taxonomy, isolation, physico-chemical properties and biological activities. *J Antibiot (Tokyo)* 54:997–1003.

90 Loidl, G., M. Groll, H.J. Musiol, L. Ditzel, R. Huber, and L. Moroder. 1999. Bifunctional inhibitors of the trypsin-like activity of eukaryotic proteasomes. *Chem Biol* 6:197–204.

91 Omura, S., K. Matsuzaki, T. Fujimoto, K. Kosuge, T. Furuya, S. Fujita, and A. Nakagawa. 1991. Structure of lactacystin, a new microbial metabolite which induces differentiation of neuroblastoma cells. *J Antibiot (Tokyo)* 44:117–118.

92 Omura, S., T. Fujimoto, K. Otoguro, K. Matsuzaki, R. Moriguchi, H. Tanaka, and Y. Sasaki. 1991. Lactacystin, a novel microbial metabolite, induces neuritogenesis of neuroblastoma cells. *J Antibiot (Tokyo)* 44:113–116.

93 Wang, E.W., B.M. Kessler, A. Borodovsky, B.F. Cravatt, M. Bogyo, H.L. Ploegh, and R. Glas. 2000. Integration of the ubiquitin–proteasome pathway with a cytosolic oligopeptidase activity. *Proc Natl Acad Sci U S A* 97:9990–9995.

94 Feling, R.H., G.O. Buchanan, T.J. Mincer, C.A. Kauffman, P.R. Jensen, and W. Fenical. 2003. Salinosporamide A: a highly cytotoxic proteasome inhibitor from a novel microbial source, a marine bacterium of the new genus salinospora. *Angew Chem Int Ed Engl* 42:355–357.

95 Reddy, L.R.S., P. Corey, E.J. 2004. A simple stereocontrolled synthesis of salinosporamide A. *Journal of the American Chemical Society* in press.

96 Meng, L., R. Mohan, B.H. Kwok, M. Elofsson, N. Sin, and C.M. Crews. 1999. Epoxomicin, a potent and selective proteasome inhibitor, exhibits *in vivo* antiinflammatory activity. *Proc Natl Acad Sci U S A* 96:10403–10408.

97 Sin, N., K.B. Kim, M. Elofsson, L. Meng, H. Auth, B.H. Kwok, and C.M. Crews. 1999. Total synthesis of the potent proteasome inhibitor epoxomicin: a useful tool for understanding proteasome biology. *Bioorg Med Chem Lett* 9:2283–2288.

98 Kim, K.B., J. Myung, N. Sin, and C.M. Crews. 1999. Proteasome inhibition by the natural products epoxomicin and dihydroeponemycin: insights into specificity and potency. *Bioorg Med Chem Lett* 9:3335–3340.

99 Groll, M., Y. Koguchi, R. Huber, and J. Kohno. 2001. Crystal structure of the 20 S proteasome:TMC-95A complex: a non-covalent proteasome inhibitor. *J Mol Biol* 311:543–548.

100 Kaiser, M., M. Groll, C. Renner, R. Huber, and L. Moroder. 2002. The core structure of TMC-95A is a promising lead for reversible proteasome inhibition. *Angew Chem Int Ed Engl* 41:780–783.

101 Koguchi, Y., J. Kohno, S. Suzuki, M. Nishio, K. Takahashi, T. Ohnuki, and S. Komatsubara. 1999. TMC-86A, B and TMC-96, new proteasome inhibitors from Streptomyces sp. TC 1084 and Saccharothrix sp. TC 1094. I. Taxonomy, fermentation, isolation, and biological activities. *J Antibiot (Tokyo)* 52:1069–1076.

102 Koguchi, Y., M. Nishio, S. Suzuki, K. Takahashi, T. Ohnuki, and S. Komatsubara. 2000. TMC-89A and B, new proteasome inhibitors from streptomyces sp. TC 1087. *J Antibiot (Tokyo)* 53:967–972.

103 Kohno, J., Y. Koguchi, M. Nishio, K. Nakao, M. Kuroda, R. Shimizu, T. Ohnuki, and S. Komatsubara. 2000. Structures of TMC-95A-D: novel proteasome inhibitors from Apiospora montagnei sacc. TC 1093. *J Org Chem* 65:990–995.

104 Koguchi, Y., J. Kohno, M. Nishio, K. Takahashi, T. Okuda, T. Ohnuki, and S. Komatsubara. 2000. TMC-95A, B, C, and D, novel proteasome inhibitors produced by Apiospora montagnei Sacc. TC 1093. Taxonomy, production, isolation, and biological activities. *J Antibiot (Tokyo)* 53:105–109.

105 Koguchi, Y., J. Kohno, S. Suzuki, M. Nishio, K. Takahashi, T.

Ohnuki, and S. Komatsubara. 2000. TMC-86A, B and TMC-96, new proteasome inhibitors from Streptomyces sp. TC 1084 and Saccharothrix sp. TC 1094. II. Physico-chemical properties and structure determination. *J Antibiot (Tokyo)* 53:63–65.

106 Ma, M.H., H.H. Yang, K. Parker, S. Manyak, J.M. Friedman, C. Altamirano, Z.Q. Wu, M.J. Borad, M. Frantzen, E. Roussos, J. Neeser, A. Mikail, J. Adams, N. Sjak-Shie, R.A. Vescio, and J.R. Berenson. 2003. The Proteasome Inhibitor PS-341 Markedly Enhances Sensitivity of Multiple Myeloma Tumor Cells to Chemotherapeutic Agents. *Clin Cancer Res* 9:1136–1144.

107 Kaiser, M., C. Siciliano, I. Assfalg-Machleidt, M. Groll, A.G. Milbradt, and L. Moroder. 2003. Synthesis of a TMC-95A ketomethylene analogue by cyclization via intramolecular Suzuki coupling. *Org Lett* 5:3435–3437.

108 Lin, S., and S.J. Danishefsky. 2002. The total synthesis of proteasome inhibitors TMC-95A and TMC-95B: discovery of a new method to generate cis-propenyl amides. *Angew Chem Int Ed Engl* 41:512–515.

109 Lin, S., and S.J. Danishefsky. 2001. Synthesis of the Functionalized Macrocyclic Core of Proteasome Inhibitors TMC-95A and B This work was supported by grants from the National Institutes of Health (grant CA28824). We thank Dr. George Sukenick and Sylvi Rusli of the MSKCC NMR Core Facility for NMR and mass spectral analyses (NIH Grant CA08748). *Angew Chem Int Ed Engl* 40:1967–1970.

110 Hideshima, T., P.G. Richardson, and K.C. Anderson. 2003. Targeting proteasome inhibition in hematologic malignancies. *Rev Clin Exp Hematol* 7:191–204.

111 Palmer, J.T., D. Rasnick, J.L. Klaus, and D. Bromme. 1995. Vinyl sulfones as mechanism-based cysteine protease inhibitors. *J Med Chem* 38:3193–3196.

112 Bogyo, M., M. Gaczynska, and H.L. Ploegh. 1997. Proteasome inhibitors and antigen presentation. *Biopolymers* 43:269–280.

113 Bogyo, M., J.S. McMaster, M. Gaczynska, D. Tortorella, A.L. Goldberg, and H. Ploegh. 1997. Covalent modification of the active site threonine of proteasomal beta subunits and the Escherichia coli homolog HslV by a new class of inhibitors. *Proc Natl Acad Sci U S A* 94:6629–6634.

114 Richardson, P.G., B. Barlogie, J. Berenson, S. Singhal, S. Jagannath, D. Irwin, S.V. Rajkumar, G. Srkalovic, M. Alsina, R. Alexanian, D. Siegel, R.Z. Orlowski, D. Kuter, S.A. Limentani, S. Lee, T. Hideshima, D.L. Esseltine, M. Kauffman, J. Adams, D.P. Schenkein, and K.C. Anderson. 2003. A phase 2 study of bortezomib in relapsed, refractory myeloma. *N Engl J Med* 348:2609–2617.

115 Otto, H.H., and T. Schirmeister. 1997. Cysteine Proteases and Their Inhibitors. *Chem Rev* 97:133–172.

116 Bogyo, M., S. Verhelst, V. Bellingard-Dubouchaud, S. Toba, and D. Greenbaum. 2000. Selective targeting of lysosomal cysteine proteases with radiolabeled electrophilic substrate analogs. *Chem Biol* 7:27–38.

117 Glas, R., M. Bogyo, J.S. McMaster, M. Gaczynska, and H.L. Ploegh. 1998. A proteolytic system that compensates for loss of proteasome function. *Nature* 392:618–622.

118 Saxon, E., J.I. Armstrong, and C.R. Bertozzi. 2000. A "traceless" Staudinger ligation for the chemoselective synthesis of amide bonds. *Org Lett* 2:2141–2143.

119 Saxon, E., and C.R. Bertozzi. 2000. Cell surface engineering by a modified Staudinger reaction. *Science* 287:2007–2010.

120 Ovaa, H., B.M. Kessler, U. Rolen, P.J. Galardy, H.L. Ploegh, and M.G. Masucci. 2003. Activity-based ubiquitin-specific protease (USP) profiling of virus-infected and malignant human cells. *Proc Natl Acad Sci U S A* 101 (8):2253–2258.

121 Ovaa, H., P.F. Van Swieten, B.M. Kessler, M.A. Leeuwenburgh, E. Fiebiger, A.M. Van Den Nieuwendijk, P.J. Galardy, G.A. Van Der Marel, H.L. Ploegh, and H.S. Overkleeft. 2003. Chemistry in Living Cells: Detection of Active Proteasomes by a Two-Step Labeling Strategy. *Angew Chem Int Ed Engl* 42:3626–3629.

122 Nazif, T., and M. Bogyo. 2001. Global analysis of proteasomal substrate specificity using positional-scanning libraries of covalent inhibitors. *Proc Natl Acad Sci U S A* 98:2967–2972.

123 Geier, E., G. Pfeifer, M. Wilm, M. Lucchiari-Hartz, W. Baumeister, K. Eichmann, and G. Niedermann. 1999. A giant protease with potential to substitute for some functions of the proteasome. *Science* 283:978–981.

124 Kessler, B., X. Hong, J. Petrovic, A. Borodovsky, N.P. Dantuma, M. Bogyo, H.S. Overkleeft, H. Ploegh, and R. Glas. 2003. Pathways accessory to proteasomal proteolysis are less efficient in major histocompatibility complex class I antigen production. *J Biol Chem* 278:10013–10021.

125 Reits, E., J. Neijssen, C. Herberts, W. Benckhuijsen, L. Janssen, J.W. Drijfhout, and J. Neefjes. 2004. A Major Role for TPPII in Trimming Proteasomal Degradation Products for MHC Class I Antigen Presentation. *Immunity* 20:495–506.

126 Seifert, U., C. Maranon, A. Shmueli, J.F. Desoutter, L. Wesoloski, K. Janek, P. Henklein, S. Diescher, M. Andrieu, H. de la Salle, T. Weinschenk, H. Schild, D. Laderach, A. Galy, G. Haas, P.M. Kloetzel, Y. Reiss, and A. Hosmalin. 2003. An essential role for tripeptidyl peptidase in the generation of an MHC class I epitope. *Nat Immunol* 4:375–379.

127 Kloetzel, P.M., and F. Ossendorp. 2004. Proteasome and peptidase function in MHC-class-I-mediated antigen presentation. *Curr Opin Immunol* 16:76–81.

128 Gavioli, R., T. Frisan, S. Vertuani, G.W. Bornkamm, and M.G. Masucci. 2001. c-myc overexpression activates alternative pathways for intracellular proteolysis in lymphoma cells. *Nat Cell Biol* 3:283–288.

129 Ganellin, C.R., P.B. Bishop, R.B. Bambal, S.M. Chan, J.K. Law, B. Marabout, P.M. Luthra, A.N. Moore, O. Peschard, P. Bourgeat, C. Rose, F. Vargas, and J.C. Schwartz. 2000. Inhibitors of tripeptidyl peptidase II. 2. Generation of the first novel lead inhibitor of cholecystokinin-8-inactivating peptidase: a strategy for the design of peptidase inhibitors. *J Med Chem* 43:664–674.

130 Facchinetti, P., C. Rose, P. Rostaing, A. Triller, and J.C. Schwartz. 1999. Immunolocalization of tripeptidyl peptidase II, a cholecystokinin-inactivating enzyme, in rat brain. *Neuroscience* 88:1225–1240.

131 Hampton, R.Y. 2002. ER-associated degradation in protein quality control and cellular regulation. *Curr Opin Cell Biol* 14:476–482.

132 Breslin, H.J., T.A. Miskowski, M.J. Kukla, W.H. Leister, H.L. De Winter, D.A. Gauthier, M.V. Somers, D.C. Peeters, and P.W. Roevens. 2002. Design, synthesis, and tripeptidyl peptidase II inhibitory activity of a novel series of (S)-2,3-dihydro-2-(4-alkyl-1H-imidazol-2-yl)-1H-indoles. *J Med Chem* 45:5303–5310.

133 Hampton, R.Y. 2000. ER stress response: getting the UPR hand on misfolded proteins. *Curr Biol* 10: R518–521.

134 Tortorella, D., B. Gewurz, D. Schust, M. Furman, and H. Ploegh. 2000. Down-regulation of MHC class I antigen presentation by HCMV; lessons for tumor immunology. *Immunol Invest* 29:97–100.

135 Tortorella, D., B.E. Gewurz, M.H. Furman, D.J. Schust, and H.L. Ploegh. 2000. Viral subversion of the immune system. *Annu Rev Immunol* 18:861–926.

136 Gewurz, B.E., E.W. Wang, D. Tortorella, D.J. Schust, and H.L. Ploegh. 2001. Human cytomegalovirus US2 endoplasmic reticulum-lumenal domain dictates association with major histocompatibility complex class I in a locus-specific manner. *J Virol* 75:5197–5204.

137 Gewurz, B.E., R. Gaudet, D. Tortorella, E.W. Wang, H.L. Ploegh, and D.C. Wiley. 2001. Antigen presentation subverted: Structure of the human cytomegalovirus protein US2 bound to the class I molecule HLA-A2. *Proc Natl Acad Sci U S A* 98:6794–6799.

138 Blom, D., C. Hirsch, P. Stern, D. Tortorella, and H.L. Ploegh. 2004. A glycosylated type I membrane protein becomes cytosolic when peptide: *N*-glycanase is compromised. *Embo J* 23:650–658.

139 Suzuki, T., H. Park, N.M. Hollingsworth, R. Sternglanz, and W.J. Lennarz. 2000. PNG1, a yeast gene encoding a highly conserved peptide: *N*-glycanase. *J Cell Biol* 149:1039–1052.

140 Suzuki, T., and W.J. Lennarz. 2003. Hypothesis: a glycoprotein-degradation complex formed by protein-protein interaction involves cytoplasmic peptide:*N*-glycanase. *Biochem Biophys Res Commun* 302:1–5.

141 Suzuki, T., H. Park, and W.J. Lennarz. 2002. Cytoplasmic peptide:*N*-glycanase (PNGase) in eukaryotic cells: occurrence, primary structure, and potential functions. *Faseb J* 16:635–641.

142 Katiyar, S., T. Suzuki, B.J. Balgobin, and W.J. Lennarz. 2002. Site-directed mutagenesis study of yeast peptide:*N*-glycanase. Insight into the reaction mechanism of deglycosylation. *J Biol Chem* 277:12953–12959.

143 Park, H., T. Suzuki, and W.J. Lennarz. 2001. Identification of proteins that interact with mammalian peptide:*N*-glycanase and implicate this hydrolase in the proteasome-dependent pathway for protein degradation. *Proc Natl Acad Sci U S A* 98:11163–11168.

144 Dempski, R.E., Jr., and B. Imperiali. 2002. Oligosaccharyl transferase: gatekeeper to the secretory pathway. *Curr Opin Chem Biol* 6:844–850.

145 Imperiali, B., and T.L. Hendrickson. 1995. Asparagine-linked glycosylation: specificity and function of oligosaccharyl transferase. *Bioorg Med Chem* 3:1565–1578.

146 Misaghi, S., M.E. Pacold, D. Blom, H.L. Ploegh, G.A. Korbel. 2004. Using a small molecule inhibition of peptide: N-glycanase to probe its role in glycoprotein turnover. *Chem. Biol.* 12:1677–87.

147 Glickman, M.H., and N. Adir. 2004. The Proteasome and the Delicate Balance between Destruction and Rescue. *PLoS Biol* 2:E13.

148 Glickman, M.H., and V. Maytal. 2002. Regulating the 26S proteasome. *Curr Top Microbiol Immunol* 268:43–72.

149 Glickman, M.H., and A. Ciechanover. 2002. The ubiquitin–proteasome proteolytic pathway: destruction for the sake of construction. *Physiol Rev* 82:373–428.

150 Wing, S.S. 2003. Deubiquitinating enzymes – the importance of driving in reverse along the ubiquitin–proteasome pathway. *Int J Biochem Cell Biol* 35:590–605.

151 Wilkinson, K.D. 2000. Ubiquitination and deubiquitination: targeting of proteins for degradation by the proteasome. *Semin Cell Dev Biol* 11:141–148.

152 Balakirev, M.Y., S.O. Tcherniuk, M. Jaquinod, and J. Chroboczek. 2003. Otubains: a new family of cysteine proteases in the ubiquitin pathway. *EMBO Rep* 4:517–522.

153 Evans, P.C., H. Ovaa, M. Hamon, P.J. Kilshaw, S. Hamm, S. Bauer, H.L. Ploegh, and T.S. Smith. 2004. Zinc-finger protein A20, a regulator of inflammation and cell survival, has deubiquitinating activity. *Biochem J* 378:727–734.

154 Evans, P.C., T.S. Smith, M.J. Lai, M.G. Williams, D.F. Burke, K. Heyninck, M.M. Kreike, R. Beyaert, T.L. Blundell, and P.J. Kilshaw.

2003. A novel type of deubiquitinating enzyme. *J Biol Chem* 278:23180–23186.

155 Wang, Y., A. Satoh, G. Warren, and H.H. Meyer. 2004. VCIP135 acts as a deubiquitinating enzyme during p97-p47-mediated reassembly of mitotic Golgi fragments. *J Cell Biol* 164:973–978.

156 Ovaa, H., B.M. Kessler, U. Rolen, P.J. Galardy, H.L. Ploegh, and M.G. Masucci. 2004. Activity-based ubiquitin-specific protease (USP) profiling of virus-infected and malignant human cells. *Proc Natl Acad Sci U S A* 101:2253–2258

157 Hemelaar, J., A. Borodovsky, B.M. Kessler, D. Reverter, J. Cook, N. Kolli, T. Gan-Erdene, K.D. Wilkinson, G. Gill, C.D. Lima, H.L. Ploegh, and H. Ovaa. 2004. Specific and covalent targeting of conjugating and deconjugating enzymes of ubiquitin-like proteins. *Mol Cell Biol* 24:84–95.

158 Borodovsky, A., H. Ovaa, N. Kolli, T. Gan-Erdene, K.D. Wilkinson, H.L. Ploegh, and B.M. Kessler. 2002. Chemistry-based functional proteomics reveals novel members of the deubiquitinating enzyme family. *Chem Biol* 9:1149–1159.

159 Guterman, A., and M.H. Glickman. 2004. Complementary roles for Rpn11 and Ubp6 in deubiquitination and proteolysis by the proteasome. *J Biol Chem* 279:1729–1738.

160 Wilson, S.M., B. Bhattacharyya, R.A. Rachel, V. Coppola, L. Tessarollo, D.B. Householder, C.F. Fletcher, R.J. Miller, N.G. Copeland, and N.A. Jenkins. 2002. Synaptic defects in ataxia mice result from a mutation in Usp14, encoding a ubiquitin-specific protease. *Nat Genet* 32:420–425.

4
MEKK1: Dual Function as a Protein Kinase and a Ubiquitin Protein Ligase

Zhimin Lu and Tony Hunter

4.1
Introduction

Protein kinases are important regulators of intracellular signal transduction pathways, which mediate the development and regulation of diverse eukaryotic cellular activities, including cellular metabolism, transcription, cytoskeletal rearrangement and cell movement, apoptosis, cell-cycle progression, and differentiation. Through phosphorylation of substrates, protein kinases also play an important role in intercellular communication during development, in physiological responses and homeostasis, and in the functioning of the nervous and immune systems [2, 27, 28]. The 518 putative protein kinase genes identified in the human genome sequence comprise approximately 2% of all human genes, making them one of the largest families of eukaryotic genes [28]. In comparing the kinase gene chromosomal map with known disease loci, 164 kinases have been mapped to amplicons that are frequently found in tumors, and 80 kinases have been mapped to loci that are associated with major diseases such as diabetes, obesity, and hypertension [23, 28]. Perturbations of protein kinase function caused by mutation, overexpression, and dysregulation have causal roles in diverse human illnesses [2, 18].

4.2
Types of Protein Kinases

Based on their catalytic specificity, protein kinases can be subdivided into two major categories, tyrosine kinases and serine/threonine kinases. They function primarily by phosphorylating tyrosine or serine/threonine residues, respectively, either their own via autophosphorylation or those of their substrates, whose activity is consequently modulated. The activation of protein kinases and the phosphorylation of their substrates can play a role in regulating protein expression levels via a ubiquitin/proteasome-mediated degradation pathway through which target proteins are covalently tagged with ubiquitin and marked for degradation. The process of conjugating ubiquitin to substrate proteins depends upon three enzymes:

Protein Degradation, Vol. 2: The Ubiquitin-Proteasome System.
Edited by R. J. Mayer, A. Ciechanover, M. Rechsteiner

ISBN: 3-527-31130-0

a ubiquitin-activating enzyme (E1), a ubiquitin-conjugating enzyme (E2), and a ubiquitin ligase (E3). Intriguingly, an increasing number of protein kinases are known to be rapidly degraded via the ubiquitin/proteasome pathway following their activation. Such protein kinase deactivation through downregulation at the protein level provides an additional feedback mechanism, along with phosphorylation and dephosphorylation, that controls protein kinase activity. Protein kinases can also initiate ubiquitin/proteasome-mediated degradation of their protein substrates through direct phosphorylation. For instance, phosphorylation of β-catenin at serines 33 and 37 by GSK 3β creates a binding motif for the β-TrCP/HOS F-box protein, which is the substrate recognition subunit of the SCF$^{\beta\text{-TrCP}}$ E3 ubiquitin ligase [15, 33]. The cyclin-dependent kinase (Cdk) inhibitor p27 is phosphorylated directly by cyclin E/Cdk2 at Thr187 [30, 38], which then interacts and is ubiquitinated by the SCF$^{\text{SKP2}}$ E3 ligase [8, 35, 36]. In addition to protein kinase phosphorylation by catalytic domains, 83 different types of domains are found in 258 protein kinases [28]. These domains regulate kinase activity, localize proteins to subcellular compartments, interact with various signaling molecules, and are involved in protein degradation. Recent reports have shown that MEK kinase 1 (MEKK1), a serine/threonine protein kinase that has an important regulatory role in mitogen-activated protein (MAP) kinase cascades, functions both as a serine/threonine kinase through its kinase domain and as an E3 ubiquitin ligase via its N-terminal cysteine-rich domain [26, 40].

E3 ligases interact with both a ubiquitin-charged E2 molecule and the targeted substrate protein, facilitating polyubiquitination and directing substrate specificity. Thus, ubiquitination is primarily controlled by regulating E3 ligase activity and E3-substrate interactions [12]. There are two distinct types of E3 ligases: enzymatic HECT (homologous to E6-AP C-terminus) domain E3s and adaptor E3s containing a RING finger domain [19]. The ~350-residue HECT domain E3s forms a thioester with ubiquitin and transfers ubiquitin to substrates. On the other hand, the ~50-residue RING finger does not form a thioester with ubiquitin. Instead, it functions as an adaptor and facilitates the interaction between substrates and the E2. The RING finger is a zinc-binding domain with an octet of cysteines and histidines with a defined spacing configuration, which function as molecular scaffolds to conjoin proteins [4]. There are two varieties of RING finger E3 ligases. In one case the RING finger is part of a single polypeptide E3 ligase, such as Cbl, whereas in the other the RING finger protein is a subunit of a multi-subunit E3 complex, such as the small RING finger proteins present in the SCF (Skp1, cullin, and F-box) and APC (anaphase-promoting complex) E3 ligases [20, 24].

MEKK1, a 195-kDa protein with a C-terminal protein kinase domain and a large non-catalytic N-terminus, acts as a MAP kinase kinase kinase (MAPKKK or MAP3K) [25, 42]. The N-terminal non-catalytic region of MEKK1 contains a 48-residue region (aa 433–488) that has seven cysteines and a histidine that are linearly arranged in a C4HC3 consensus sequence [26] (Figure 4.1). Based on classification by the order of cysteine and histidine residues arrayed in a domain, this consensus sequence is categorized as a plant homeodomain (PHD) domain (also called a leukemia-associated protein (LAP) domain) rather than a RING finger do-

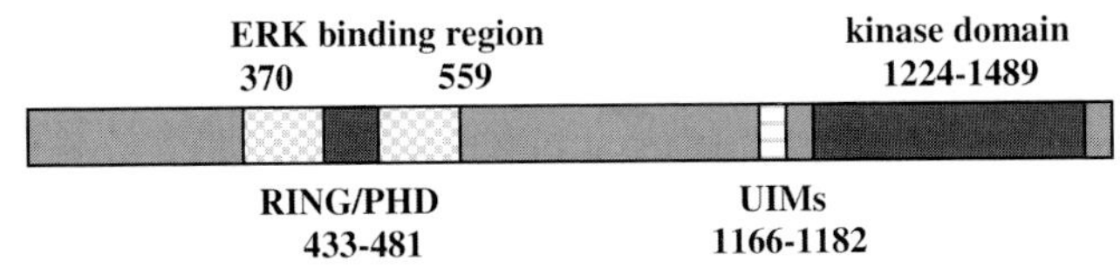

Fig. 4.1. RING/PHD domain (aa 433–481), ERK binding region (aa 370–559), UIMs (aa 1166–1182), and kinase domain (aa 1224–1489) of MEKK1 in a schematic structure.

main [6, 10, 26, 31]. The PHD domain is an approximately 50-residue C4HC3 zinc finger–binding motif, whereas the classical RING finger has a C3HC4 zinc finger–binding motif. The PHD domain structurally resembles the RING finger domain, with eight similarly spaced conserved metal-binding ligands [5, 7, 32]. Based on a different classification, which uses sequence profile or Hidden Markov models, a search of the PROSITE and Pfam databases or the non-redundant database of protein alignment shows that the cysteine-rich domain of MEKK1 retrieves NFX1- and H2-type variant RING domains [1, 34]. Given the close structural similarity between the RING and PHD domains and because the cysteine-rich domain of MEKK1 has features of both the RING and PHD domains, further studies are needed to determine whether the MEKK1 domain properly belongs to the RING finger or PHD domain family. For now, this domain will be called a RING/PHD domain, and this issue will be considered further in the discussion of whether a subset of PHD domains may, like many RING domains, have E3 ligase activity.

MEKK1 is one member of a family of related serine/threonine protein kinases that regulate three-tiered MAP kinase cascades. The three-tiered cascades are composed of a MAP3K, a MAP kinase kinase (MAPKK or MAP2K), and a MAP kinase (MAPK) (Figure 4.2A). MAP3Ks transduce signals received at the cell surface into

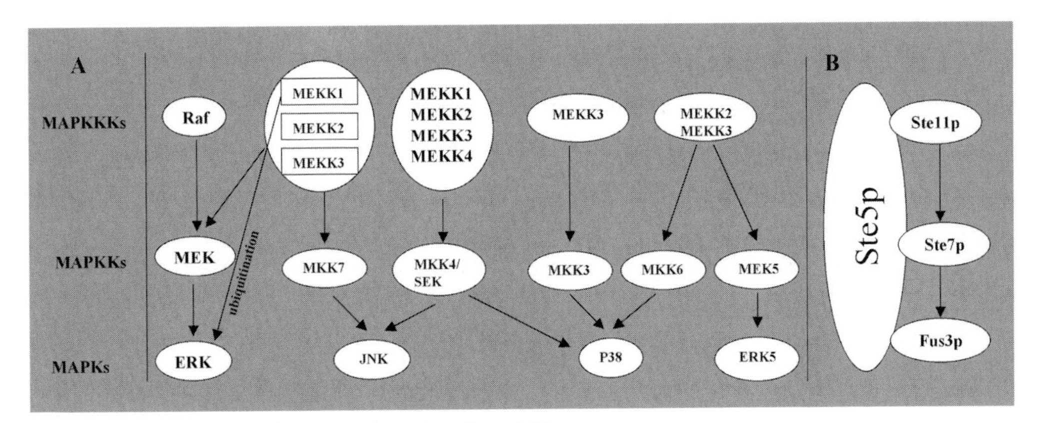

Fig. 4.2. MAP kinase cascade in (A) eukaryotic cells and (B) *Saccharomyces cerevisiae* (modified from [17]).

the nucleus by activating MAPK family members, including extracellular signal-regulated kinases (ERKs), the c-Jun NH2-terminal kinase (JNK), and p38. MAPK activation occurs in response to growth factor stimulation, cellular stress (e.g., UV and irradiation, osmotic stress, heat shock, and protein synthesis inhibitors), inflammatory cytokines (e.g., tumor necrosis factor [TNF] and interleukin-1 [IL1]), and G protein–coupled receptor agonists (e.g., thrombin) [17]. The activation of MAPK cascades has been implicated in cell growth and differentiation, apoptosis, oncogenesis, and inflammatory responses. Upon stimulation, MAP3Ks phosphorylate and activate their substrates, MAP2Ks, which in turn phosphorylate critical threonine and tyrosine residues in the activation loop of MAPK, thereby activating them.

4.3 Functions of Protein Kinases

The MEKK1 kinase domain phosphorylates several different MAP2K/MEKs and can regulate both the ERK and JNK pathways in response to specific stimuli [22, 41, 43] (Figure 4.2A). MEKK1 is activated in response to cellular stresses, including hyperosmolarity. When triggered by a mitogenic stimulus, ERK1/2 MAPKs are phosphorylated and activated by the MEK1/2 MAP2Ks. Nevertheless, this activated state is only transient in response to EGF and serum treatment, as ERK1/2 can be dephosphorylated by MAPK phosphatases (MKPs). In contrast, a hyperosmotic stimulus (sorbitol) results in sustained activation of ERK1/2, which is downregulated by ubiquitin/proteasome-mediated ERK1/2 protein degradation rather than by MKP activity. This illustrates that at least two different mechanisms downregulate MAPK activity after its initial activation: (1) dephosphorylation mediated by MKPs without a decrease in ERK1/2 protein levels upon mitogenic stimuli and (2) ubiquitination-mediated degradation of ERK1/2 without detectable dephosphorylation in response to hyperosmotic stimulation. Interestingly, ERK activity per se is not required for its own degradation since blocking its activation by inhibition of MEK does not block ERK degradation. Given that MEK1/2 can be activated by phosphorylation on serines 218 and 222 either by Raf or MEKK1, and that sorbitol-induced MEKK1-MEK-ERK but not mitogenically induced RAF-MEK-ERK activation involves protein degradation, MEK activation and the interaction between MEK and ERK are probably not important regulatory factors of ERK degradation [26].

MEKK1, like Raf, interacts with MEK via its catalytic domain. In addition, it also binds ERK2 through residues 370–559 in its N-terminal domain [21]. The ability of MEKK1 to interact with ERK and the fact that some RING domains possess E3 ubiquitin ligase activity suggested that the MEKK1 RING/PHD domain might play a role in ERK degradation in response to a hyperosmotic stimulus. Consistent with this idea, the MEKK1 RING/PHD domain exhibits ubiquitin ligase activity toward ERK both *in vivo* and *in vitro* [26]. In the presence of recombinant E1, E2 (Ubc4), and ubiquitin, the GST-MEKK1 RING/PHD fusion protein autoubiquitinates and

2001. MHC class I ubiquitination by a viral PHD/LAP finger protein. Immunity **15**:627–36.

4 Borden, K. L. 2000. RING domains: master builders of molecular scaffolds? J Mol Biol **295**:1103–12.

5 Borden, K. L., and P. S. Freemont. 1996. The RING finger domain: a recent example of a sequence-structure family. Curr Opin Struct Biol **6**:395–401.

6 Capili, A. D., D. C. Schultz, I. F. Rauscher, and K. L. Borden. 2001. Solution structure of the PHD domain from the KAP-1 corepressor: structural determinants for PHD, RING and LIM zinc-binding domains. Embo J **20**:165–77.

7 Capili, A. D., D. C. Schultz, I. F. Rauscher, and K. L. Borden. 2001. Solution structure of the PHD domain from the KAP-1 corepressor: structural determinants for PHD, RING and LIM zinc-binding domains. Embo J **20**:165–77.

8 Carrano, A. C., E. Eytan, A. Hershko, and M. Pagano. 1999. SKP2 is required for ubiquitin-mediated degradation of the CDK inhibitor p27. Nat Cell Biol **1**:193–9.

9 Coscoy, L., and D. Ganem. 2003. PHD domains and E3 ubiquitin ligases: viruses make the connection. Trends Cell Biol **13**:7–12.

10 Coscoy, L., and D. Ganem. 2003. Reply to Scheel and Hofmann: Progressing towards a better definition of PHD domains. Trends Cell Biol **13**:287–8.

11 Coscoy, L., D. J. Sanchez, and D. Ganem. 2001. A novel class of herpesvirus-encoded membrane-bound E3 ubiquitin ligases regulates endocytosis of proteins involved in immune recognition. J Cell Biol **155**:1265–73.

12 Deshaies, R. J. 1999. SCF and Cullin/Ring H2-based ubiquitin ligases. Annu Rev Cell Dev Biol **15**:435–67.

13 Dohlman, H. G., and J. W. Thorner. 2001. Regulation of G protein-initiated signal transduction in yeast: paradigms and principles. Annu Rev Biochem **70**:703–54.

14 Esch, R. K., and B. Errede. 2002. Pheromone induction promotes Ste11 degradation through a MAPK feedback and ubiquitin-dependent mechanism. Proc Natl Acad Sci U S A **99**:9160–5.

15 Fuchs, S. Y. 2002. The role of ubiquitin-proteasome pathway in oncogenic signaling. Cancer Biol Ther **1**:337–41.

16 Goto, E., S. Ishido, Y. Sato, S. Ohgimoto, K. Ohgimoto, M. Nagano-Fujii, and H. Hotta. 2003. c-MIR, a human E3 ubiquitin ligase, is a functional homolog of herpesvirus proteins MIR1 and MIR2 and has similar activity. J Biol Chem **278**:14657–68.

17 Hagemann, C., and J. L. Blank. 2001. The ups and downs of MEK kinase interactions. Cell Signal **13**:863–75.

18 Hunter, T. 2000. Signaling – 2000 and beyond. Cell **100**:113–27.

19 Jackson, P. K., A. G. Eldridge, E. Freed, L. Furstenthal, J. Y. Hsu, B. K. Kaiser, and J. D. Reimann. 2000. The lore of the RINGs: substrate recognition and catalysis by ubiquitin ligases. Trends Cell Biol **10**:429–39.

20 Johnson, E. S., P. C. Ma, I. M. Ota, and A. Varshavsky. 1995. A proteolytic pathway that recognizes ubiquitin as a degradation signal. J Biol Chem **270**:17442–56.

21 Karandikar, M., S. Xu, and M. H. Cobb. 2000. MEKK1 binds raf-1 and the ERK2 cascade components. J Biol Chem **275**:40120–7.

22 Karandikar, M., S. Xu, and M. H. Cobb. 2000. MEKK1 binds raf-1 and the ERK2 cascade components. J Biol Chem **275**:40120–7.

23 Knuutila, S., A. M. Bjorkqvist, K. Autio, M. Tarkkanen, M. Wolf, O. Monni, J. Szymanska, M. L. Larramendy, J. Tapper, H. Pere, W. El-Rifai, S. Hemmer, V. M. Wasenius, V. Vidgren, and Y. Zhu. 1998. DNA copy number amplifications in human neoplasms: review of comparative genomic hybridization studies. Am J Pathol **152**:1107–23.

24 Koegl, M., T. Hoppe, S. Schlenker, H. D. Ulrich, T. U. Mayer, and S. Jentsch. 1999. A novel ubiquitination factor, E4, is involved in multi-ubiquitin chain assembly. Cell **96**: 635–44.

25 Lange-Carter, C. A., C. M. Pleiman, A. M. Gardner, K. J. Blumer, and G. L. Johnson. 1993. A divergence in the MAP kinase regulatory network defined by MEK kinase and Raf. Science **260**:315–9.

26 Lu, Z., S. Xu, C. Joazeiro, M. H. Cobb, and T. Hunter. 2002. The PHD domain of MEKK1 acts as an E3 ubiquitin ligase and mediates ubiquitination and degradation of ERK1/2. Mol Cell **9**:945–56.

27 Manning, G., G. D. Plowman, T. Hunter, and S. Sudarsanam. 2002. Evolution of protein kinase signaling from yeast to man. Trends Biochem Sci **27**:514–20.

28 Manning, G., D. B. Whyte, R. Martinez, T. Hunter, and S. Sudarsanam. 2002. The protein kinase complement of the human genome. Science **298**:1912–34.

29 Mansouri, M., E. Bartee, K. Gouveia, B. T. Hovey Nerenberg, J. Barrett, L. Thomas, G. Thomas, G. McFadden, and K. Fruh. 2003. The PHD/LAP-domain protein M153R of myxomavirus is a ubiquitin ligase that induces the rapid internalization and lysosomal destruction of CD4. J Virol **77**:1427–40.

30 Montagnoli, A., F. Fiore, E. Eytan, A. C. Carrano, G. F. Draetta, A. Hershko, and M. Pagano. 1999. Ubiquitination of p27 is regulated by Cdk-dependent phosphorylation and trimeric complex formation. Genes Dev **13**:1181–9.

31 Pascual, J., M. Martinez-Yamout, H. J. Dyson, and P. E. Wright. 2000. Structure of the PHD zinc finger from human Williams-Beuren syndrome transcription factor. J Mol Biol **304**:723–9.

32 Pascual, J., M. Martinez-Yamout, H. J. Dyson, and P. E. Wright. 2000. Structure of the PHD zinc finger from human Williams-Beuren syndrome transcription factor. J Mol Biol **304**:723–9.

33 Polakis, P. 2000. Wnt signaling and cancer. Genes Dev **14**:1837–51.

34 Scheel, H., and K. Hofmann. 2003. No evidence for PHD fingers as ubiquitin ligases. Trends Cell Biol **13**:285–7; author reply 287–8.

35 Sutterluty, H., E. Chatelain, A. Marti, C. Wirbelauer, M. Senften, U. Muller, and W. Krek. 1999. p45SKP2 promotes p27Kip1 degradation and induces S phase in quiescent cells. Nat Cell Biol **1**:207–14.

36 Tsvetkov, L. M., K. H. Yeh, S. J. Lee, H. Sun, and H. Zhang. 1999. p27(Kip1) ubiquitination and degradation is regulated by the SCF(Skp2) complex through phosphorylated Thr187 in p27. Curr Biol **9**:661–4.

37 Uchida, D., S. Hatakeyama, A. Matsushima, H. Han, S. Ishido, H. Hotta, J. Kudoh, N. Shimizu, V. Doucas, K. I. Nakayama, N. Kuroda, and M. Matsumoto. 2004. AIRE Functions As an E3 Ubiquitin Ligase. J Exp Med **199**:167–72.

38 Vlach, J., S. Hennecke, and B. Amati. 1997. Phosphorylation-dependent degradation of the cyclin-dependent kinase inhibitor p27. Embo J **16**:5334–44.

39 Wang, Y., Q. Ge, D. Houston, J. Thorner, B. Errede, and H. G. Dohlman. 2003. Regulation of Ste7 ubiquitination by Ste11 phosphorylation and the Skp1-Cullin-F-box complex. J Biol Chem **278**:22284–9.

40 Witowsky, J. A., and G. L. Johnson. 2003. Ubiquitylation of MEKK1 inhibits its phosphorylation of MKK1 and MKK4 and activation of the ERK1/2 and JNK pathways. J Biol Chem **278**:1403–6.

41 Xu, S., D. Robbins, J. Frost, A. Dang, C. Lange-Carter, and M. H. Cobb. 1995. MEKK1 phosphorylates MEK1 and MEK2 but does not cause activation of mitogen-activated protein kinase. Proc Natl Acad Sci U S A **92**:6808–12.

42 Xu, S., D. J. Robbins, L. B.

CHRISTERSON, J. M. ENGLISH, C. A. VANDERBILT, and M. H. COBB. 1996. Cloning of rat MEK kinase 1 cDNA reveals an endogenous membrane-associated 195-kDa protein with a large regulatory domain. Proc Natl Acad Sci U S A **93**:5291–5.

43 YUJIRI, T., S. SATHER, G. R. FANGER, and G. L. JOHNSON. 1998. Role of MEKK1 in cell survival and activation of JNK and ERK pathways defined by targeted gene disruption. Science **282**:1911–4.

5
Proteasome Activators

Andreas Förster and Christopher P. Hill

Abstract

In this chapter we discuss the possible roles and mechanisms of protein complexes that bind and stimulate 20S proteasomes, the primary proteases of the cytosol and nucleus of eukaryotic cells. We review structural and biochemical studies of 11S/PA28 activators and PA200, two protein complexes that are known to activate proteasomes. Our discussion of biological functions will be brief, since these are currently quite speculative and have been addressed elsewhere (Rechsteiner and Hill 2005). Instead our focus will be on structural studies and biochemical mechanisms. We start by briefly reviewing salient features of 20S proteasome architecture and mechanism. We will emphasize the role of N-terminal residues of 20S proteasome α subunits in restricting substrate access and their activator-induced reorganization to an open conformation. We do not discuss the 19S/PA700 activator in detail, since this topic has been discussed in this series (DeMartino and C. Wojcik 2005). Nor do we address reports of protein inhibitors of the 20S proteasome, since these have been discussed elsewhere (Rechsteiner and Hill 2005).

5.1
Introduction

5.1.1
20S Proteasomes

20S proteasomes are abundant proteases in all eukaryotic cells examined, where they are found in the cytosol and nucleus, and appear to perform the majority of proteolysis that occurs in these compartments (Coux et al. 1996). Many proteins have been identified as proteasome substrates, generally as polyubiquitylated substrates of the 26S proteasome, which is comprised of the 20S proteasome and 19S/PA700 (Pickart and Cohen 2004). Substrates include short-lived regulatory proteins (Hershko and Ciechanover 1998) and proteins that are damaged, denatured, or misfolded (Goldberg 2003). Given the fundamental importance of protein turnover,

Protein Degradation, Vol. 2: The Ubiquitin-Proteasome System.
Edited by R. J. Mayer, A. Ciechanover, M. Rechsteiner

ISBN: 3-527-31130-0

it is not surprising that 20S proteasomes are essential in yeast (Emori et al. 1991; Heinemeyer et al. 1994; Velichutina et al. 2004). 20S proteasomes are also found in archaea and in a few prokaryotes (De Mot et al. 1999), although prokaryotes generally make use of mechanistically related but distinct protein complexes (Gottesman 2003).

20S proteasomes are barrel-shaped structures comprised of four rings that each contain seven subunits, with α subunits forming the two end rings and β subunits forming the two central rings (Figure 5.1). For reviews of the structural studies, see Baumeister et al. (1998), Groll and Clausen (2003), and Groll and Huber (2003). Whereas archaeal 20S proteasomes are built from multiple copies of identical α and β subunits, eukaryotic proteasomes have seven different α subunits ($\alpha1$–$\alpha7$) and seven different β subunits ($\beta1$–$\beta7$), with each subunit occupying a precise location in the appropriate ring.

The mechanism by which 20S proteasomes avoid indiscriminate degradation of folded proteins was explained by the crystal structure of the proteasome from *T. acidophilum* (Löwe et al. 1995) and later confirmed with crystal structures of the 20S proteasomes from yeast (Groll et al. 1997) and cow (Unno et al. 2002). The proteolytically active sites (Seemüller et al. 1995) are sequestered within the central catalytic chamber formed by the β subunits. Access to the proteasome interior is through a pore in the middle of the α-subunit ring that permits passage of unfolded substrates (Wenzel and Baumeister 1995). This aperture (α-annulus; green in Figure 5.1) is defined by the main-chain atoms of the short loops in the middle of the α-subunit sequences and appears to have a fixed diameter of $\sim$17 Å between atomic nuclei.

A wealth of structural data on proteasome–inhibitor complexes has illuminated the mechanism of proteolysis at the active sites, which are located at the N-termini of some (eukaryotes) or all (archaea) proteasome β subunits (Groll and Clausen 2003; Seemüller et al. 1995). The 20S proteasome active sites are fully formed in the unliganded proteasome. The naturally repressed state of isolated 20S proteasomes results, therefore, entirely from sequestration of the active sites within the hollow structure, with the α-annulus preventing entrance of folded proteins and a closed-gate structure (see next section) blocking smaller substrates. This mechanism for preventing hydrolysis of inappropriate substrates is in marked contrast to the analogous bacterial HslV protease, for which binding of the HslU activator induces formation of an active conformation at the proteolytic active sites (Ramachandran et al. 2002; Sousa et al. 2002; Sousa et al. 2000; Wang et al. 2001).

5.1.2
The 20S Proteasome Gate

Eukaryotic 20S proteasomes seal their entrance/exit port through the α-annulus by a gate structure formed by the N-terminal residues of their α subunits (Groll et al. 1997; Unno et al. 2002) (Figures 5.1c and 5.1d). In particular, the N-terminal residues of subunits $\alpha2$, $\alpha3$, and $\alpha4$ adopt unique ordered conformations that are stabilized by an extensive network of hydrogen-bonding and van der Waals interactions. The other four subunits make less extensive contributions to the closed conformation. Rather than crossing the central gate area, their N-terminal residues project away from the proteasome surface. This asymmetric arrangement results from the unique amino acid sequences of the proteasome α-subunit N-termini, which are well conserved between equivalent subunits of different species but differ significantly between paralogous subunits. In pairwise comparisons between equivalent yeast and human α subunits, the residues prior to residue 13 (archaeal *T. acidophilum* proteasome numbering) are 56–100% identical between species. In contrast, only Tyr8 and Asp9 are highly conserved between different subunits of the same species.

In contrast to the ordered closed gate of eukaryotic proteasomes, the N-terminal 12 residues of isolated archaeal 20S proteasomes from *T. acidophilum* (Löwe et al. 1995) and *Archaeoglobus fulgidus* (Groll et al. 2003a) are disordered and presumably flexible (Löwe et al. 1995). The inability of archaeal 20S proteasomes to form the ordered, closed-gate conformation is explained by the symmetric configuration in which all seven α subunits have the same sequences and therefore are unable to form the asymmetric closed state. In contrast to the eukaryotic enzymes, the archaeal 20S proteasome degrades small peptides efficiently, since they are apparently able to diffuse through the "curtain" of flexible α-subunit N-termini with little hindrance. The flexible tails do, however, provide a significant barrier to passage of unfolded protein substrates, since a variant in which the N-terminal tails have been deleted degrades unfolded proteins, whereas the wild-type *T. acidophilum* 20S proteasome does not (Benaroudj et al. 2003).

Although the two reported crystal structures of intact archaeal 20S proteasomes show the α-subunit N-terminal residues to be unstructured (Groll et al. 2003a; Löwe et al. 1995), there is one example in which these gate residues adopt an ordered, open conformation. The structure of an isolated ring of α subunits from the archaeon *A. fulgidus* reveals a conformation essentially identical to that observed for the yeast 20S proteasome in complex with the activator PA26 (Förster et al. 2003), which is discussed below. The original motivation for determining the α-subunit ring structure was to understand the process of proteasome assembly (Groll et al. 2003a). However, because of the similarity to the activator complex and the apparent absence of a role for the open conformation in assembly, we favor the possibility that crystallization here fortuitously captured a conformation that is functionally important but not highly populated in the absence of an activator. As discussed below, formation of this ordered, open conformation appears to be important for efficient entry and degradation of protein substrates.

It is not entirely clear why eukaryotic 20S proteasomes require an ordered, closed-gate structure, since the α-annulus and flexible N-terminal residues of archaeal 20S proteasomes are sufficient to restrict passage of folded protein substrates. Indeed, the closed-gate conformation does not appear to be important for logarithmic growth of yeast under favorable conditions, although a defect in release from the stationary phase is revealed in yeast when the gate is disrupted by mutagenesis

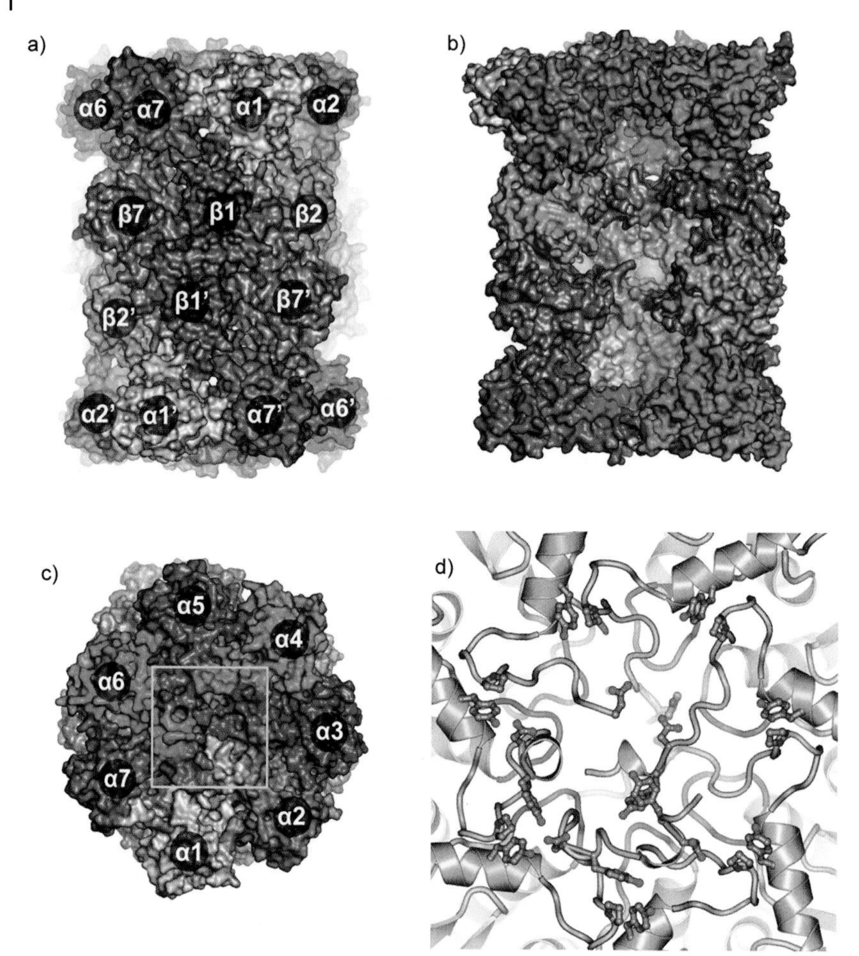

Fig. 5.1. Structure of the yeast 20S proteasome (Groll et al. 1997). (a) Space-filling representation, side view. (b) Same as panel a, with subunits closest to the viewer removed to reveal the hollow interior and proteolytic active sites (yellow). The α-annulus is colored green. (c) Same as panel a, top view. (d) Ribbon representation showing the central boxed region of panel c. Tyr8, Asp9, Pro17, and Tyr26 side chains from each α subunit are shown explicitly. These residues form clusters that stabilize the open conformation (Förster et al. 2003) (Figure 5.5).

(Bajorek et al. 2003). This mutant proteasome also displayed an accelerated rate of protein turnover *in vitro* and *in vivo*. One possibility is that eukaryotes contain more natively unfolded proteins or functionally important oligopeptides that are able to pass a flexible gate. It is clear, however, that disruption of the stable gate structure is necessary for proteolysis to occur, and that even the flexible gate of archaeal proteasomes provides a significant barrier to passage of protein substrates.

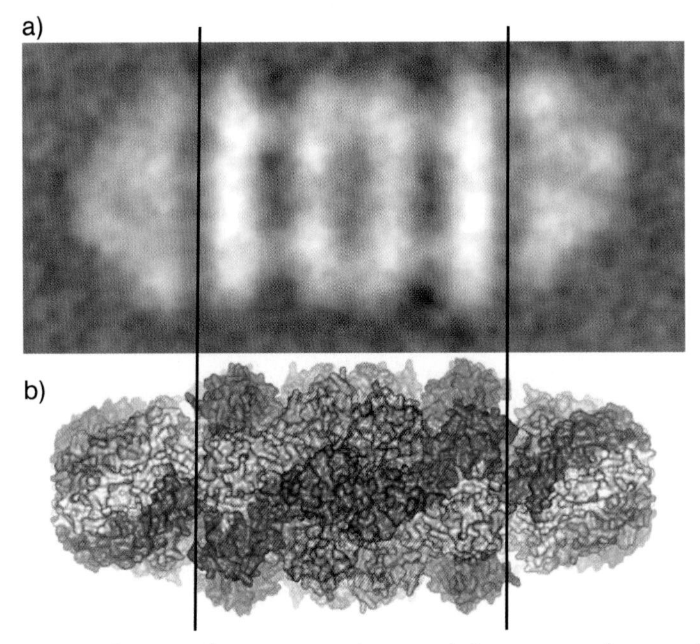

Fig. 5.2. Structure of proteasome–activator complexes. (a) Averaged negative-stain electron micro-graph of bovine 20S proteasome–PA200 complex (A. Steven and J. Ortega, personal communication). (b) Crystal structure of yeast 20S proteasome in complex with *T. brucei* PA26 (yellow) (Förster et al. 2003).

5.1.3 Proteasome Activators

Proteasomes are activated by protein complexes that bind to one or both rings of α subunits (Figure 5.2). The best known of these activators is the 19S activator, also known as proteasome activator MW 700 (PA700) and regulatory complex (RC). PA700 has a well-defined biological role, namely, the degradation of polyubiquitylated protein substrates. It is a remarkable machine, comprised of at least 17 stoichiometric subunits (Glickman et al. 1998) and a number of other transient or weakly associated components (Leggett et al. 2002; Verma et al. 2000). It contains subunits that recognize polyubiquitin chains, edit the chains, remove chains from substrates, unfold the substrate, open the proteasome gate, and translocate substrate into the 20S proteasome interior for degradation. We do not discuss 19S further here, since it is the focus of the chapter in this series by George DeMartino and Cezary Wojcik. Rather, our focus is on the other characterized activators, 11S (also called PA28, REG, PA26; reviewed in DeMartino and Slaughter 1999, Hill et al. 2002, and Kuehn and Dahlmann 1997) and PA200 (Ustrell et al. 2002). Unlike 19S, 11S and PA200 do not recognize ubiquitin or utilize ATP and have unknown *in vivo* substrates but, at least in the case of 11S, are better understood from a biochemical and structural perspective.

5.2
11S Activators: Sequence and Structure

5.2.1
Amino Acid Sequences

Members of the 11S family were first identified as protein complexes that stimulate the peptidase activity of 20S proteasomes (Dubiel et al. 1992; Ma et al. 1992). The three homologues of higher eukaryotes, PA28α, -β, and -γ (REGα, -β, and -γ) each have a subunit mass of ~28 kDa and share 35–50% sequence identity in pairwise comparisons. More primitive species than jawed vertebrates appear to have only one 11S activator, which is most closely related to PA28γ, and yeasts and plants appear to lack an 11S homologue (Masson et al. 2001; Paesen and Nuttall 1996; Murray et al. 2000). Sequence analyses indicate that duplication and divergence of the PA28γ gene produced the PA28α gene, which duplicated in turn to produce PA28β (Kim et al. 2003). A very distantly related homologue, PA26, has been identified in *Trypanosoma brucei* and found to share only ~14% identity with other 11S activators (Yao et al. 1999).

5.2.2
Oligomeric State

Following some initial confusion, 11S activators are now known to be assembled as ~200-kDa heptamers (Johnston et al. 1997; Knowlton et al. 1997; Li et al. 2000, 2001a; Yao et al. 1999; Zhang et al. 1999) (see Hill et al. 2002 for a full discussion). This is an important point because, as discussed later, structural studies have shown that the sevenfold assembly is central to the mechanism of binding and gate opening by 11S (Förster et al. 2003; Whitby et al. 2000). Whereas PA28γ forms a homoheptamer (Li et al. 2001a; Realini et al. 1997; Tanahashi et al. 1997), PA28α and PA28β preferentially assemble as hetero-oligomers with a stochastic distribution of α and β subunits (Zhang et al. 1999). Consistent with this, PA28α and PA28β copurify as a single complex from tissues (Kuehn and Dahlmann 1996a, 1996b, 1997; Mott et al. 1994). PA28α and PA28β can also each assemble into functional heptamers, although PA28β is monomeric and inactive at low concentrations (Realini et al. 1997; Song et al. 1997; Wilk et al. 2000a; Zhang et al. 1998b). The distant relative from *T. brucei*, PA26, is also known to be heptameric (Whitby et al. 2000; Yao et al. 1999).

5.2.3
PA28α Crystal Structure

The crystal structure of human PA28α (Figure 5.3) revealed that each subunit forms an elongated bundle of four helices and that subunits assemble to form a doughnut-shaped heptamer that has a central channel of 20–30 Å diameter (between atomic nuclei) (Knowlton et al. 1997). The base of the heptamer provides a sevenfold symmetric array of two functionally important motifs; the C-terminal tails, which are important for proteasome binding (Li et al. 2000; Ma et al. 1993; Song et al. 1997), and the activation loops, which are required to stimulate the proteasome's peptidase activity (Zhang et al. 1998a). The C-terminal eight residues of human PA28α are disordered and apparently flexible in the crystal structure. As described below, these residues provide a flexible tether that becomes partially ordered upon binding proteasome.

5.2.4
Activation Loop

The activation loop was identified from a random mutagenesis screen as a segment of nine residues, located in the turn between helices 2 and 3, that is important for proteasome activation (Zhang et al. 1998a). Interestingly, mutations in this region have been identified that bind 20S proteasome with the same affinity as wild type yet fail to stimulate the peptidase activity, thereby indicating that binding and activation are to some extent separable. As discussed below, PA26–proteasome complex crystal structures have revealed contacts with the activation loops that explain how PA26 opens the proteasome entrance/exit gate. Remarkably, however, whereas residues of the activation loop are almost universally conserved among 11S activators, PA26 is an exception that has very different residues in this functionally important part of the structure. A structure-based alignment of PA28α and PA26 sequences is shown in Figure 5.4. The basis for how the different activation loops might stabilize the same activated proteasome conformation is discussed below.

5.2.5
Homologue-specific Inserts

The sequences of PA28α, -β, and -γ are primarily distinguished by 15–30 residue segments known as "homologue-specific inserts" (Song et al. 1997; Zhang et al. 1998a, 1998c). As shown by the structure of human PA28α (Knowlton et al. 1997) (Figure 5.3), the PA28α homologue-specific insert is disordered but located at the end of the activator distant from the proteasome-binding surface. The PA28α insert sequences are rich in lysine and glutamate residues and define the so-called KEKE motif, which has been suggested to mediate protein–protein interactions (Realini et al. 1994). The PA28β insert has a similar amino acid composition but is shorter, whereas the PA28γ insert has a mixed composition with a larger number of hydrophobic residues. Notably, PA26 lacks homologue-specific insert sequences but possesses tight turns of just a few ordered residues between helices 1 and 2 (Whitby et al. 2000). One attractive possibility is that the homologue-specific inserts function by binding specific partner(s), although no such partners have been convincingly demonstrated to date.

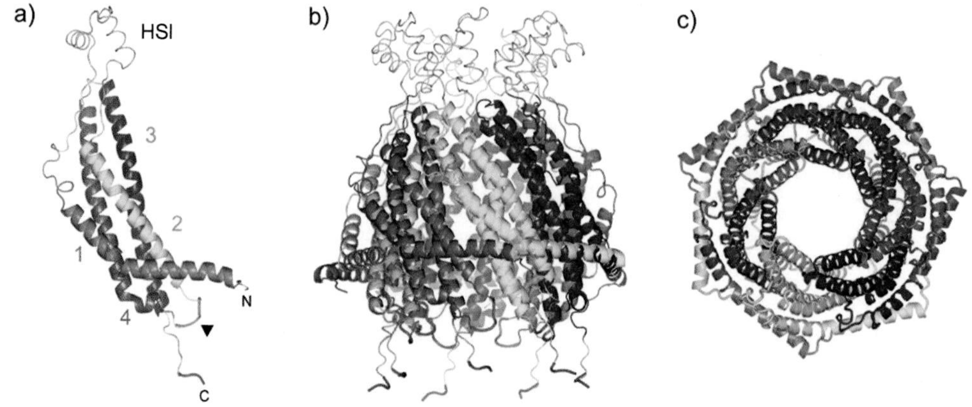

Fig. 5.3. PA28α crystal structure (Knowlton et al. 1997). (a) Structure of an isolated subunit. N- and C-termini are labeled. The homologue-specific insert (HSI) is disordered in the structure and is included here in an arbitrary conformation. The activation loop is colored red and indicated with a triangle. The C-terminal tails, which are disordered in the isolated PA28α structure, have been included in the conformation observed for a high-resolution archaeal 20S proteasome–PA26 structure (Förster and Hill, unpublished), with residues that contact proteasome colored magenta. (b) Side view of the PA28α heptamer colored by subunit. Activation loops and C-terminal tails are colored as in panel a. (c) Top view of panel b. For clarity, the disordered HSI has been omitted.

5.3 PA26–Proteasome Complex Structures

The mechanism of proteasome activation by 11S activators has been explained, in part, by the crystal structure (Förster et al. 2003; Whitby et al. 2000) of a complex

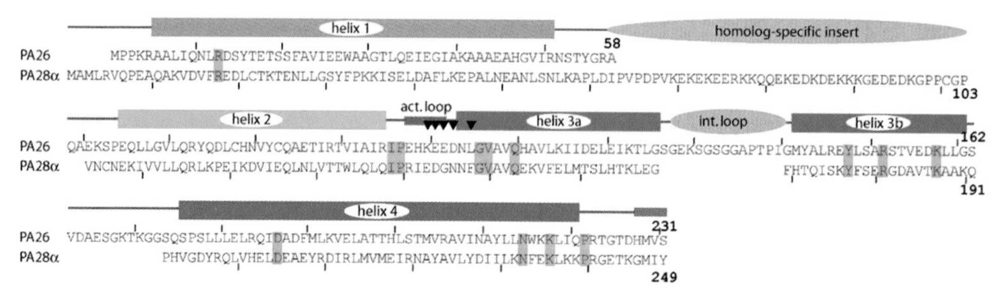

Fig. 5.4. Structure-based sequence alignment of PA26 and PA28α. The structural alignment is clear for residues from the beginning of helix 2 through to the C-terminus, but is ambiguous for helix 1. Residue identities that are conserved in PA26 and all three of the human PA28 homologues (αβγ) are shown on a yellow background. Residues of the PA26 activation loops that are within Van der Waals contact distance of a proteasome atom are indicated with black triangles. Residues at the PA26 C-terminus that contact the proteasome are indicated with a purple box.

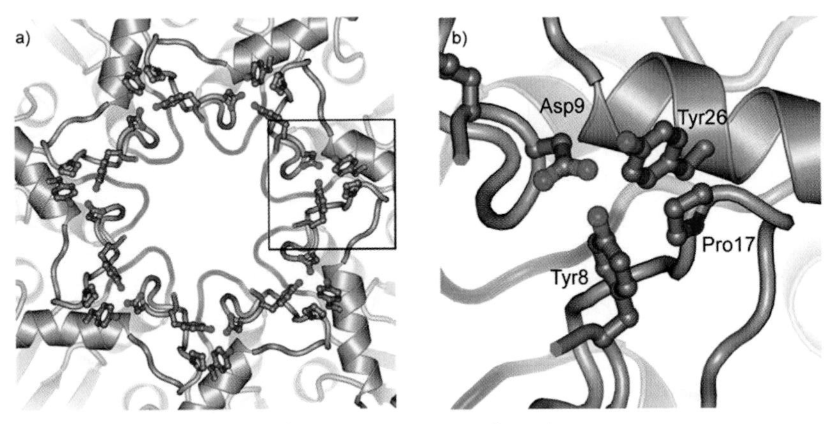

Fig. 5.5. Open conformation of yeast 20S proteasome formed in complex with PA26 (Förster et al. 2003). (a) Same view as closed conformation of Figure 5.1d. (b) Close-up of the cluster of invariant residues (Tyr8, Asp9, Pro17, Tyr26) boxed in panel a.

between the yeast 20S proteasome and PA26 (Figure 5.2b). Binding of PA26 induces the entrance/exit gate that is closed in isolated yeast 20S proteasomes to adopt an open conformation (Figure 5.5). This allows peptide substrates to diffuse freely into the proteasome interior. It is important to note that the structure was solved for a highly non-cognate complex, especially considering that yeast do not appear to posses 11S activators. Nevertheless, 11S activates seem to stimulate essentially any 20S proteasome, regardless of source. For example, PA26 activates 20S proteasome from rat (Yao et al. 1999) and yeast (Whitby et al. 2000), human PA28α activates proteasome from cow (Eugene Masters, personal communication) and yeast (Martin Rechsteiner, personal communication), and cow PA28 activates 20S proteasome from lobster (Mykles 1996). It therefore seems likely that the yeast 20S proteasome–PA26 complex reveals conformational changes that underlie activation of cognate complexes. The structural analysis has recently been advanced by determination of complexes between PA26 and an archaeal 20S proteasome that diffract to relatively high resolution (Förster and Hill, unpublished).

5.3.1
Binding

The mechanisms of binding and activation depend in large part upon the symmetry of the PA26 heptamer. PA26, like human PA28α (Knowlton et al. 1997), is exactly sevenfold symmetric. As indicated by the earlier biochemical observations (Li et al. 2000; Song et al. 1997), binding is mediated by the C-terminal tails of PA28 subunits, which project into pockets that are formed between neighboring α subunits on the 20S proteasome surface. The exact match in spacing of the seven PA26 C-terminal tails with the seven pockets on the proteasome surface explains how the individual interactions, which are probably quite weak, can sum to provide a significant binding affinity.

Details of this interaction were obscure in the medium-resolution yeast 20S proteasome complex. Recently, a structure of a complex between PA26 and an archaeal 20S proteasome (Förster and Hill, unpublished) has revealed that the interaction is largely mediated by main chain–main chain contacts. This explains why the different 11S homologues can all bind the same proteasome and why most 11S activators bind 20S proteasomes from most species, even though the 11S C-terminal residues are highly variable between the homologues. It also explains why activation is tolerant of many mutations in the C-terminal residues of PA28 (Song et al. 1997).

5.3.2
Symmetry Mismatch Mechanism of Gate Opening

The PA26–proteasome complex structure revealed a symmetry-mismatch mechanism of gate opening. As illustrated in Figure 5.1d, the closed-gate structure of unbound proteasomes is asymmetric, with the N-terminal residues of subunits $\alpha2$, $\alpha3$, and $\alpha4$ adopting unique, ordered conformations that make a large number of specific hydrogen-bonding and van der Waals interactions. As the C-terminal residues of the seven PA26 subunits bind into the appropriately spaced pockets between each of the seven proteasome α subunits, the symmetric surface of activation loops is pressed against the reverse turns containing Pro17 of proteasome α subunits. This induces the proteasome to follow the symmetry of PA26 by moving Pro17 of individual subunits by as much as 2.5 Å. This displacement in a subset of the proteasome Pro17 turns appears to be the trigger that leads to gate opening.

Repositioning of the Pro17 turns appears to induce gate opening for two reasons. Firstly, consequent displacement of more N-terminal residues disrupts the many van der Waals and hydrogen-bonding interactions formed by the N-terminal residues of subunits $\alpha2$, $\alpha3$, and $\alpha4$. This destabilization explains why the closed-gate conformation is no longer maintained, although it does not explain why a specific open-gate conformation is assumed rather than the disordered state observed in crystal structures of isolated archaeal proteasomes. As described below, the second contribution to gate opening provided by displacement of Pro17 turns is to allow formation of stabilizing interactions between conserved proteasome residues.

It is remarkable that only one of the PA26 activation loop residues, Glu102, appears to make significant contact with the proteasome. Surprisingly, this residue is not present in any of the other known 11S activators. The equivalent residue in all PA28 sequences is an invariant glycine. It is not obvious how this Gly145 (PA28α numbering) would contact 20S proteasome to reposition the Pro17 turn. One possibility, suggested from the structural overlap of PA26 and PA28α (Figure 5.6), is that the previous residue, Asp144 of PA28α, which is also invariant among PA28 sequences, might be functionally equivalent to PA26 Glu102. This unusual shift of a functionally critical residue along an amino acid sequence might explain why the Cα trace of the activation loops is relatively divergent between PA28α and PA26.

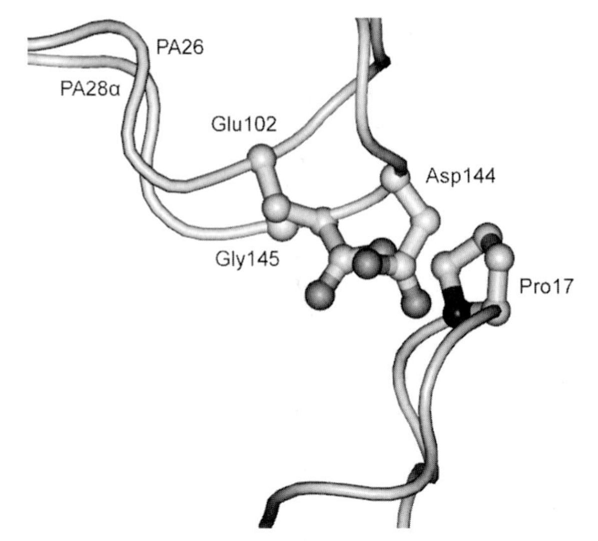

Fig. 5.6. Superposition of PA26 and PA28α activation loops after global overlap. The primary activating contact to proteasome is through Glu102 of PA26. Based simply on structural overlap, the equivalent residue in PA28α is Gly145. We speculate that the adjacent residue, Asp144, might mediate contacts equivalent to those of PA26 Glu102. The Asp144 conformation shown here is a preferred rotamer that differs from the rotamer seen in the crystal structure of isolated PA28α. No other adjustments were made to the crystal structures for these figures.

(Following global overlap of the PA28α and PA26 structures, the six residues N- and C-terminal to the activation loop have equivalent Cα atoms separated by an average of less than 1.0 Å, whereas activation loop (PA26 99–104) Cα atoms show an average deviation of 2.0 Å in this structural alignment.) Resolution of this point will require determination of a crystal structure of 20S proteasome bound to a PA28 activator.

5.3.3
Open-gate Stabilization by Conserved Proteasome Residues

Because of the limited resolution of the yeast 20S proteasome–PA26 complex, it was initially not appreciated that the proteasome's open conformation is ordered, rather than being comprised of disordered α-subunit N-terminal residues (Whitby et al. 2000). Closer inspection revealed that the open conformation is in fact ordered and appears to be stabilized by interactions between four highly conserved proteasome residues; Tyr8, Asp9, Pro17, and Try26 (Förster et al. 2003). Although these residues do not make direct contacts with the activator, they are allowed to form stabilizing clusters between adjacent α subunits mediated by the repositioning of Pro17 induced by PA26 (Figure 5.5). The importance of these residues for stabilizing the symmetric, open conformation explains why they are conserved between the different α subunits, whereas other residues N-terminal to residue 15 are variable between subunits.

The four cluster residues are not absolutely conserved between all proteasome α subunits. The few deviations from perfect conservation seen for these cluster residues are limited to the cluster between α1 and α2 (based on sequences of man, rat, mouse, *Arabidopsis*, fruit fly, worm, and fission and baker's yeasts). This cluster, which we call the non-canonical cluster, has a slightly different structure from the other six. In particular, residue 9 of α2 is always a serine or smaller side chain, rather than the invariant aspartate of the other α subunits. This substitution is explained by the requirement that the adjacent residue 10 invariably be a phenylalanine for subunit α2, whereas it is never a phenylalanine for the other subunits. Phe10 of α2 occupies a critical buried position in the closed conformation, and its location in the open conformation appears to be incompatible with an aspartate at position 9. As described more fully in Förster et al. (2003), these two substitutions in α2 explain the occasional additional substitution at position 8 of α1 and position 26 of α2.

The model that the cluster residues are required for formation of the ordered, open conformation is supported by the recently determined structures of PA26 with wild-type and mutant (Asp9Ser) archaeal proteasome (Förster and Hill, unpublished). These structures have been refined against ~2-Å data. The wild-type archaeal 20S proteasome complex shows the same ordered, open conformation as seen for the yeast 20S proteasome complex. In contrast, α subunits in the Asp9Ser mutant complex are disordered before residue 12. This indicates that whereas eukaryotic proteasomes can accommodate one non-cognate cluster, the open conformation is not stable when all seven α subunits substitute the aspartate at position 9 for a serine.

5.3.4
Do Other Activators Induce the Same Open Conformation?

Not only are the four cluster residues conserved in the different subunits of the yeast proteasome, they are also conserved in all known proteasome sequences, including eukaryotes from yeast to human and 18 archaeal species. It was initially surprising to realize that conservation of Tyr8, Asp9, Pro17, and Tyr26 extends to species, such as yeast and archaea, that do not appear to possess 11S activators. This observation implies that these residues are important for a function that does not involve 11S activators. Our preferred explanation is that proteasomes have just one open conformation and that different activators, such as 19S, function, in part, by inducing the same open conformation as seen in the PA26–proteasome crystal structure.

Support for this hypothesis has been obtained using a mutagenic/biochemical approach with an archaeal 20S proteasome and PAN, an archaeal analogue of the 19S activator. In this study (Förster et al. 2003), degradation of PAN-dependent model substrates was impaired in a number of mutant proteasomes that had disruptions in the cluster residues. In particular, the proteasome in which Tyr8 and Asp9 were both mutated to glycine was inactive in this assay. This is especially noteworthy because the model that activators induce a disordered gate structure would predict that this mutant, with inherently much more flexible amino-terminal tails, would be even more active than wild type. The observed impaired activity against protein substrates supports the hypothesis that these residues adopt an ordered, open conformation, which is necessary for efficient degradation of protein substrates.

Further indication that proteasomes have an inherent ability to adopt this open conformation, independent of binding by an 11S activator, is demonstrated by observation of the open conformation in an isolated ring of α subunits from the archaeon *Archaeoglobus fulgidus* (Groll et al. 2003b). Based upon the available structural and biochemical data, we find the hypothesis that other activators induce the same open conformation attractive, although resolution of this point will require structure determination of other proteasome–activator complexes.

5.3.5
Differential Stimulation of Proteasome Peptidase Activities

20S proteasome active sites have been characterized by their ability to stimulate hydrolysis of small fluorogenic peptides. The three distinct active sites at the N-termini of $\beta1$, $\beta2$, and $\beta5$ are referred to, respectively, as peptidyl-glutamyl peptide hydrolytic (PGPH), trypsin-like, and chymotrypsin-like because they preferentially cleave following acidic, basic, and hydrophobic residues (see Bochtler et al. 1999 and references therein). Although other determinants are also important for specificity (Bogyo et al. 1998; Groll et al. 2002; Harris et al. 2001; Wang et al. 2003), the simplest interpretation of the structural data is that stimulation of proteasome peptidase activity by 11S activators results simply from opening of the gate, since no conformational changes are evident at the active sites in the β subunits. Presumably, therefore, the different extents to which hydrolysis of different peptide substrates is enhanced results from the relative rates at which the peptides can pass through the activator channel and into the proteasomes antechamber and catalytic chamber.

Although the structural data support the simple model outlined above, another possibility has been suggested by studies on recombinant human PA28γ (Li et al. 2001a). Whereas PA28α stimulates the hydrolysis of all three fluorogenic peptide substrates that are diagnostic for hydrolysis at the three distinct 20S proteasome active sites, PA28γ stimulates hydrolysis of the "trypsin site" peptide but not of the chymotrypsin or PGPH peptides. This difference in specificity exists despite identical activation loop sequences for PA28α and PA28γ, and substitution of residues close to the activation loop does not alter specificities (Li et al. 2000). Curiously, substitution of PA28γ Lys188 with either Asp or Glu changes the specificity of PA28γ to that of PA28α (Li et al. 2001a). Since Lys188 is thought, based upon homology modeling with the PA28α structure, to face the central channel at the end distant from the proteasome, one possibility is that this residue performs a gating function by restricting passage of the chymotrypsin and PGPH substrates. This explanation is challenged, however, by the observation that positively charged substrates are processed rapidly when residue 188 is lysine, whereas these are the type of substrates that are expected to be excluded by a simple gating mode. Li et al. (2001a) therefore proposed that binding of PA28γ not only opens the substrate entrance gate but also induces long-range conformational changes that repress activity at the chymotryptic and PGPH active sites. Lys188Asp/Glu PA28α forms less stable heptamers, and Li et al. (2001a) proposed this mutant to be more flexible than PA28γ and therefore unable to impose the same conformational changes on the proteasome. PA28γ has also been shown by a second group to activate all three active sites (Wilk et al. 2000b). In that case, however, protein purification included ammonium sulfate precipitation, which yielded PA28γ much less stable than activator purified the traditional way (Gao et al. 2004). Reconciliation of these observations with the currently available structural data will require further study. In particular, it will be important to determine structures at higher resolution and for cognate complexes with mammalian proteasomes and activators.

5.3.6
Hybrid Proteasomes

Immune precipitation with monoclonal antibodies specific for 19S or PA28 demonstrated that both of these activators are present in the same complex, which was presumed to be comprised of 19S and P28 complexes bound to the rings of α subunits at opposite ends of the same 20S proteasome (Hendil et al. 1998). Analysis of HeLa cell extracts determined that these "hybrid" proteasomes account for about a fourth of all proteasomes and that induction of PA28 expression and hybrid proteasome formation with γINF appreciably enhanced degradation of a PA700-dependent substrate (Tanahashi et al. 2000).

Two studies have reported reconstitution of hybrid proteasomes from singly capped 19S–20S complexes and recombinant PA28. Kopp et al. (2001) reported that hybrid proteasomes have peptidase activities similar to those of singly capped PA28 or 19S complexes, and also reported negative-stain electron micrographs. The expected complex is formed from its components without apparent structural distortion. Cascio et al. (2002) performed a similar analysis and showed similar electron micrographs. They found enhanced hydrolysis of small peptides in hybrid proteasomes, but no significant acceleration of protein breakdown. They also demonstrated that hybrid proteasomes generate a pattern of peptide products different from those generated by 26S proteasomes, without altering mean product length. This observation suggests that the change in peptides produced accounts for the capacity of PA28 to enhance antigen presentation and argues against the proposal of Whitby et al. (2000) that binding of PA28 facilitates release of larger product peptides.

5.4 Biological Roles of 11S Activators

Our discussion of *in vivo* function will be brief because the biological roles of 11S activators are not yet precisely defined and because potential roles have been recently discussed at more length elsewhere (Rechsteiner and Hill 2005). Several observations suggest that PA28α and PA28β function in the immune system (reviewed in Rechsteiner et al. 2000). These activators appear to have arisen during evolution at roughly the same time as vertebrate cellular immunity. They are particularly enriched in immune tissues and virtually absent in brain. Finally, PA28$\alpha\beta$, but not PA28γ, is induced by interferon-γ and infection (Khan et al. 2001; Maksymowych et al. 1998; Tanahashi et al. 1997), and the presence of PA28$\alpha\beta$ influences production of some class I epitopes. Evidence of the involvement of PA28$\alpha\beta$ in antigen presentation is reviewed in Kloetzel and Ossendorp (2004), Rechsteiner et al. (2000), and Rock et al. (2002).

The role of PA28$\alpha\beta$ in antigen presentation is confused by the contradictory findings of two independent knockout mouse studies. Preckel et al. (1999) found a general impairment in CTL responses and concluded that that PA28 functions in immunoproteasome assembly. (Immunoproteasomes are the same as constitutive 20S proteasomes except that the three catalytic subunits are replaced with inducible counterparts [Rock and Goldberg 1999]). In contrast, Murata et al. (2001) found a phenotype that was almost wild type, including normal immunoproteasome assembly, although these mice were unable to process a specific epitope. Also arguing against the immunoproteasome assembly model is the finding that upregulation of immunoproteasome subunits occurs early in dendritic cell maturation, whereas PA28$\alpha\beta$ subunits are expressed later (Li et al. 2001b). Thus, although PA28$\alpha\beta$ appears to function in antigen presentation, the mechanistic basis for this activity is currently unknown. Alternative proposed mechanisms include substrate channeling, facilitating release of longer products (Whitby et al. 2000), and alteration of cleavage sites (Murata et al. 2001). Currently, the importance of these possible mechanisms is unclear. Several non-immune functions of PA28$\alpha\beta$ have also been proposed (Rechsteiner and Hill 2005), but these also lack validation and a convincing mechanism.

The biological role of PA28γ is even more obscure than that of PA28$\alpha\beta$. The knockout mice have mild phenotypes (Murata et al. 1999), including defects in processing some specific antigens (Barton et al. 2004). It is likely, however, that PA28γ has at least one important role that does not involve antigen presentation by MHC class I molecules, since simple eukaryotes that lack an MHC class I system generally contain a single PA28 molecule that is most closely related to the γ homologue of higher eukaryotes. Unlike PA28$\alpha\beta$, which most reports describe as being mainly cytoplasmic, PA28γ is largely confined to the nucleus (Soza et al. 1997; Wojcik 1999; Wojcik et al. 1998). Also, whereas PA28$\alpha\beta$ is most heavily expressed in immune tissues such as spleen, PA28γ is most heavily expressed in brain. One possibility is that PA28γ functions in apoptosis and/or cell-cycle progression (Masson et al. 2003). The physiological role of PA26 in *T. brucei* is unknown.

5.5 PA200/Blm10p

PA200 is a large nuclear protein that stimulates the proteasome's peptidase activity (Ustrell et al. 2002). It is a single-chain activator with a molecular weight of ~200 kDa, which is similar to that of the PA28 heptamer. Like PA28, bovine PA200 stimulates proteasomal hydrolysis of peptides, but not proteins, and it does not utilize ATP. The likely homologue of PA200 in yeast is Blm10p, (previously known as Blm3p, Doherty et al. 2004) with which PA200 shares just 13% sequence identity. Blm10p is also a nuclear protein, although, surprisingly, Blm10p does not appear to stimulate the 20S proteasome's peptidase activity (Fehlker et al. 2003).

It was suggested that PA200 functions in DNA repair because gamma irradiation of HeLa cells resulted in alteration of the usual uniform nuclear distribution of PA200 to a punctate pattern, a behavior characteristic of many DNA-repair proteins (Ustrell et al. 2002). Further support for this proposal was provided by earlier work that described Blm10p as complementing the bleomycin hypersensitivity of the blm3-1 mutation, thereby indicating a role in DNA repair (Febres et al. 2001; Moore 1991). This proposal has been weakened, however, by a subsequent analysis that concluded that Blm10p does not in fact contribute to bleomycin resistance in yeast (Aouida et al. 2004).

An alternative function for Blm10p in the assembly of nuclear proteasomes has been suggested based upon the finding that Blm10p associated with nascent proteasomes (Fehlker et al. 2003). Because the Blm10p-deleted strain was found to display an increased rate of processing of a proteasome subunit precursor, it was proposed that Blm10p functions to regulate late stages of proteasome assembly and maturation in the nucleus. Given the fundamental importance of proteasome maturation, however, it is surprising that the phenotypes associated with Blm10p deletion and overexpression are quite mild. At the current time we consider the functional role(s) of PA200/Blm10p to be an open question. The proposed roles in DNA repair and proteasome assembly require further clarification and support. Indeed, in view of their highly diverged amino acid sequences, it is possible that PA200 and Blm10p perform different *in vivo* roles.

Biochemical and structural analysis of PA200 activity is at a relatively preliminary stage (Figure 5.2). Based upon analysis of amino acid sequences, it has been concluded that PA200 is comprised of multiple HEAT/ARM repeats (Kajava et al. 2004). This implies that PA200 is comprised almost entirely of helices, a property shared with 11S activators, although the role of multiple PA28 C-termini in proteasome binding indicates that PA200 must bind via a different arrangement. Similarly, the amino acid sequence does not give obvious clues about the mechanism of gate opening. Because HEAT repeat proteins generally function in protein–protein interactions, it is attractive to speculate that PA200 forms hybrid proteasomes and functions to localize 26S proteasome activity to specific intracellular locations.

5.6 Concluding Remarks and Future Challenges

In contrast to the 19S activator, whose biological role is well established, it is not clear what physiological roles are performed by 11S activators or by PA200. Numerous publications link PA28$\alpha\beta$ to production of peptide ligands for MHC class I molecules, although, arguably, definitive data are lacking. The two reports of knockout mice, which might have provided conclusive data on this point, are largely contradictory. PA28γ has been linked to apoptosis and cell-cycle progression, although a convincing direct connection is elusive. Studies of PA200/Blm10p are similarly inconclusive, and the true role of PA200/Blm10p is still an open question. We believe that PA28 and PA200 will be found to perform important biological functions, since evolution is unlikely to have preserved 200-kDa proteins/complexes that bind and activate 20S proteasomes unless they provide a functional advantage. It is an urgent priority for the field to firmly establish the physiological roles of these proteasome activators.

Biochemical analysis of PA28 and PA200 is generally more advanced than the biological studies. Important mechanistic questions for the future include whether activation results simply from opening of the entrance/exit gate. A possible role for channels through PA28 and PA200 in defining substrate preference through filtering mechanisms should be tested explicitly. Similarly, the possibility of long-range allosteric changes that selectively repress specific active sites needs to be confirmed or refuted.

Structural analysis of 11S activators (PA26) is relatively advanced. We know how they bind in a manner that is insensitive to a specific amino acid sequence. We also know how PA26 repositions the proteasome Pro17 turn, and thereby induces formation of the open-gate conformation. PA26 is only distantly related to other PA28 activators, however, with no residues conserved in the activation loop. Therefore, questions still remain about the mechanism of activation by other 11S activators, and it would be interesting to visualize the structure of a cognate mammalian PA28–proteasome complex. Obtaining higher-resolution structural information on PA200/Blm10p and 19S remains an important priority.

An attractive possibility is that PA28 and PA200 might primarily function in the context of hybrid proteasomes as adaptors that tether 20S–19S complexes to specific intracellular locations or substrate complexes. In this model, opening of the entrance/exit gate might be of some advantage, but the primary role would be to define a location or association. The idea that PA28 and PA200 are primarily adaptors, rather than activators, is quite general and has a number of possibilities. For example, mediating interactions with components of the ER have been suggested as a possible mechanism for delivery of product peptides to the TAP transporter and hence to nascent MHC class I molecules (Realini et al. 1994; Rechsteiner et al. 2000). It is also conceivable that PA28 or PA200 interacts with substrate complexes or with chaperones that deliver substrates for degradation. There is still much to be resolved concerning the biology and biochemistry of proteasome activators.

References

Aouida, M., Page, N., Leduc, A., Peter, M. and Ramotar, D. (2004) A genome-wide screen in Saccharomyces cerevisiae reveals altered transport as a mechanism of resistance to the anticancer drug bleomycin. *Cancer Res*, **64**, 1102–1109.

Bajorek, M., Finley, D. and Glickman, M.H. (2003) Proteasome disassembly and downregulation is correlated with viability during stationary phase. *Curr Biol*, **13**, 1140–1144.

Barton, L.F., Runnels, H.A., Schell, T.D., Cho, Y., Gibbons, R., Tevethia, S.S., Deepe, G.S., Jr. and Monaco, J.J. (2004) Immune defects in 28-kDa proteasome activator gamma-deficient mice. *J Immunol*, **172**, 3948–3954.

Baumeister, W., Walz, J., Zuhl, F. and Seemuller, E. (1998) The proteasome: paradigm of a self-compartmentalizing protease. *Cell*, **92**, 367–380.

Benaroudj, N., Zwickl, P., Seemuller, E., Baumeister, W. and Goldberg, A.L. (2003) ATP hydrolysis by the proteasome regulatory complex PAN serves multiple functions in protein degradation. *Mol Cell*, **11**, 69–78.

Bochtler, M., Ditzel, L., Groll, M., Hartmann, C. and Huber, R. (1999) The proteasome. *Ann. Rev. Biophys. Struct.*, **28**, 295–317.

Bogyo, M., Shin, S., McMaster, J.S. and Ploegh, H.L. (1998) Substrate binding and sequence preference of the proteasome revealed by active-site-directed affinity probes. *Chem Biol*, **5**, 307–320.

Cascio, P., Call, M., Petre, B.M., Walz, T. and Goldberg, A.L. (2002) Properties of the hybrid form of the 26S proteasome containing both 19S and PA28 complexes. *Embo J*, **21**, 2636–2645.

Coux, O., Tanaka, K. and Goldberg, A.L. (1996) Structure and Functions of the 20S and 26S Proteasomes. *Ann. Rev. Biochem.*, **65**, 801–847.

De Mot, R., Nagy, I., Walz, J. and Baumeister, W. (1999) Proteasomes and other self-compartmentalizing proteases in prokaryotes. *Trends Microbiol*, **7**, 88–92.

DeMartino, G.N. and Slaughter, C.A. (1999) The proteasome, a novel protease regulated by multiple mechanisms. *J. Biol. Chem.*, **274**, 22123–22126.

DeMartino, G.N. and Wojcik C. (2005) Proteasome Regulator, PA700 (19s Regulatory Particle in *Protein Degradation* (Eds: R.J. Mayer, A. Ciechanover, M. Rechsteiner), Vol. 1, Wiley-VCH, Weinheim, 288–317.

Doherty, K., Pramanik, A., Pride, L., Lukose, J. and Moore, C.W. (2004) Expression of the expanded YFL007w ORF and assignment of the gene name BLM10. *Yeast*, *21*, 1021–1023.

Dubiel, W., Pratt, G., Ferrell, K. and Rechsteiner, M. (1992) Purification of an 11S regulator of the multicatalytic protease. *J.Biol.Chem.*, **267**, 22369–22377.

Emori, Y., Tsukahara, T., Kawasaki, H., Ishiura, S., Sugita, H. and Suzuki, K. (1991) Molecular cloning and functional analysis of three subunits of yeast proteasome. *Mol Cell Biol*, **11**, 344–353.

Febres, D.E., Pramanik, A., Caton, M., Doherty, K., McKoy, J., Garcia, E., Alejo, W. and Moore, C.W. (2001) The novel BLM3 gene encodes a protein that protects against lethal effects of oxidative damage. *Cell Mol Biol (Noisy-le-grand)*, **47**, 1149–1162.

Fehlker, M., Wendler, P., Lehmann, A. and Enenkel, C. (2003) Blm3 is part of nascent proteasomes and is involved in a late stage of nuclear proteasome assembly. *EMBO Rep*, **4**, 959–963.

Förster, A., Whitby, F.G. and Hill, C.P. (2003) The pore of activated 20S proteasomes has an ordered 7-fold symmetric conformation. *EMBO J*, **22**, 4356–4364.

Gao, X., Li, J., Pratt, G., Wilk, S. and Rechsteiner, M. (2004) Purification procedures determine the proteasome activation properties of REG gamma (PA28 gamma). *Arch Biochem Biophys*, **425**, 158–164.

Glickman, M.H., Rubin, D.M., Fried, V.A. and Finley, D. (1998) A regulatory particle of the Saccharomyces cerevisiae proteasome. *Mol. Cell. Biol.*, **18**, 3149–3162.

Goldberg, A.L. (2003) Protein degradation and protection against misfolded or damaged proteins. *Nature*, **426**, 895–899.

Gottesman, S. (2003) Proteolysis in bacterial regulatory circuits. *Annu Rev Cell Dev Biol*, **19**, 565–587.

Groll, M. and Clausen, T. (2003) Molecular shredders: how proteasomes fulfill their role. *Curr Opin Struct Biol*, **13**, 665–673.

Groll, M. and Huber, R. (2003) Substrate access and processing by the 20S proteasome core particle. *Int J Biochem Cell Biol*, **35**, 606–616.

Groll, M., Ditzel, L., Löwe, J., Stock, D., Bochtler, M., Bartunik, H.D. and Huber, R. (1997) Structure of 20S proteasome from yeast at 2.4 Å resolution. *Nature*, **386**, 463–471.

Groll, M., Nazif, T., Huber, R. and Bogyo, M. (2002) Probing structural determinants distal to the site of hydrolysis that control substrate specificity of the 20S proteasome. *Chem Biol*, **9**, 655–662.

Groll, M., Brandstetter, H., Bartunik, H., Bourenkow, G. and Huber, R. (2003a) Investigations on the maturation and regulation of archaebacterial proteasomes. *J Mol Biol*, **327**, 75–83.

Groll, M., Brandstetter, H., Bartunik, H., Bourenkow, G. and Huber, R. (2003b) Investigations on the maturation and regulation of archaebacterial proteasomes. *J Mol Biol*, **327**, 75–83.

Harris, J.L., Alper, P.B., Li, J., Rechsteiner, M. and Backes, B.J. (2001) Substrate specificity of the human proteasome. *Chem Biol*, **8**, 1131–1141.

Heinemeyer, W., Trondle, N., Albrecht, G. and Wolf, D.H. (1994) PRE5 and PRE6, the last missing genes encoding 20S proteasome subunits from yeast? Indication for a set of 14 different subunits in the eukaryotic proteasome core. *Biochemistry*, **33**, 12229–12237.

Hendil, K.B., Khan, S. and Tanaka, K. (1998) Simultaneous binding of PA28 and PA700 activators to 20 S proteasomes. *Biochem J*, **332 (Pt 3)**, 749–754.

Hershko, A. and Ciechanover, A. (1998) The ubiquitin system. *Annu Rev Biochem*, **67**, 425–479.

Hill, C.P., Masters, E.I. and Whitby, F.G. (2002) The 11S regulators of 20S proteasome activity. *Curr Top Microbiol Immunol*, **268**, 73–89.

Johnston, S.C., Whitby, F.G., Realini, C., Rechsteiner, M. and Hill, C.P. (1997) The proteasome 11S regulator subunit REG alpha (PA28 alpha) is a heptamer. *Protein Sci*, **6**, 2469–2473.

Kajava, A.V., Gorbea, C., Ortega, J., Rechsteiner, M. and Steven, A.C. (2004) New HEAT-like repeat motifs in proteins regulating proteasome structure and function. *J Struct Biol*, **146**, 425–430.

Khan, S., van den Broek, M., Schwarz, K., de Giuli, R., Diener, P.A. and Groettrup, M. (2001) Immunoproteasomes largely replace constitutive proteasomes during an antiviral and antibacterial immune response in the liver. *J Immunol*, **167**, 6859–6868.

Kim, D.H., Lee, S.M., Hong, B.Y., Kim, Y.T. and Choi, T.J. (2003) Cloning and sequence analysis of cDNA for the proteasome activator PA28-beta subunit of flounder (Paralichthys olivaceus). *Mol Immunol*, **40**, 611–616.

Kloetzel, P.M. and Ossendorp, F. (2004) Proteasome and peptidase function in MHC-class-I-mediated antigen presentation. *Curr Opin Immunol*, **16**, 76–81.

Knowlton, J.R., Johnston, S.C., Whitby, F.G., Realini, C., Zhang, Z., Rechsteiner, M. and Hill, C.P. (1997) Structure of the proteasome activator REGalpha (PA28alpha). *Nature*, **390**, 639–643.

Kopp, F., Dahlmann, B. and Kuehn, L. (2001) Reconstitution of hybrid proteasomes from purified PA700-20 S complexes and PA28alphabeta activator: ultrastructure and peptidase activities. *J Mol Biol*, **313**, 465–471.

Kuehn, L. and Dahlmann, B. (1996a) Proteasome activator PA28 and its interaction with 20 S proteasomes. *Arch. Biochem. Biophys.*, **329**, 87–96.

Kuehn, L. and Dahlmann, B. (1996b) Reconstitution of proteasome activator

PA28 from isolated subunits: optimal activity is associated with an alpha,beta-heteromultimer. *FEBS Lett.*, **394**, 183–186.

Kuehn, L. and Dahlmann, B. (1997) Structural and functional properties of proteasome activator PA28. *Mol. Biol. Rep.*, **24**, 89–93.

Leggett, D.S., Hanna, J., Borodovsky, A., Crosas, B., Schmidt, M., Baker, R.T., Walz, T., Ploegh, H. and Finley, D. (2002) Multiple associated proteins regulate proteasome structure and function. *Molecular Cell*, **10**, 495–507.

Li, J., Gao, X., Joss, L. and Rechsteiner, M. (2000) The proteasome activator 11 S REG or PA28: chimeras implicate carboxyl-terminal sequences in oligomerization and proteasome binding but not in the activation of specific proteasome catalytic subunits. *J. Mol. Biol.*, **299**, 641–654.

Li, J., Gao, X., Ortega, J., Nazif, T., Joss, L., Bogyo, M., Steven, A.C. and Rechsteiner, M. (2001a) Lysine 188 substitutions convert the pattern of proteasome activation by REGγ to that of REGs α and β. *EMBO J.*, **20**, 3359–3369.

Li, J., Schuler-Thurner, B., Schuler, G., Huber, C. and Seliger, B. (2001b) Bipartite regulation of different components of the MHC class I antigen-processing machinery during dendritic cell maturation. *Int Immunol*, **13**, 1515–1523.

Löwe, J., Stock, D., Jap, B., Zwickl, P., Baumeister, W. and Huber, R. (1995) Crystal structure of the 20S proteasome from the archaeon T. acidophilum at 3.4 A resolution. *Science*, **268**, 533–539.

Ma, C.P., Slaughter, C.A. and DeMartino, G.N. (1992) Identification, purification, and characterization of a protein activator (PA28) of the 20 S proteasome (macropain). *J. Biol. Chem.*, **267**, 10515–10523.

Ma, C.P., Willy, P.J., Slaughter, C.A. and DeMartino, G.N. (1993) PA28, an activator of the 20 S proteasome, is inactivated by proteolytic modification at its carboxyl terminus. *J Biol Chem*, **268**, 22514–22519.

Maksymowych, W.P., Ikawa, T., Yamaguchi, A., Ikeda, M., McDonald, D., Laouar, L., Lahesmaa, R., Tamura, N., Khuong, A., Yu, D.T.Y. and Kane, K.P. (1998) Invasion by Salmonella typhimurium induces increased expression of the LMP, MECL, and PA28 proteasome genes and changes in the peptide repertoire of HLA-B27. *Infect. Immun.*, **66**, 4624–4632.

Masson, P., Andersson, O., Petersen, U.M. and Young, P. (2001) Identification and characterization of a Drosophila nuclear proteasome regulator. A homolog of human 11 S REGgamma (PA28gamma). *J Biol Chem*, **276**, 1383–1390.

Masson, P., Lundgren, J. and Young, P. (2003) Drosophila proteasome regulator REGgamma: transcriptional activation by DNA replication-related factor DREF and evidence for a role in cell cycle progression. *J Mol Biol*, **327**, 1001–1012.

Moore, C.W. (1991) Further characterizations of bleomycin-sensitive (blm) mutants of Saccharomyces cerevisiae with implications for a radiomimetic model. *J Bacteriol*, **173**, 3605–3608.

Mott, J.D., Pramanik, B.C., Moomaw, C.R., Afendis, S.J., DeMartino, G.N. and Slaughter, C.A. (1994) PA28, an activator of the 20 S proteasome, is composed of two nonidentical but homologous subunits. *J. Biol. Chem.*, **269**, 31466–31471.

Murata, S., Kawahara, H., Tohma, S., Yamamoto, K., Kasahara, M., Nabeshima, Y., Tanaka, K. and Chiba, T. (1999) Growth retardation in mice lacking the proteasome activator PA28gamma. *J Biol Chem*, **274**, 38211–38215.

Murata, S., Udono, H., Tanahashi, N., Hamada, N., Watanabe, K., Adachi, K., Yamano, T., Yui, K., Kobayashi, N., Kasahara, M., Tanaka, K. and Chiba, T. (2001) Immunoproteasome assembly and antigen presentation in mice lacking both PA28alpha and PA28beta. *Embo J*, **20**, 5898–5907.

Murray, B.W., Sultmann, H. and Klein, J. (2000) Identification and linkage of the proteasome activator complex PA28 subunit genes in zebrafish. *Scand. J. Immunol.*, **51**, 571–576.

Mykles, D.L. (1996) Differential effects of bovine PA28 on six peptidase activities of the lobster muscle proteasome (multicatalytic proteinase). *Arch Biochem Biophys*, **325**, 77–81.

Paesen, G.C. and Nuttall, P.A. (1996) A tick homolog of the human Ki nuclear autoantigen. *Biochim. Biophys. Acta*, **1309**, 9–13.

Pickart, C.M. and Cohen, R.E. (2004)

Proteasomes and their kin: proteases in the machine age. *Nat Rev Mol Cell Biol*, **5**, 177–187.

Preckel, T., Fung-Leung, W.P., Cai, Z., Vitiello, A., Salter-Cid, L., Winqvist, O., Wolfe, T.G., Von Herrath, M., Angulo, A., Ghazal, P., Lee, J.D., Fourie, A.M., Wu, Y., Pang, J., Ngo, K., Peterson, P.A., Fruh, K. and Yang, Y. (1999) Impaired immunoproteasome assembly and immune responses in PA28–/– mice. *Science*, **286**, 2162–2165.

Ramachandran, R., Hartmann, C., Song, H.K., Huber, R. and Bochtler, M. (2002) Functional interactions of HslV (ClpQ) with the ATPase HslU (ClpY). *Proc Natl Acad Sci U S A*, **99**, 7396–7401.

Realini, C., Rogers, S.W. and Rechsteiner, M. (1994) KEKE motifs: Proposed roles in protein-protein association and presentation of peptides by MHC Class I receptors. *FEBS Lett.*, **348**, 109–113.

Realini, C., Jensen, C.C., Zhang, Z., Johnston, S.C., Knowlton, J.R., Hill, C.P. and Rechsteiner, M. (1997) Characterization of recombinant REGalpha, REGbeta, and REGgamma proteasome activators. *J Biol Chem*, **272**, 25483–25492.

Rechsteiner, M. and Hill, C.P. (2005) Mobilizing the cell's proteolytic machine: biological roles of proteasome activators and inhibitors. *Trends Cell Biol*, **15**, 27–33.

Rechsteiner, M., Realini, C. and Ustrell, V. (2000) The proteasome activator 11 S REG (PA28) and class I antigen presentation. *Biochem J*, **345 (Pt 1)**, 1–15.

Rock, K.L. and Goldberg, A.L. (1999) Degradation of cell proteins and the generation of MHC class I-presented peptides. *Annu. Rev. Immunol.*, **17**, 739–779.

Rock, K.L., York, I.A., Saric, T. and Goldberg, A.L. (2002) Protein degradation and the generation of MHC class I-presented peptides. *Adv Immunol*, **80**, 1–70.

Seemüller, E., Lupas, A., Stock, D., Löwe, J., Huber, R. and Baumeister, W. (1995) Proteasome from Thermoplasma acidophilum: a threonine protease. *Science*, **268**, 579–582.

Song, X., von Kampen, J., Slaughter, C.A. and DeMartino, G.N. (1997) Relative functions of the α and β subunits of the proteasome activator, PA28. *J. Biol. Chem.*, **272**, 27994–28000.

Sousa, M.C., Trame, C.B., Tsuruta, H., Wilbanks, S.M., Reddy, V.S. and McKay, D.B. (2000) Crystal and solution structures of an HslUV protease-chaperone complex. *Cell*, **103**, 633–643.

Sousa, M.C., Kessler, B.M., Overkleeft, H.S. and McKay, D.B. (2002) Crystal structure of HslUV complexed with a vinyl sulfone inhibitor: corroboration of a proposed mechanism of allosteric activation of HslV by HslU. *J Mol Biol*, **318**, 779–785.

Soza, A., Knuehl, C., Groettrup, M., Henklein, P., Tanaka, K. and Kloetzel, P.M. (1997) Expression and subcellular localization of mouse 20S proteasome activator complex PA28. *FEBS Lett.*, **413**, 27–34.

Tanahashi, N., Yokota, K., Ahn, J.Y., Chung, C.H., Fujiwara, T., Takahashi, E., DeMartino, G.N., Slaughter, C.A., Toyonaga, T., Yamamura, K., Shimbara, N. and Tanaka, K. (1997) Molecular properties of the proteasome activator PA28 family proteins and gamma-interferon regulation. *Genes Cells*, **2**, 195–211.

Tanahashi, N., Murakami, Y., Minami, Y., Shimbara, N., Hendil, K.B. and Tanaka, K. (2000) Hybrid proteasomes. Induction by interferon-gamma and contribution to ATP-dependent proteolysis. *J Biol Chem*, **275**, 14336–14345.

Unno, M., Mizushima, T., Morimoto, Y., Tomisugi, Y., Tanaka, K., Yasuoka, N. and Tsukihara, T. (2002) The structure of the mammalian 20S proteasome at 2.75 Å resolution. *Structure*, **10**, 609–618.

Ustrell, V., Hoffman, L., Pratt, G. and Rechsteiner, M. (2002) PA200, a nuclear proteasome activator involved in DNA repair. *EMBO J*, **21**, 3516–3525.

Velichutina, I., Connerly, P.L., Arendt, C.S., Li, X. and Hochstrasser, M. (2004) Plasticity in eucaryotic 20S proteasome ring assembly revealed by a subunit deletion in yeast. *EMBO J*, **23**, 500–510.

Verma, R., Chen, S., Feldman, R., Schieltz, D., Yates, J., Dohmen, J. and Deshaies, R.J. (2000) Proteasomal proteomics: identification of nucleotide-sensitive proteasome-interacting proteins by mass spectrometric analysis of affinity-purified proteasomes. *Mol Biol Cell*, **11**, 3425–3439.

Wang, J., Song, J.J., Franklin, M.C.,

Kamtekar, S., Im, Y.J., Rho, S.H., Seong, I.S., Lee, C.S., Chung, C.H. and Eom, S.H. (2001) Crystal structures of the HslVU peptidase-ATPase complex reveal an ATP-dependent proteolysis mechanism. *Structure (Camb)*, **9**, 177–184.

Wang, C.C., Bozdech, Z., Liu, C.L., Shipway, A., Backes, B.J., Harris, J.L. and Bogyo, M. (2003) Biochemical analysis of the 20 S proteasome of Trypanosoma brucei. *J Biol Chem*, **278**, 15800–15808.

Wenzel, T. and Baumeister, W. (1995) Conformational constraints in protein degradation by the 20S proteasome. *Nature Structural Biology*, **2**, 199–204.

Whitby, F.G., Masters, E.I., Kramer, L., Knowlton, J.R., Yao, Y., Wang, C.C. and Hill, C.P. (2000) Structural basis for the activation of 20S proteasomes by 11S regulators. *Nature*, **408**, 115–120.

Wilk, S., Chen, W.E. and Magnusson, R.P. (2000a) Properties of the beta subunit of the proteasome activator PA28 (11S REG). *Arch Biochem Biophys*, **384**, 174–180.

Wilk, S., Chen, W.E. and Magnusson, R.P. (2000b) Properties of the nuclear proteasome activator PA28gamma (REGgamma). *Arch Biochem Biophys*, **383**, 265–271.

Wojcik, C. (1999) Proteasome activator subunit PA28 alpha and related Ki antigen (PA28gamma) are absent from the nuclear fraction purified by sucrose gradient centrifugation. *Int. J. Biochem. Cell. Biol.*, **31**, 273–276.

Wojcik, C., Tanaka, K., Paweletz, N., Naab, U. and Wilk, S. (1998) Proteasome activator (PA28) subunits, alpha, beta and gamma (Ki antigen) in NT2 neuronal precursor cells and HeLa S3 cells. *Eur. J. Cell. Biol.*, **77**, 151–160.

Yao, Y., Huang, L., Krutchinsky, A., Wong, M.L., Standing, K.G., Burlingame, A.L. and Wang, C.C. (1999) Structural and functional characterization of the proteasome-activating protein PA26 from Trypanosoma brucei. *J. Biol. Chem.*, **274**, 33921–33930.

Zhang, Z., Clawson, A., Realini, C., Jensen, C.C., Knowlton, J.R., Hill, C.P. and Rechsteiner, M. (1998a) Identification of an activation region in the proteasome activator REGalpha. *Proc Natl Acad Sci U S A*, **95**, 2807–2811.

Zhang, Z., Clawson, A. and Rechsteiner, M. (1998b) The proteasome activator or PA28. Contribution by both α and β subunits to proteasome activation. *J. Biol. Chem.*, **273**, 30660–30668.

Zhang, Z., Realini, C., Clawson, A., Endicott, S. and Rechsteiner, M. (1998c) Proteasome activation by REG molecules lacking homolog-specific inserts. *J. Biol. Chem.*, **273**, 9501–9509.

Zhang, Z., Kruchinsky, A., Endicott, S., Realini, C., Rechsteiner, M. and Standing, K.G. (1999) Proteasome activator 11S REG or PA28. Recombinant REGα/REGβ hetero-oligomers are heptamers. *Biochemistry*, **38**, 5651–5658.

6
The Proteasome Portal and Regulation of Proteolysis

Monika Bajorek and Michael H. Glickman

Abstract

The proteolytic active sites of the 26S proteasome are sequestered within the central chamber of its 20S catalytic core particle. Access to this chamber is through a narrow channel defined by the outer alpha subunits. An intricate lattice of interactions anchors the N-termini of these alpha subunits, blocking access to the channel in free 20S core particles of eukaryotes. Entry of substrates can be enhanced by attachment of activators or regulatory particles to the proteolytic 20S core. Regulatory particles rearrange the blocking residues to form an open pore and promote substrate entry into the proteolytic chamber. Channel gating is apparently partially rate limiting for proteasome activity, as facilitating substrate entry in the open channel state leads to enhanced overall proteolysis rates. Interestingly, some substrates, particularly hydrophobic ones, can activate gate opening themselves, thus facilitating their own destruction. Properties of channel gating and the interactions required to maintain stable closed and open conformations and their consequences for proteasome function are discussed.

6.1
Background

In eukaryotes, the 26S proteasome hydrolyzes most nuclear, cytoplasmic, and endoreticulum (ER) proteins into peptides of varying lengths. Normally, substrates destined for elimination are first covalently attached to multiple molecules of ubiquitin (Ub) – a process that is executed by a cascade of ubiquitinating enzymes specific for each class of substrate – and then recognized by the 26S proteasome, unfolded, translocated into the proteolytic chamber, and irreversibly degraded [1–3]. The somewhat simpler 20S proteasomes found in archaea and some bacteria (actinomycetes) can degrade and remove non-ubiquitinated proteins in a very similar manner, though recognition of substrates and anchoring them to the proteasome while they are prepared for degradation probably differ due to lack of the ubiquitin tag in these organisms [4, 5]. In either case, substrates are threaded through a nar-

Protein Degradation, Vol. 2: The Ubiquitin-Proteasome System.
Edited by R. J. Mayer, A. Ciechanover, M. Rechsteiner

ISBN: 3-527-31130-0

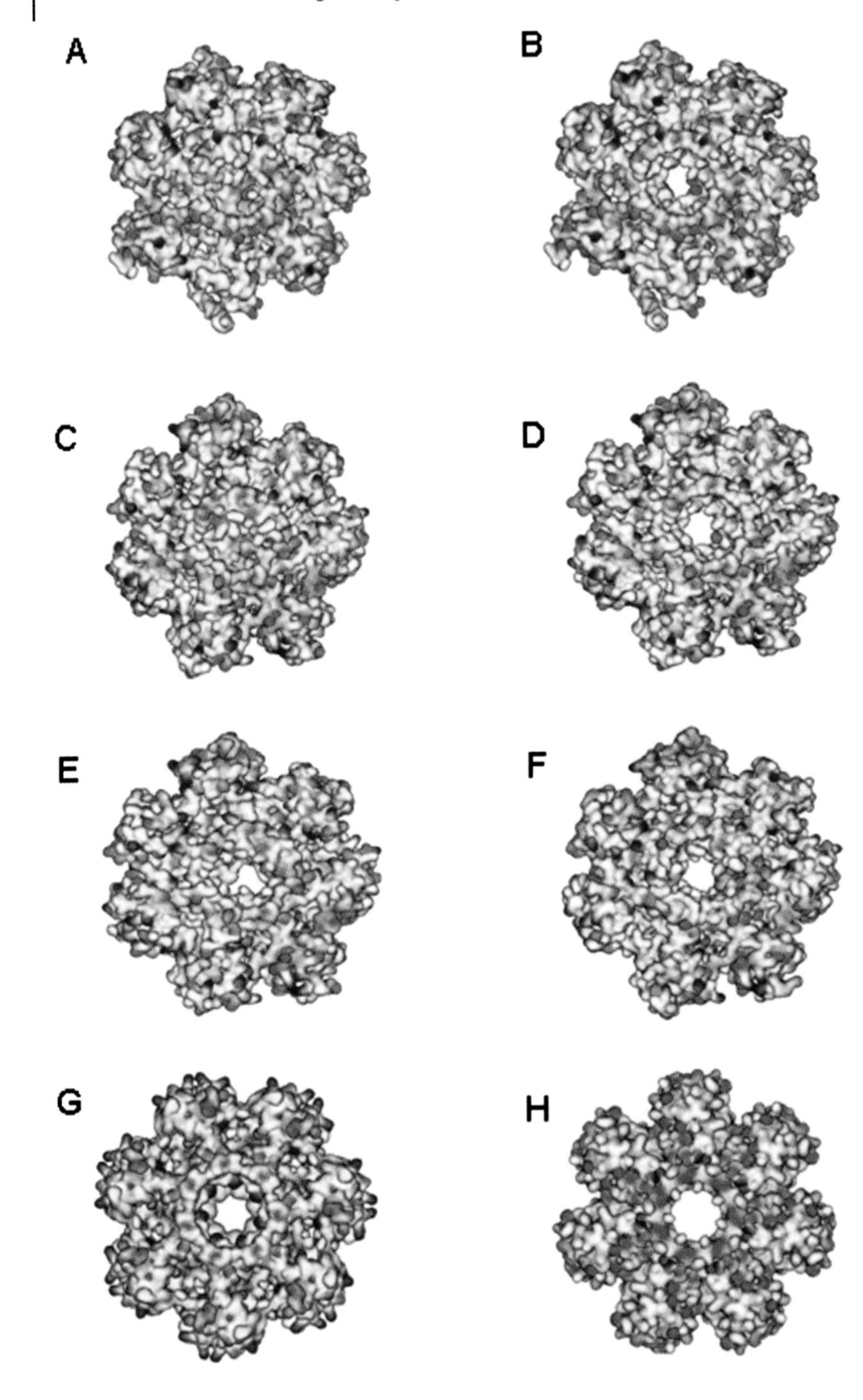
A
B
C
D
E
F
G
H

row channel leading into the isolated internal chamber where the proteolytic active sites are located. A gated porthole at the entrance to this channel may play a defined role in controlling both the nature of preferred substrates and the rate at which they enter the proteasome lumen where they are irreversibly hydrolyzed (Figure 6.1). This chapter will focus on this gate and mechanisms of maintaining opened or closed states. Other chapters in the book series will cover in depth the myriad steps leading up to, or those following, this process.

The 26S proteasome is composed of two sub-complexes: the 20S core particle of the proteasome (CP) where proteolysis takes place and a 19S regulatory particle (RP) that prepares substrates for entry into the CP. A detailed description of proteasome structure and associated activities can be found in other chapters in this volume, as well as in many detailed reviews [6–15]. Pertinent to understanding regulation of substrate entry, the 20S CP is a cylindrical structure composed of four stacked heptameric rings engendering a sequestered proteolytic chamber. Each of the two outer rings is composed of seven α subunits, and each of the two identical inner rings is formed from seven β subunits. The β rings contain the proteolytic active sites, while the outer α rings define the channel leading into the internal pro-

Fig. 6.1. A porthole into the proteasome. Top view presentations showing the surface structure of the α ring of free 20S CPs from various preparations: (A) mammalian, (B) mammalian in theoretical open state, (C) yeast (*S. cerevisiae*), (D) yeast in theoretical fully open state, (E) the $\alpha3\Delta N$ mutant from yeast, (F) yeast in open conformation imposed by attachment of the PA26S activator, (G) archaea (*T. acidophilum*), (H) bacteria (*Rhodococcus erythropolis*). Structures depicted in A, C, E, F, G, and H are the actual 2D determinations extracted from the crystal structure deposited in the Protein Data Bank (PDB) and visualized as a surface view with the Viewerlite program. Acidic residues are colored in red, alkali in blue, and hydrophobic in white. Note the dominance of acidic residues in the pore region of the closed state in yeast and mammalian 20S CPs. These structures indicate that the gross surface structure of eukaryotic proteasomes is remarkably similar, with slight structural divergence apparent in the archaeal and bacterial versions. The N-terminus upstream of threonine 13 in the archaeal complex (the gray region shown in Figure 6.2) is disordered and thus does not show up in the electron density map, giving the appearance of an open conformation (G). Structures shown in B and D are models in which the electron density of the corresponding tail residues in each α subunit was deleted to mimic an open conformation. Complete removal of these tail segments unveils a pore in the open state. The actual crystal structure determination of the $\alpha3\Delta N$ 20S CP indicates that $\alpha3$ is a pivotal subunit in controlling channel gating. Deletion of the $\alpha3$ tail alone (E) causes disordering in neighboring subunits (up to the red arrow in Figure 6.2); however, some obstruction remains when compared with the theoretical fully open state (D). Deletion of the two opposing tails from the $\alpha3$ and $\alpha7$ subunits generates a fully activated complex, apparently due to removal of this residual obstruction [22]. Interestingly, attachment of PA26 to the α-ring surface rearranges the tail regions of each subunit, coupled with significant disordering (up to the blue arrow in Figure 6.2), thus imposing an open conformation (F). By deleting the electron density of the PA26 chains and focusing only on the structure of the 20S subunits, we depict in panel F the actual situation that occurs upon PA26 binding to 20S from yeast. Additional details can be found in the original publications [16–19, 55, 59].

teolytic chamber [16–19]. Overall, the 20S CP forms a $\alpha_7\beta_7\beta_7\alpha_7$ barrel structure [6]. The α and β rings of archaeal 20S proteasomes are made up of seven copies each of a single α or β subunit, giving the structure as a whole a sevenfold symmetry along the central axis. In contrast, there are seven distinct subunits in each ring of the 20S CP from eukaryotes. When visualized from outside, these subunits are labeled counterclockwise $\alpha1$ through $\alpha7$ and $\beta1$ through $\beta7$, respectively. Although the seven α or β subunits within the 20S CP of each species are structurally almost superimposable, their sequence identities are usually in the 20–40% range (see Ref. [7] and references therein). These differences are probably of significance as they are well maintained; sequences of orthologous subunits from different species are more than 55% identical [20, 21].

The purified 20S CP in its so-called latent form can slowly hydrolyze short or unstructured polypeptides as well as some proteins with hydrophobic or misfolded patches [22–25]. There is increasing evidence that degradation of unstructured non-ubiquitinated proteins by the 20S CP may play biological roles *in vivo* as well [22, 26–31]. To degrade ubiquitinated substrates, attachment of the ATPase-containing 19S RP to the surface of the α ring of the 20S CP is required [25, 32]. Attachment of 19S RP enhances the basal peptidase activity of proteasomes as well [33, 34]. In archaea and select prokaryotes, ATPase rings such as the proteasome-activating nucleotidase (PAN) complex or the AAA ATPase ring complex (ARC) serve as rudimentary regulatory particles that enhance proteolysis by 20S proteasomes but probably do not affect peptidase rates [35–38]. One manner by which such ATPase-containing regulatory particles activate proteolysis is by unfolding substrates and translocating them into the proteolytic chamber [36, 39–45]. However, other regulatory particles can also influence the proteolytic activity of the proteasome in an ATP-independent manner. For example, a number of non-ATPase activators – such as PA28, PA26, and 11S Reg complexes – attach to the 20S CP and enhance peptidase, but not proteolysis, rates [46–53]. Other conditions such as exposure to low ionic strength, sodium dodecyl sulfate (SDS), or small hydrophobic peptides can also activate hydrolysis rates and give rise to an activated 20S CP [18, 22, 24, 34, 54, 55]. These results (together with other observations) suggest that the latent 20S CP is found in a self-imposed repressed state and that it is possible to activate the intrinsic peptidase activity in an ATP-independent manner. Structural analysis revealed the nature of this auto-inhibition by identifying a gated channel leading substrates into the proteolytic chamber [17–19]. Alleviation of inhibition by opening the substrate channel for traffic is a basic feature of proteasomes, a key property necessary to understand their function.

6.2 The Importance of Channel Gating

In order to protect cells from mistaken degradation of random proteins, entry into the proteasome lumen must be strictly regulated. Indeed, access to the channel in the latent 20S CP of eukaryotes is restricted by the N-termini of the seven α sub-

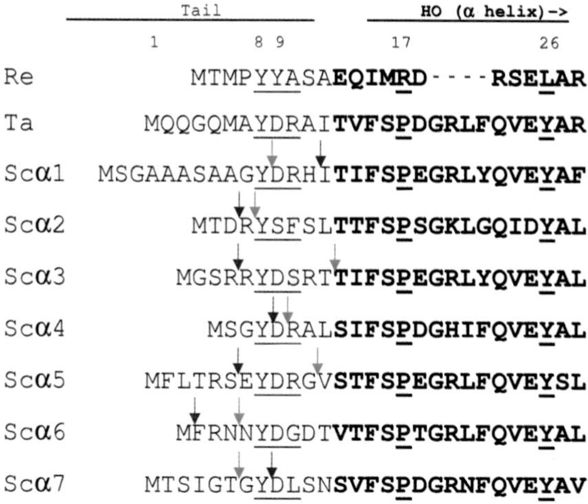

Fig. 6.2. Sequence alignment of α subunit N-termini. N-termini of the seven different α subunits from S. cerevisiae compared to archaeal and bacterial homologues. The seven α subunits (α1–α7) of yeast (Sc) are shown along with the single α subunit found in bacteria (Re) and archaea (Ta). In normal text, to the left of the bold region lie the tail sequences, which differ from one α subunit to the other. Residue numbers for the yeast and bacteria α subunits are assigned based on the alignment to archaea. Residues in the N-terminus to the arrow are disordered in α3ΔN mutant (red arrow) and upon attachment of PA26S (blue arrow). The conserved residues forming the so-called canonical cluster (YDR-P-Y) documented to play a role in stabilizing the closed or open conformations are underlined. The N-terminal sequence up to the first α helical structure of each CP α subunit (as determined by the crystal structure of the yeast CP [17, 18]) is shown in light gray. This region is homologous to the disordered segments in the N-terminal regions of the α subunit from *T. acidophilum* [16, 57]. The YD(R) motif in each tail is underlined. Note that in all subunits from yeast a tyrosine residue is present in the same location. In six of the seven subunits, an aspartate follows, and in three, an arginine completes the YDR motif. The remainder of the tail region is divergent between subunits.

units [17–19]. Each tail assumes a unique conformation while pointing inwards to the center of the ring [7, 9, 10]. Figures 6.1A and 6.1C describe the surface of the latent 20S CP as it may appear to an approaching substrate; the interlaced residues in latent proteasomes form a sealed surface obstructing passage of proteins and even short peptides. This property is a common feature of eukaryotic proteasomes and is well conserved between yeast and mammalian complexes. The blocking residues at the entrance to the proteolytic channel probably account for the repressed proteolytic activity of the latent 20S CP of eukaryotes [18, 56].

The behavior of archaeal proteasomes is somewhat different from that of the proteasomes of eukaryotes. In contrast to the sealed chamber in the eukaryotic complex, crystal structures of 20S proteasomes obtained from archaeal organisms distinctly find the proteasome in an open state. Disorder in the conformation of the first 12 amino acids of each α subunit creates a pore in the center of the α ring contiguous with the channel leading into the lumen (Figure 6.1G). This open state accounts for enhanced peptidase rates measured for this complex [16, 36, 57]. Despite the appearance of an open pore in the archaeal 20S CP (Figure 6.1), dynamic conformations of the α-subunit tails may partially restrict passage of intact proteins through the pore, necessitating mechanisms for activation of protease activity. For instance, by locking these residues in a stable, open conformation, regulatory complexes found in these organisms (such as PAN) are able to accelerate proteolysis of full-sized proteins [36, 43, 55, 57, 58]. Certain conditions might promote switching of the N-termini from an unstructured into a structured, open conformation without requiring a regulatory complex [57]. Apparently, a similar situation occurs for 20S proteasomes present in bacteria [59]. The N-termini encompassing the first eight amino acids of the subunits in the α ring of the *Rhodococcus* proteasome are disordered (Figure 6.2), creating the appearance of an open pore (Figure 6.1H). The ARC regulatory complex found in these organisms probably stimulates proteolytic activity by anchoring these tails into a static, open conformation, similarly to the role of PAN in archaeal organisms.

Eukaryotic 20S CPs can also be found in an open conformation. Repeated freeze-thawing of purified 20S CPs; mild chemical treatments such as exposure to low ionic strength, to low levels of sodium dodecyl sulfate (SDS), or to short hydrophobic peptides; and attachment of regulatory complexes and even mutations in residues near the central pore all activate hydrolysis rates [18, 22, 24, 34, 54, 55]. Presumably, all these treatments lead to disordering in pore residues involved in gating the channel, thus opening up a porthole into the 20S CP (Figure 6.1B,D). Rearrangement of channel-blocking residues results in facilitated substrate access into the proteolytic chamber and activation of proteasomes. Activated 20S CPs can easily hydrolyze unfolded or hydrophobic proteins, in some instances more rapidly even than intact 26S holoenzymes [22, 23]. It should be noted that free 20S CPs can spontaneously switch between the latent "closed" and activated "open" conformations [60]; however, under physiological conditions it appears that the majority of free 20S CPs from eukaryotes are found in a closed and latent state.

Similar to the situation described above for archaeal proteasomes, a distinction can be made between the appearance of an open channel due to disordering in residues lining the pore region and one in which the blocking tails are secured in an open structure. Structural determination of a 20S CP–PA26 complex depicts the α subunit N-termini pointing away from the center of the ring, opening up an unobstructed porthole into the channel [55] (see also Figure 6.1). The conformation differences between the closed and open states of N-termini could explain how regulatory particles activate proteolytic activity by rearranging the blocking residues to facilitate substrate entry [43, 54, 55].

Regulatory complexes such as the 19S RP participate in channel gating and facilitate substrate entry. This activates proteolysis and allows the resulting proteasome holoenzyme to fulfill its role in regulated protein degradation [7]. Indeed, under standard growth conditions, the majority of proteasomes in yeast cells are found as 26S holoenzymes. That said, 20S CPs are abundant in certain cases and are found regularly in mammalian tissue. Given that 20S core particles are slower than 26S holoenzymes at hydrolyzing most test substrates and are unable to proteolyze polyubiquitinated proteins, it is unclear what the biological function of the 20S CP is, or whether it is an important player in cellular protein breakdown. Abating bulk protein degradation upon proteasome disassembly may be a requirement for survival under certain stress conditions. For instance, prolonged starvation, oxidative stress, and severe heat-induced damage result in dissociation of proteasome holoenzymes and elevated levels of 20S CPs [22, 28, 29, 61]. There is some evidence that the catalytically repressed 20S CP can serve as a reservoir of proteasome components for reassembly when resumption of proteolysis is needed. Nevertheless, free 20S CPs may play limited roles in degradation of unstructured or non-ubiquitinated proteins under both normal growth and stress conditions [22, 26, 28–31].

6.3 A Porthole into the Proteasome

6.3.1 The Closed State

The first 12 residues in the archaeal complex, right up to the first α helix in the protein (Figure 6.2), are naturally disordered, giving the impression of an open channel (Figure 6.1G). Similar disordering is found in the equivalent sections of α subunits in the bacterial 20S proteasome purified from *Rhodococcus* (Figure 6.1H). In comparison, the corresponding N-termini of the seven α subunits in eukaryotic complexes point towards the center of the ring, sealing the entrance to the proteolytic channel (Figure 6.1A,C). In eukaryotes, the paralogous subunits within the α ring show structural and sequence similarities over the bulk of the protein, yet diverge at their amino-terminal region in both sequence and relative length (Figure 6.2). Precisely at the center of the ring, each tail accepts a unique conformation, stabilizing a well-defined and non-symmetric closed configuration [7] (see also Figure 6.3). The tail of $\alpha3$ is somewhat distinct from the others in that it points directly across the surface of the α ring towards the center, maintaining close contact to every other α subunit. The importance of these tail regions is highlighted by their extreme conservation across eukaryotes; while each tail is highly conserved in different species, the corresponding regions are divergent from one subunit to another [7]. These properties suggest that the N-termini play a critical structural role that has been maintained in core particles in all eukaryotes, and it is precisely their differences that are integral to their function.

Low-energy bonds formed between specific tail residues are critical for stabilizing the open or closed conformations. For example, in latent 20S CPs, aspartate at position 9 in the N-terminus of $\alpha3$ contacts both tyrosine 8 and arginine 10 in neighboring $\alpha4$ [18]. A salt bridge is formed between the carboxylate group of aspartate 9 in $\alpha3$ and the guanidinium group of arginine 10 in $\alpha4$, simultaneous with a hydrogen bond linking aspartate 9 of $\alpha3$ with tyrosine 8 of $\alpha4$ [18]. Similar bonds probably link the analogous residues in mammalian 20S CPs [19]. Embedded

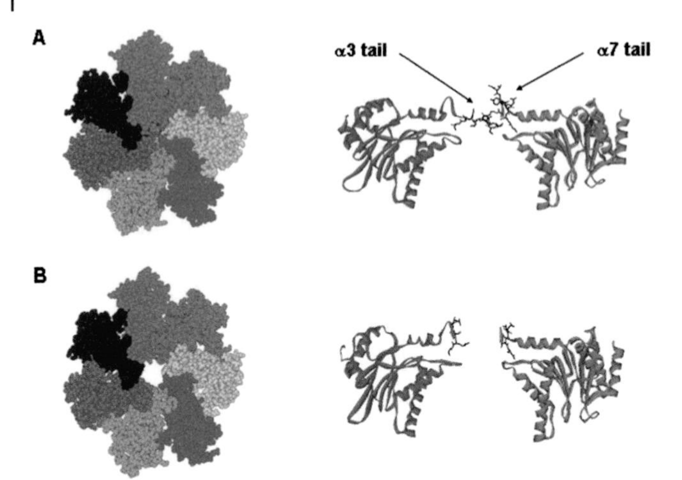

Fig. 6.3. A gate-and-latch system determines open and closed states of the substrate channel. Yeast 20S CP in a latent closed conformation (A) and in an open state imposed by the PA26S activator (B). At left a top view of the seven-member α ring of the 20S CP is shown. All α subunits are color coded, starting with α1 (light blue) on top and running counterclockwise to α7 (red). On the right, a side view focusing only on two opposing subunits – α3 (pink) and α7 (red) – highlights the conformational switch that occurs in the N-termini (black) between the closed state (A) to an open state (B) upon binding of PA26. Note that the N-terminal tails of α3 and α7 adopt different conformations in the closed state and point inwards to block access through the channel (A). In contrast, in the open state induced by PA26S, the tails adopt a new ordered conformation, pointing away from the channel region. A portion of the N-terminus (up to the blue arrow in Figure 6.2) is disordered and invisible in the crystal structure.

within the N-terminal segments of most α subunits is a short consensus sequence: Tyr8-Asp9-Arg10 or "the YDR motif" (Figure 6.2). Conservation of tyrosine at position 8 is absolute among subunits in yeast (and is invariable at this location in most subunits of other organisms as well); aspartate at position 9 is present in the tail of all yeast subunits except for α2, while conservation of arginine as residue number 10 is less strict. The direct contacts formed between these residues in adjacent subunits may explain their correlated evolutionary conservation. Interactions involving YDR residues could be critical for maintaining distinct open and closed conformations of 20S proteasomes.

Finding YDR residues in all subunits that form the α ring makes it somewhat puzzling that crystal structure determination did not pick out similar contacts between other neighboring subunits in the pore region. This raises the possibility that the interaction between α3 and α4 plays a unique and central role in maintaining the closed conformation of the proteasome. Support for the pivotal role of α3 in gating the 20S CP channel was provided upon truncation of the tail region of α3. Truncation of the N-terminus of the α3 subunit in yeast (the α3ΔN mutant) resulted in a purified 20S CP that was found in the open pore conformation (Figures

6.1 and 6.3). Furthermore, removal of the N-terminus of α3 caused disordering in neighboring subunits concomitant with stimulation of 20S CP peptidase activity. These results indicate that the tail of α3 is important for stabilizing neighboring tails in the closed conformation. Moreover, an aspartate-to-alanine substitution in the YDR motif of α3 (the α3 D9A mutant) appears to increase peptidase activity of purified 20S CP *in vitro*, on par with the activation observed upon deletion of the entire tail region in α3ΔN [18]. Both mutations associated with α3 point to a functional significance of the YDR motif in stabilization of the closed state of the gate. Interestingly, the YDR sequence is found intact in the α subunits from various archaea, even though the tail regions are not anchored in the latent state of 20S proteasomes from these organisms [16, 57]. This observation points to a wider role for α-subunit tail interactions in defining proteasome conformation.

6.3.2 The Open State

In order for substrates to enter the proteolytic chamber, and most likely for products to exit as well, the blocking N-terminal residues of the α subunits in the closed state must be rearranged. Rearrangement obviously necessitates breaking of the interactions that anchor the tails in the closed conformation, while forming competing interactions to stabilize them in an open conformation. For example, removal of the first nine residues at the N-terminus of the α3 subunit in yeast breaks stabilizing interactions with neighboring subunits, causing significant disordering in neighboring tails (Figure 6.2) and leading to an open pore roughly 13 Å across at the center of the α-ring surface (Figure 6.1F). Disordering alone is insufficient to allow unobstructed entry through the pore, as no enhancement of protein degradation rates has been observed so far with proteasomes purified from this mutant [18, 22]. Each α subunit plays a unique role in gating. Deletion of the equivalent N-terminal tail of α7 does not significantly increase the peptidase activity compared to wild type, pointing to a significant role for α3 [22]. Because of its peripheral location at the α-ring surface, truncation of α7 alone may not result in sufficient loss of order in neighboring tails to generate an opening wide enough for entry of small peptides. However, deletion of tails of two opposing α subunits (the α3α7ΔN strain; [22]) may act synergistically to relieve hindrance of entry of proteins into the proteolytic chamber. Thus, gating of the proteolytic channel emerges as a pivotal property in regulating proteolysis rates.

In fact, activated proteasomes may require more than just a disordered state of channel gating residues. Blocking residues may need to be removed (as in the truncation studies described in the preceding paragraph) or anchored into a stable, open conformation to relieve obstruction of the channel. Studies on archaeal proteasomes point to interactions involving YDR residues as influencing channel opening rather than stabilizing the closed state. While the N-terminal tails (12 amino acids) of each subunit of the α ring are disordered, they nevertheless occupy the pore region and impose a partial barrier to passage of protein substrates [16, 57]. The lack of defined electron density reflects that under the experimental conditions mutual interactions were not strong enough to anchor these tails into a single stable conformation, resulting in multiple possible conformations of these segments between individual molecules in the sample. Evidence that the YDR motif may play a role in stabilizing the open state was provided by a combined mutagenesis and biochemical study of archaeal proteasomes. Purified archaeal 20S proteasomes slowly hydrolyze GFP tagged with a short C-terminal extension, GFP-ssrA. The rate is accelerated upon addition of the archaeal proteasome activator PAN. However, PAN was unable to activate proteolysis of GFP-ssrA by 20S proteasomes from the archaeon *T. acidophilum* that were mutated at positions tyrosine 8 or arginine 9 (of the YDR motif in the N-tail regions) of their α subunit [55]. It was suggested that these residues are required to stabilize the tails in an open conformation that is preferred upon attachment of PAN.

A stably open conformation was observed in proteasome assembly precursors as well. During proteasome biogenesis, the seven α subunits form an intermediate homomeric ring, the α_7 ring, which only then interacts with β subunits to yield the mature complex with $\alpha_7\beta_7$ composition [62]. In contrast to mature archaeal 20S CPs, structure determination of such an α_7 ring precursor from the archaeon *Archaeoglobus fulgidus* found the N-terminal segments anchored in a stable, open state [57]. In this conformation, the tail regions that contain the YDR motif adopted a helical structure motif and pointed away from the ring surface: tyrosine 8 of each subunit made a hydrogen bond with aspartate 9 of the preceding α subunit, whereas arginine 10 pointed inwards towards the central channel and did not partake in anchoring neighboring tails. The region N-terminal to tyrosine 8 of each subunit was disordered, creating the appearance of a pore roughly 13 Å in diameter contiguous with the substrate channel. In mature archaeal 20S CPs, the interactions between the N-termini of the α subunits appear to be broken, causing disordering in a greater portion of the tail regions up to residue number 12 (inclusive). These disordered tails, which are not stably anchored, partially block the pore and interfere with passage of proteins into the proteasome lumen.

Eukaryotic proteasomes can be found in an open state as well. Crystallography analysis of a PA26–20S CP complex depicted the α ring in an ordered, open, symmetric conformation [55]. Attachment of PA26 induces all seven of the α subunit N-terminal tails to adopt an ordered conformation for residues 7–12 away from the center of the ring. A cluster of four highly conserved residues, Tyr8 and Asp9 (part of the YDR motif in the tail region) together with downstream residues Pro17 and Tyr26 (in the first stable alpha helix; HO), is critical for the open state. Attachment of PA26 to the surface of the α ring repositions Pro17, which in turn induces a conformational change in tail segments to lift up and away from the center of the ring (Figure 6.2). In stark contrast to the closed state, the open state is remarkably symmetric, with all tails conforming to a similar structure held in place by similar interactions. In this open state, Asp9 forms a hydrogen bond with Tyr26 of the same subunit. Cooperativity between subunits is communicated via an additional hydrogen bond linking Asp9 of one subunit and Tyr8 of the preceding (clockwise) tail [55]. Consequently, the hydrogen bond that holds Asp9 and Tyr8 of $\alpha 3$ and $\alpha 4$, respectively, in the closed state must be broken and rearranged to allow for the open state in which Asp9 of $\alpha3$ now interacts with Tyr8 of $\alpha2$. The term "proteasome gating" refers to switching between these two conformations. A principal role of regulatory particles is to promote channel opening by stabilizing one conformation over the other.

The interactions occurring between different subunits in the pore region of the eukaryotic PA26–20S CP complex are quite similar to those found in the α_7-ring precursor complex of archaea, which is also found in a stable, open conformation [57]. Asp9 in each subunit interacts with Tyr8 of the preceding (clockwise) tail in α_7 rings, leading to a symmetric, open structure. Asp9 of each α subunit also interacts with the downstream Tyr26 residue of the same subunit, repositioning the tail away from the center of the ring and up into the cavity of the docking proteasome activator complex (PA26 in this case). This conformation can be seen clearly in a side view of the 20S CP complex shown in Figure 6.3.

The mesh of internal interactions among α subunits described in the preceding paragraph and the apparent lack of stable interactions between α-subunit tail residues with the proteasome activator suggest that a stable, open conformation is an intrinsic property of proteasomes and may explain how, under some conditions, 20S core particles can spontaneously adopt the activated form without need for attachment of an activator complex [60]. Nevertheless, comparative studies show that the 20S CP from *Rhodococcus* is also found in an open conformation [59], even though the α subunits in this organism do not contain any of the signature YD or PY residues in the tail or HO regions (Figures 6.1, 6.2). Apparently, the semblance of an open state does not absolutely require interactions among these residues, suggesting additional mechanisms for stabilizing the closed or open states and switching between them. Additional studies will have to be completed for a clear understanding of gating mechanisms.

6.4
Facilitating Traffic Through the Gated Channel

6.4.1
Regulatory Complexes

As mentioned above, a number of ATP-independent activators are known to attach to the 20S CP and activate its peptidase and protease activities. These include the 11S Reg/PA28, PA26, and PA200 [46–53]. Attachment of PA28, for example, increases V_{max} for hydrolysis of certain peptides by the 20S CP by up to 100-fold, but in contrast to ATPase-containing activators (such as the 19S RP or PAN), PA28 does not promote protein degradation by the 20S CP [46, 63]. Activation of peptidase activity by non-ATPase-containing regulators can be attributed to imposing an open-channel conformation and facilitated substrate entry upon attachment of the regulatory complex. For example, activation has been documented for the PA26–20S hybrid complex, formed *in vitro* from 20S CP from *S. cerevisiae* and PA26 from *T. brucei* [55]. *S. cerevisiae* apparently lacks natural homologues of this class of activators, but it is assumed that attachment of other symmetric activators induces a similar conformational change. For example, the symmetric PAN complex found in archaea appears to drive formation of the open conformation of archaeal proteasomes in much the same manner [55]. Attachment of PAN expedites proteolysis by wild-type archaeal 20S proteasomes, yet is unable to activate proteasomes mutant in any of the channel cluster residues (Tyr8, Asp9, Pro17 and Tyr26; see Section 6.3.2). This result suggests that each residue in the conserved YD-P-Y cluster is critical for stabilizing the open conformation in activated proteasomes. Mutations in these residues may lead to disordered tails that are unable to "lock" into an open conformation even upon PAN attachment, thus impinging on substrate entry. Activation is not achieved merely by realigning the residues that obstruct traffic through the alpha ring in the closed state, but necessitates locking the α tails into a stable, open conformation.

So far, it is unclear whether asymmetric activators, such as the 19S RP, open the channel in a manner similar to that of the symmetric examples given by PA26 and PAN. However, evidence linking the 19S RP to gating can be deduced from a substitution mutation in the ATP-binding site of a single ATPase (*RPT2*) that severely affects peptidase activity of the proteasome, probably due to hampering the ability of the RP to properly gate the channel into the CP [56, 64]. This observation indicates that even the entry of small peptides – which do not need to be unfolded – can be controlled by the RP. Furthermore, a constitutively open-channel 20S CP generated upon deletions of tail residues in the $\alpha3$ and $\alpha7$ subunits ($\alpha3\alpha7\Delta N$ mutant) exhibits activated peptidase activity, similar to that measured for 26S proteasome holoenzymes [22]. The implication is that attachment of the 19S RP realigns the α-subunit tails to facilitate passage of substrates, thus enhancing proteolysis rates. It should be emphasized, however, that in contrast to the homomeric rings found in PA26 or PAN, the 19S RP is a heterogeneous complex that contains six different ATPases. The two classes of regulators may induce different conformational changes on the sevenfold symmetry of the α ring. Furthermore, it has not yet been verified that the six ATPases indeed form a six-member ring at the base of the 19S RP. A limited subset of Rpt subunits have been found to come in direct contact with an α subunit ($\alpha2$–Rpt4, $\alpha2$–Rpt5, $\alpha4$–Rpt4, $\alpha7$–Rpt4, $\alpha1$–Rpt6, $\alpha2$–Rpt6, $\alpha4$–Rpt2, $\alpha6$–Rpt4; [65–68]). The pair Rpt2–$\alpha3$ has been shown to be involved in gating the channel into the CP [18, 56, 64], though gating may be controlled by additional Rpt–α subunit interactions. As there does not appear to be simple pairing of each Rpt with a single α subunit, it is possible that the conformation adopted by α-subunit tails in the 26S proteasome holoenzyme will differ from the "symmetric" open state observed for the PA26–20S CP hybrid.

6.4.2
Substrate-facilitated Traffic

Recent studies show that some natively disordered proteins can enter the proteasome without assistance of ATP-dependent activators [26–28, 30]. The ability of latent 20S CP to catalyze cleavage of some unfolded proteins suggests that they may directly interact with the α ring to promote gating to facilitate their own entry. This mechanism is not general: latent 20S CPs degrade most substrates slower than activated proteasomes [22]. Nevertheless, some substrates with unfolded domains or hydrophobic patches are degraded rapidly by latent 20S CP, faster even than by 26S holoenzymes [23]. Presumably, sequence motifs in the substrate interact with channel-gating residues in α subunits and aid in channel opening. For example, p21 and α-synuclein facilitate their own degradation and, when fused to stable and hard-to-degrade proteins such as GFP, promote their degradation as well indicating that gating sequences are transferable. Support for such a mechanism can be deduced from certain peptides that interact with the proteasome in a noncompetitive way to modulate the proteolytic activity of the proteasome [24, 69]. This stimulation was not observed for open-channel proteasomes (such as $\alpha 3\Delta N$ or the PA26–20S CP complex), suggesting that they specifically interact with channel-gating residues and promote channel opening. Whether these interactions involve the YDR motif or the YD-P-Y cluster in α subunits, similar to the manner by which regulatory particles activate proteasome activity, has not been elucidated. Interestingly, the pore region of the 20S CP, in the closed state, exposes only hydrophobic or negatively charged side chains. No positively charged groups are present on the surface of the α ring in eukaryotes (Figure 6.1). Thus, as a substrate approaches the surface of the α ring, it "sees" a predominantly negatively charged surface. This may explain how the latent 20S CP discriminates between substrates and why certain unstructured substrates with hydrophobic or positively charge stretches may interact with gating residues in the pore region and facilitate their own translocation inwards, whereas others do not [23]. For example, hydrolysis of casein – which is highly phosphorylated and carries multiple negatively charged groups – is remarkably slow by latent 20S CP, yet can be dramatically accelerated upon channel opening possibly by removal of tail residues as in Figure 6.1E [22].

6.5 Summary: Consequences for Regulated Proteolysis

It appears that proteasome-dependent proteolysis is a regulated process that can be enhanced or inhibited under certain conditions. There are reports that the proteasome itself can be a target of such regulation [22, 29, 61, 70]. Indeed, enhancement of overall *in vivo* proteolysis rates observed in the open-channel mutant indicates that the proteasome may be partially rate limiting in the overall cascade of ubiquitin-dependent protein degradation [22]. Polyubiquitinated substrates must be stable enough, even if only transiently, to allow for competition between degradation and reversal of fate. Channel gating within 26S holoenzymes may participate in the delicate balance between proteolysis and rescue.

A function of a gated channel leading into the CP is to impose inhibition during assembly of the mature CP. In the final stage of CP assembly, self-compartmentalization is achieved by the association of two $\alpha_7\beta_7$ half-CPs at the β–β interface. These half-CPs are inactive due to propeptides in the critical β

subunits that mask their active site. As these half-CPs are joined, inhibition by β-subunit N-termini is relieved by autolysis [17, 62], while inhibition by the blocking N-termini of the α subunits is imposed. Binding of the regulatory particles relieves this inhibition by opening the channel and thus activating proteolysis. There is increasing evidence that (at least in yeast) certain stress constitutions such as prolonged starvation or severe heat shock naturally promote proteasome dissociation into separate 20S CP and 19S RP subcomponents [22, 29, 61, 70]. These conditions may require repressed proteasome-dependent degradation for survival. One manner by which proteasome activity could be downregulated is by reinstating autoinhibition of the dissociated 20S CP. Indeed, the open-channel mutant that lacks the ability to enter the closed conformation exhibits low viability under conditions that promote proteasome disassembly [22].

An additional reason for a gated channel could be to regulate exit of products from the proteasome. It is possible that under normal conditions product release is slowed down by a gated channel in order to increase processivity or to decrease average peptide length. Most of these short peptides are quickly removed from the cytoplasm. Under certain conditions (such as during immune response) it might be beneficial to produce peptides with other lengths or properties. For example, upon interferon-γ induction, attachment of PA28/11S Reg plays a role in antigen processing by altering the makeup of peptides generated by the hybrid proteasome 19S RP–20S CP–11S Reg complexes [52, 53, 71, 72]. In analogy to the distantly related PA26, PA28 probably attaches to the α-ring surface and rearranges the blocking N-termini, promoting the open-channel conformation. It is possible that the open state increases the exit rate of peptides generated in the proteolytic chamber and alters their makeup to fit better antigen-presentation requirements.

References

1 Glickman, M.H. and A. Ciechanover, *The Ubiquitin-proteasome Proteolytic Pathway: Destruction for the sake of construction*. Physiol. Rev., 2002. **82**: p. 373–428.

2 Pickart, C.M., *MECHANISMS UNDERLYING UBIQUITINATION*. Annu. Rev. Biochem., 2001. **70**: p. 503–533.

3 Weissman, A.M., *Themes and variations on ubiquitylation*. Nature Rev. Cell Mol. Biol., 2001. **2**(3): p. 169–179.

4 Maupin-Furlow, J.A., et al., *Proteasomes in the archaea: from structure to function*. Front. Biosci., 2000. **5**: p. D837–D865.

5 De Mot, R., et al., *Proteasomes and other self-compartmentalizing proteases in prokaryotes*. Trends Microbiol., 1999. **7**(2): p. 88–92.

6 Pickart, C.M. and R.E. Cohen, *PROTEASOMES AND THEIR KIN: PROTEASES IN THE MACHINE AGE*. Nat. Rev. Mol. Cell Biol., 2004. **5**(3): p. 177–187.

7 Bajorek, M. and M.H. Glickman, *Proteasome regulatory particles: Keepers of the gates*. Cell. Mol. Life Sci., 2004. **61**(13): p. 1579–1588.

8 Guterman, A. and M.H. Glickman, *Deubiquitinating enzymes are IN(trinsic to proteasome function)*. Curr. Prot. Pep. Sci., 2004. **5**(3): p. 201–210.

9 Groll, M. and T. Clausen, *Molecular shredders: how proteasomes fulfill their role*. Current Opinion in Structural Biology, 2003. **13**(6): p. 665–673.

10 GROLL, M. and R. HUBER, *Substrate access and processing by the 20S proteasome core particle.* Int. J. Biochem. Cell Biol., 2003. **35**(5): p. 606–616.

11 GLICKMAN, M.H. and V. MAYTAL, *Regulating the 26S proteasome.*, in *Curr. Top. Microbiol. and Immunol.*, P. ZWICKL and W. BAUMEISTER, Editors. 2002, Springer-Verlag: Germany. p. 43–72.

12 ZWICKL, P. and E. SEEMUELLER, *20S proteasomes*, in *Curr. Topics Microbiol. Immunol.*, P. ZWICKL and W. BAUMEISTER, Editors. 2002, Springer-Verlag: Germany. p. 23–41.

13 HEINEMEYER, W. and D.H. WOLF, *Active sites and assembly of the 20S proteasome.*, in *Proteasomes: The world of regulatory proteolysis*, D.H. WOLF and W. HILT, Editors. 2000, Eurekah.com/ LANDES BIOSCIENCE Publishing Company: Georgetown, Texas USA. p. 48–70.

14 BOCHTLER, M., et al., *The Proteasome.* Annu. Rev. Biophys. Biomol. Struct., 1999. **28**: p. 295–317.

15 VOGES, D., P. ZWICKL, and W. BAUMEISTER, *THE 26S PROTEASOME: A Molecular Machine Designed for Controlled Proteolysis. Annu. Rev. Biochem*, 1999. **68**: p. 1015–1068.

16 LOEWE, J., et al., *Crystal structure of the 20S proteasome from the archeon T. acidophilum at 3.4 Angstrom resolution.* Science, 1995. **268**: p. 533–539.

17 GROLL, M., et al., *Structure of 20S proteasome from yeast at a 2.4 Angstrom resolution.* Nature, 1997. **386**: p. 463–471.

18 GROLL, M., et al., *A gated channel into the core particle of the proteasome.* Nat. Struct. Biol., 2000. **7**: p. 1062–1067.

19 UNNO, M., et al., *The structure of the mammalian 20S proteasome at 2.75A resolution.* Structure, 2002. **10**: p. 609–618.

20 VOLKER, C. and A. LUPAS, *Molecular Evolution of Proteasomes*, in *Curr. Top. Microbiol. and Immunol.*, P. ZWICKL and W. BAUMEISTER, Editors. 2002, Springer-Verlag: Berlin Heidelberg. p. 1–22.

21 GILLE, C., et al., *A Comprehensive View on Proteasomal Sequences: Implications for the Evolution of the Proteasome.* Journal of Molecular Biology, 2003. **326**(5): p. 1437–1448.

22 BAJOREK, M., D. FINLEY, and M.H. GLICKMAN, *Proteasome Disassembly and Downregulation Is Correlated with Viability during Stationary Phase.* Current Biology, 2003. **13**(13): p. 1140–1144.

23 LIU, C.-W., et al., *Endoproteolytic activity of the proteasome.* Science, 2003. **299**: p. 408–411.

24 KISSELEV, A.F., D. KAGANOVICH, and A.L. GOLDBERG, *Binding of hydrophobic peptides to several non-catalytic sites promotes peptide hydrolysis by all active sites of 20 S proteasomes. Evidence for peptide-induced channel opening in the alpha-rings.* J. Biol. Chem., 2002. **277**: p. 22260–22270.

25 GLICKMAN, M.H., et al., *A subcomplex of the proteasome regulatory particle required for ubiquitin-conjugate degradation and related to the COP9/ Signalosome and eIF3.* Cell, 1998. **94**(5): p. 615–623.

26 ORLOWSKI, M. and S. WILK, *Ubiquitin-independent proteolytic functions of the proteasome.* Archives of Biochemistry and Biophysics, 2003. **415**(1): p. 1–5.

27 FORSTER, A. and C.P. HILL, *Proteasome degradation: enter the substrate.* Trends in Cell Biology, 2003. **13**(11): p. 550–553.

28 GRUNE, T., et al., *Selective degradation of oxidatively modified protein substrates by the proteasome.* Biochemical and Biophysical Research Communications, 2003. **305**(3): p. 709–718.

29 SHRINGARPURE, R., et al., *Ubiquitin Conjugation Is Not Required for the Degradation of Oxidized Proteins by Proteasome.* J. Biol. Chem., 2003. **278**(1): p. 311–318.

30 ASHER, G., et al., *Mdm-2 and ubiquitin-independent p53 proteasomal degradation regulated by NQO1.* PNAS, 2002. **99**(20): p. 13125–13130.

31 HOYT, M.A. and P. COFFINO, *Ubiquitin-free routes into the proteasome.* Cell. Mol. Life Sci., 2004. **61**(13): p. 1596–1600.

32 Guterman, A. and M.H. Glickman, *Complementary roles for Rpn11 and Ubp6 in deubiquitination and proteolysis by the proteasome.* J. Biol. Chem., 2004. **279**(3): p. 1729–1738.

33 Adams, G.M., et al., *Formation of proteasome-PA700 complexes directly correlates with activation of peptidase activity.* Biochem., 1998. **37**: p. 12927–12932.

34 Glickman, M.H., et al., *The regulatory particle of the S. cerevisiae proteasome.* Mol. Cell. Biol., 1998. **18**: p. 3149–3162.

35 Frohlich, K.U., *An AAA family tree.* J. Cell Sci., 2001. **114**: p. 1601–1602.

36 Zwickl, P., et al., *An archaebacterial ATPase, homologous to ATPases in the eukaryotic 26 S proteasome, activates protein breakdown by 20 S proteasomes.* J. Biol. Chem., 1999. **274**: p. 26008–26014.

37 Wolf, S., et al., *Characterization of ARC, a divergent member of the AAA ATPase family from Rhodococcus erythropolis.* J. Mol. Biol., 1998. **277**: p. 13–25.

38 Wollenberg, K. and J.C. Swaffield, *Evolution of proteasomal ATPases.* Mol. Biol. Evol., 2001. **18**(6): p. 962–974.

39 Braun, B.C., et al., *The base of the proteasome regulatory particle exhibits chaperone-like activity.* Nat. Cell Biol., 1999. **1**: p. 221–226.

40 Strickland, E., et al., *Recognition of misfolding proteins by PA700, the regulatory subcomplex of the 26S proteasome.* J. Biol. Chem., 2000. **275**: p. 5565–5572.

41 Navon, A. and A.L. Goldberg, *Proteins are unfolded on the surface of the ATPase ring before transport into the proteasome.* Mol. Cell, 2001. **8**(6): p. 1339–1349.

42 Liu, C.-W., et al., *Conformational remodeling of proteasomal substrates by PA700, the 19S regulatory complex of the 26S Proteasome.* J. Biol. Chem., 2002. **277**(30): p. 26815–26820.

43 Benaroudj, N., et al., *ATP hydrolysis by the proteasome regulatory complex PAN serves multiple functions in protein degradation.* Mol. Cell, 2003. **11**: p. 69–78.

44 Ogura, T. and K. Tanaka, *Dissecting various ATP-dependent steps involved in proteasomal degradation.* Mol. Cell, 2003. **11**(1): p. 3–5.

45 Kenniston, J.A., et al., *Studies with a prokaryotic protease show that the cost of translocation consumes much of the energy that is used in degradation.* Cell, 2003. **114**: p. 511–520.

46 Ma, C.P., C.A. Slaughter, and G.N. DeMartino, *Identification, purification, and characterization of a protein activator (PA28) of the 20S proteasome.* J. Biol. Chem., 1992. **267**: p. 10515–10523.

47 Realini, C., et al., *Characterization of recombinant REGα, REGβ, and REGγ proteasome activators.* J. Biol. Chem., 1997. **272**: p. 25483–25492.

48 Hendil, K.B., S. Khan, and K. Tanaka, *Simultaneous binding of PA28 and PA700 activators to 20S proteasomes.* Biochem. J., 1998. **332**: p. 749–754.

49 Yao, Y., et al., *Structural and functional characterizations of the proteasome-activating protein PA26 from Trypanosoma brucei.* J. Biol. Chem., 1999. **274**(48): p. 33921–33930.

50 McCutchen-Maloney, S.L., et al., *cDNA cloning, expression, and functional characterization of PI31, a proline-rich inhibitor of the proteasome.* J. Biol. Chem., 2000. **275**: p. 18557–18565.

51 Ustrell, V., et al., *PA200, a nuclear proteasome activator involved in DNA repair.* EMBO J., 2002. **21**(13): p. 3516–3525.

52 Rechsteiner, M., C. Realini, and V. Ustrell, *The proteasome activator 11S REG (PA28) and class I antigen presentation.* Biochem. J., 2000. **345**: p. 1–15.

53 Stohwasser, R., et al., *Kinetic evidences for facilitation of peptide channeling by the proteasomal activator PA28.* Eur. J. Biochem., 2000. **267**: p. 6221–6230.

54 Kohler, A., et al., *The substrate translocation channel of the proteasome.* Biochimie, 2001. **83**(3–4): p. 325–332.

55 Forster, A., F.G. Whitby, and C.P. Hill, *The pore of activated 20S*

proteasomes has an ordered 7-fold symmetric conformation. EMBO J., 2003. **22**(17): p. 4356–4364.

56 Koehler, A., et al., *The axial channel of the proteasome core particle is gated by the Rpt2 ATPase and controls both substrate entry and product release.* Mol. Cell, 2001. **7**: p. 1143–1152.

57 Groll, M., et al., *Investigations on the Maturation and Regulation of Archaebacterial Proteasomes+.* Journal of Molecular Biology, 2003. **327**(1 SU –): p. 75–83.

58 Wilson, H.L., et al., *Biochemical and physical properties of the M. jannaschii 20S proteasome and PAN, a homolog of the ATPase (Rpt) subunits of the eucaryal 26S proteasome.* J. Bacteriol., 2000. **186**(6): p. 1680–1692.

59 Kwon, Y.D., et al., *Crystal structures of the Rhodococcus proteasome with and without its pro-peptides: implications for the role of the pro-peptide in proteasome assembly.* J. Mol. Biol., 2004. **335**: p. 233–245.

60 Osmulski, P.A. and M. Gaczynska, *Nanoenzymology of the 20S proteasome: Proteasomal actions are controlled by the allosteric transition.* Biochem., 2002. **41**: p. 7047–7053.

61 Imai, J., et al., *The molecular chaperone Hsp90 plays a role in the assembly and maintenance of the 26S proteasome.* EMBO J., 2003. **22**(14): p. 3557–3567.

62 Chen, P. and M. Hochstrasser, *Autocatalytic subunit processing couples active site formation in the 20S proteasome to completion of assembly.* Cell, 1996. **86**: p. 961–972.

63 Dubiel, W., et al., *Purification of an 11S regulator of the multicatalytic protease.* J. Biol. Chem., 1992. **267**(31): p. 22369–22377.

64 Rubin, D.M., et al., *Active site mutants in the six regulatory particle ATPases reveal multiple roles for ATP in the proteasome.* EMBO J., 1998. **17**(17): p. 4909–4919.

65 Coux, O., *An interaction map of proteasome subunits.* Biochem. Soc. Trans., 2003. **31**: p. 465–469.

66 Davy, A., et al., *A protein-protein map of the C. elegans 26S proteasome.* EMBO Rep., 2001. **2**(9): p. 821–828.

67 Fu, H.Y., et al., *Subunit interaction maps for the regulatory particle of the 26s proteasome and the cop9 signalosome reveal a conserved core structure.* EMBO J., 2001. **20**(24): p. 7096–7107.

68 Ferrell, K., et al., *Regulatory subunit interactions of the 26S proteasome, a complex problem.* Trends Biochem. Sci., 2000. **25**(2): p. 83–88.

69 Papapostolou, D., O. Coux, and M. Reboud-Ravaux, *Regulation of the 26S proteasome activities by peptides mimicking cleavage products*1.* Biochemical and Biophysical Research Communications, 2002. **295**(5): p. 1090–1095.

70 Zhang, F., et al., *O-GlcNAc modification is an endogenous inhibitor of the proteasome.* Cell, 2003. **115**(6): p. 715–725.

71 Groettrup, M., et al., *A role for the proteasome regulator PA28a in antigen presentation.* Nature, 1996. **381**: p. 166–168.

72 Cascio, P., et al., *Properties of the hybrid form of the 26S proteasome containing both 19S and PA28 complexes.* EMBO J., 2002. **21**(11): p. 2636–2645.

7
Ubiquity and Diversity of the Proteasome System

Keiji Tanaka, Hideki Yashiroda, and Shigeo Murata

7.1
Introduction

The proteasome is an ATP-dependent protease complex known to collaborate with ubiquitin (Ub), and its polymerization acts as a marker of regulated proteolysis in eukaryotic cells [1–3]. The covalent attachment of multiple ubiquitins on the target proteins is achieved by a cascade of enzymatic reactions catalyzed by the E1 (Ub-activating), E2 (Ub-conjugating), and E3 (Ub-ligating) enzymes [4]. The resulting polyubiquitin chain serves as a signal for trapping the target protein, and consequently the substrate is destroyed after proteolytic attack by the proteasome. Numerous studies have recently emphasized the biological importance of the ubiquitin–proteasome system, which is capable of catalyzing rapidly, timely, and unidirectionally a diverse array of biological processes that are responsible for cell-cycle progression, DNA repair, cell death (e.g., apoptosis), immune response, signal transduction, transcription, metabolism, protein quality control, and developmental programs. Details of the ubiquitin–proteasome pathway have been reviewed [5–11], but the field continues to expand rapidly.

It has become clear that most cellular proteins are targeted for degradation by the proteasome. The proteasome is an unusually large protein complex, consisting of two parts: the catalytic core and the regulatory particle, both of which are composed of a set of multiple distinct subunits. Thus, the proteasome acts as a highly organized apparatus designed for efficient and exhaustive hydrolysis of proteins; in fact, it can be regarded as the protein-destroying machinery in living cells. Whereas the proteasome complex has been highly conserved during evolution due to its fundamental roles in cells, it has also acquired considerable diversity in multicellular organisms (particularly vertebrates), the purpose of which is to adapt evolutionarily to emergencies in environmental status. Indeed, the acquisition of divergent protein factors is closely linked to the development of temporal and spatial regulations driven by the proteasome in species-specific fashions. In our current knowledge, the structural and functional heterogeneity of the proteasome expands the roles of proteolysis in the cell. In this review, therefore, we focus our attention on the diversity of the proteasome system, with a special reference to its physiological roles.

Protein Degradation, Vol. 2: The Ubiquitin-Proteasome System.
Edited by R. J. Mayer, A. Ciechanover, M. Rechsteiner

ISBN: 3-527-31130-0

7.2
Catalytic Machine

7.2.1
Standard Proteasome

The 20S proteasome is the central machine with multiple catalytic sites to hydrolyze the peptide bonds of proteins. There are two types of major isoforms in cells, which include standard (alias constitutive) proteasomes and immunoproteasomes (for details, see subsequent section). The standard proteasome has been well characterized at the molecular level. It is a large protein complex with a sedimentation coefficient of 20S and a molecular mass of about 750 kDa. Electron microscopic examination revealed resemblance in the cylindrical configurations of 20S proteasomes in various sources ranging from yeast to mammal [12]. It is a barrel-like particle formed by the axial stacking of four rings made up of two outer α rings and two inner β rings, which are each made up of seven structurally similar α and β subunits (Table 7.1), respectively, being associated in the order of $\alpha_{1-7}\beta_{1-7}\beta_{1-7}\alpha_{1-7}$. The overall architectures of the high-ordered structures of yeast (*Saccharomyces cerevisiae*) and mammalian (bovine) 20S proteasomes are indistinguishable, as demonstrated by X-ray crystallography [13, 14]. The subunits of the 20S proteasome exhibit a unique location with C2 symmetry.

The three β-type subunits of each inner ring have catalytically active threonine residues at their N-termini, all of which show N-terminal nucleophile (Ntn) hydrolase activity, indicating that the proteasome is a novel threonine protease, differing from the known protease family categorized into seryl-, thiol, carboxyl, and metalloproteases. Those $\beta1$, $\beta2$, and $\beta5$ subunits correspond to caspase-like/PGPH (peptidyl glutamyl–peptide hydrolyzing), trypsin-like, and chymotrypsin-like activities, respectively, which are capable of cleaving peptide bonds at the C-terminal side of acidic, basic, and hydrophobic amino acid residues, respectively. Two pairs of these three active sites face the interior of the cylinder and reside in a chamber formed by the centers of the abutting β rings (Figure 7.1).

X-ray crystallographic analysis of the bovine 20S proteasome raises the possibility that one additional novel Ntn hydrolase activity may be present in the $\beta7$ subunit, because the functional groups that satisfy the requirement for the Ntn-hydrolase active sites are located around the N-terminal threonine of the $\beta7$ subunit [14]. Intriguingly, the hollow around this active center is much smaller than the S1 pockets of $\beta1$, $\beta2$, or $\beta5$, indicating that this active site may have a small neutral amino acid–preferring (SNAAP) activity. However, the direction of the N-terminal main chain of $\beta7$ indicates that the new active site is not in the chamber formed by the two β rings but is close to the interface formed by the α and β rings. Whether or not the $\beta7$ subunit is indeed a catalytic site requires further studies.

It is obvious that the proteasome is present in both the nucleus and cytoplasm of eukaryotic cells [15]. Indeed, it is predominantly located in the nuclei of mammalian tumor cells but dynamically moves between these two compartments. How does the proteasome alter its subcellular localization? In this regard, it is

Table 7.1. Subunits and auxiliary factors of the proteasome.

Category	*Subclassification*	*Systematic nomenclature*	*HUGO*	*Miscellaneous nomenclature*		*Human (yeast) amino acids*	*Motif*	*Lethality*
				Human	*Yeast (budding/fission)*			
20S	α-type subunits	α1	PSMA6	Iota	SCL1, YC7	246	NLS	+
		α2	PSMA2	C3	PRE8, Y7	233	NLS	+
		α3	PSMA4	C9	PRE9, Y13	261	NLS	−
		α4	PSMA7	C6	PRE6	248	NLS	+
		α5	PSMA5	Zeta	PUP2, DOA5	241		+
		α6	PSMA1	C2	PRE5	263		+
		α7	PSMA3	C8	PRE10, YC1	254		+
	β-type subunits	β1	PSMB6	Y, delta	PRE3	34 + 205	Ntn	+
		β2	PSMB7	Z	PUP1	43 + 234	Ntn	+
		β3	PSMB3	C10	PUP3	205		+
		β4	PSMB2	C7	PRE1	201		+
		β5	PSMB5	X, MB1, epsilon	PRE2, DOA3	59 + 204	Ntn	+
		β6	PSMB1	C5	PRE7	28 + 213		+
		β7	PSMB4	N3, beta	PRE4	45 + 219		+
		β1i	PSMB9	LMP2, RING12	–	20 + 199	Ntn	
		β2i	PSMB10	MECL1, LMP10	–	39 + 234	Ntn	
		β5i	PSMB8	LMP7, RING10	–	72 + 204	Ntn	
PA700 (19S)	ATPase subunits	Rpt1	PSMC2	S7, Mss1	YTA3, CIM5	433	AAA	+
		Rpt2	PSMC1	S4, p56	YTA5/mts2	440	AAA	+
		Rpt3	PSMC4	S6, Tbp7, P48	YTA2	418	AAA	+
		Rpt4	PSMC6	S10b, p42	SUG2, PCS1, CRL13	389	AAA	+
		Rpt5	PSMC3	S6′, Tbp1	YTA1	439	AAA	+
		Rpt6	PSMC5	S8, p45, Trip1	SUG1, CRL3, CIM3/let1	406	AAA	+
	Non-ATPase subunits	Rpn1	PSMD2	S2, p97	HRD2, NAS1/mts4	908	PC	+
		Rpn2	PSMD1	S1, p112	SEN3	953	PC	+
		Rpn3	PSMD3	S3, p58	SUN2	534	PCI, PAM	+

Table 7.1 *(continued)*

Category	*Subclassification*	*Systematic nomenclature*	*HUGO*	*Miscellaneous nomenclature*		*Human (yeast) amino acids*	*Motif*	*Lethality*
				Human	*Yeast (budding/fission)*			
		Rpn4		–	SON1, UFD5	(531)	Zn finger	–
		Rpn5	PSMD12	p55	NAS5	456	PCI	+
		Rpn6	PSMD11	S9, p44.5	NAS4	422	PCI, PAM	+
		Rpn7	PSMD6	S10a, p44		389	PCI	+
		Rpn8	PSMD7	S12, p40, MOV34	NAS3	324	MPN	+
		Rpn9	PSMD13	S11, p40.5	NAS7/mts1	376	PCI	–
		Rpn10	PSMD4	S5a, Mbp1	SUN1, MCB1/pus1	377	UIM, VWA	–
		Rpn11	PSMD14	S13, Poh1	MPR1/pad1, mts5	310	MPN	+
		Rpn12	PSMD8	S14, p31	NIN1/mts3	257	PCI	+
		Rpn13		–	DAQ1	(156)	ARM	–
		Rpn14		FLJ11848	YGL004C	392	WD40, G-β	–
		Rpn15		DSS1, SHFM1	SEM1	70		–
PA28 (11S)			PSME1	PA28α, REGα	–	249		
			PSME2	PA28β, REGβ	–	239		
			PSME3	PA28γ, REGγ, Ki	–	254		
PA200			PSME4	PA200, TEMO	BLM3	1843	HEAT, ARM	–
PI31			PSMF1		–	271	Proline-rich	
Others			PSMD5	S5b, p50.5	–	504	ARM	
			PSMD9	p27	NAS2	223	PDZ	–
			PSMD10	p28, gankyrin	NAS6	226	ANK	–
				KIAA0368	ECM29	1870	HEAT	–

HUGO: Human Genome Organization; Ntn: N-terminal nucleophile hydrolase; AAA: ATPase associated with diverse cellular activities; PAM: PCI-associated module; PCI: proteasome, COP9, eIF3; MPN: Mpr1, Pad1 N-terminal; UIM: ubiquitin-interacting motif; VWA: von Willebrand factor type A; NLS: nuclear localization signal; PC: proteasome/cyclosome repeat; PDZ: PSD-95/DLG/ZO-1; ANK: ankyrin repeats; ARM: Armadillo repeats; +; lethal –; non-lethal.

noteworthy that the classical nuclear localization signal (NLS), which consists of the basic amino acid cluster whose consensus sequence is X-X-K-K(R)-X-K(R), where X is any residue, is present in the four α-type subunits; i.e., $\alpha1$, $\alpha2$, $\alpha3$, and $\alpha4$ (Figure 7.2), but lacking in other subunits including seven β-type subunits (Table 7.1). Ample evidence confirms that the NLSs are functionally active, because they are able to induce complete translocation of the reporter protein into the cell nucleus, when the NLS sequence is fused to the protein [16, 17]. Moreover, structural analysis reveals that these four NLSs are at the surface of the molecule, suggesting that they all participate in the nuclear localization of the 20S proteasome [14]. Presumably, the proteasome moves as a large particle, but not as individual subunits, through nuclear membranes, because only limited subunits have the NLS sequence and free subunits are not present in the cell in general.

In the budding yeast, gene disruption analysis reveals that deletion of all the 20S subunit genes, except the $\alpha3$ subunit gene, is lethal, indicating that the proteasome is essential for cell proliferation (Table 7.1). The reason the $\alpha3$ subunit is not essential is that the $\alpha4$ subunit takes the position occupied by the $\alpha3$ subunit in $\alpha3$ subunit–deficient cells [18]. This also suggests the functional importance of molecular organization as the 20S proteasome, rather than the indispensable role of its individual subunits. Of course, the catalytic subunits themselves are of considerable importance in clarifying the biological role of the proteasome, because mutations of active threonine residues cause death of the cells. Taken together, the 20S proteasome plays a pivotal role as a basic machine for proteolysis in eukaryotes, and thereby the overall structures and functions of individual subunits are highly conserved across species, except a specialized case linked to the adaptive immune response, which will be described in the next section.

7.2.2
The Immunoproteasome

The budding yeast has seven β-type subunit genes, consistent with the configuration that the β ring of the 20S proteasome is made up of seven subunits. In contrast, mammals have 10 β-type subunit genes; this observation is puzzling, taking into consideration that the proteasomal β ring is organized by seven β subunits. On the other hand, judging from the α-ring organization, it is rational that both organisms have seven α-subunit genes. The enigma regarding the extra number of β subunits in mammals could be explained by the existence of three major immunomodulatory cytokine interferon-γ (γ-IFN) inducible subunits, $\beta1i$, $\beta2i$, and $\beta5i$, that are structurally related to $\beta1$, $\beta2$, and $\beta5$, respectively, which are regulated negatively in response to γ-IFN [19–21]. The reciprocal expression of three pairs of subunits with extremely high amino acid similarity indicates that γ-IFN may induce subunit replacement of $\beta1$, $\beta2$, and $\beta5$ by the structurally similar subunits $\beta1i$, $\beta2i$, and $\beta5i$, respectively. Based on these observations, we have proposed that γ-IFN-inducible proteasomes be called "immunoproteasomes" to emphasize their specialized functions in immune response and to distinguish them from those containing constitutively expressed subunits (see the simplified model depicted

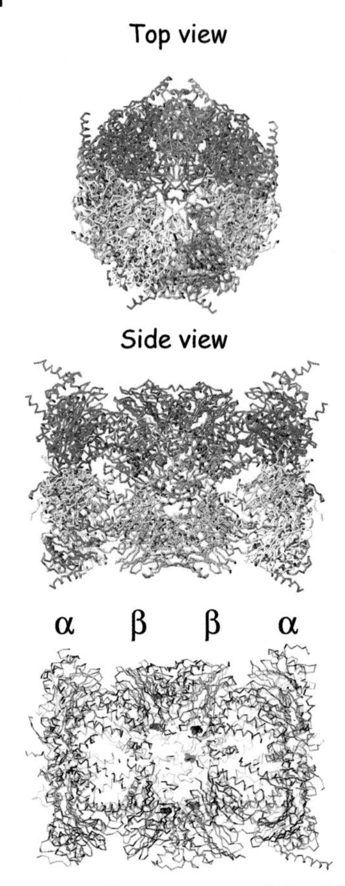

Fig. 7.1. Tertiary structure of the 20S proteasome from the bovine liver. Top panel: top view; middle and bottom panels: side view. Active threonine residues of $\beta 1$, $\beta 2$, and $\beta 5$ appear in blue, green, and red, respectively (bottom panel). For details, see [117].

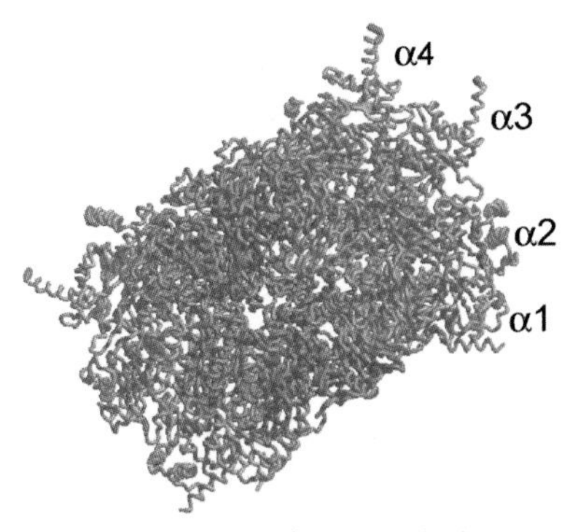

Fig. 7.2. Locations of NLSs in the bovine 20S proteasome. α subunits, β subunits, and NLSs are colored blue, green, and red, respectively.

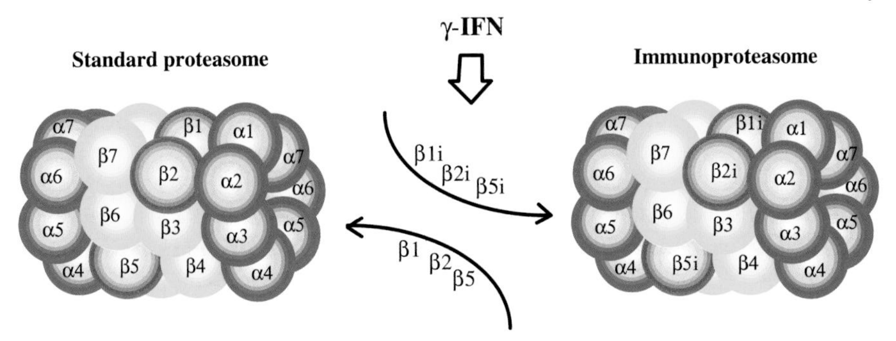

Fig. 7.3. Models of standard proteasomes (left) and immunoproteasomes (right). For details, see text.

in Figure 7.3) [19, 22]. The major histocompatibility complex (MHC) class I molecule continuously binds peptides produced by proteolysis of cytosolic proteins and displays them on the cell surface. This mechanism enables cytotoxic T lymphocytes (CTLs) to detect and destroy abnormal cells that synthesize viral or other foreign proteins [23] . Over a decade ago, the proteasome was identified as a plausible candidate-processing enzyme of intracellular antigens. To date, the roles of the immunoproteasome, which in concert contributes to the efficient production of CTL epitopes, have been highlighted in the MHC class I–restricted antigen-processing pathway [24, 25].

Of the 10 β-type subunits, three pairs of γ-IFN-regulated subunits have active threonine residues, indicating that the subunit exchanges induced by γ-IFN are likely to confer functional alterations upon the proteasome. In fact, γ-IFN alters the proteolytic specificities of the proteasomes, increasing their trypsin- and chymotrypsin-like activities for cleavage of peptide bonds on the carboxyl side of basic and hydrophobic amino acid residues of fluorogenic substrates, respectively, but decreasing their caspase-like activities for peptides containing acidic amino acid residues [21, 24, 26]. Comparison of the tertiary structures of the standard proteasome and the immunoproteasome constructed by computer-assisted modeling suggests that the caspase-like activity would be reduced and chymotryptic and tryptic activities would be enhanced in the immunoproteasome [14]. These changes of peptidase activities suggest that the immunoproteasome of γ-IFN-treated cells should generate more peptides that have hydrophobic or basic carboxyl termini and fewer peptides with acidic carboxyl termini. The peptides generated by the immunoproteasome favor settlement into the peptide-binding pocket of MHC class I molecules, because hydrophobic or basic carboxyl terminal residues normally serve as anchors for binding to MHC class I molecules. Thus, γ-IFN produces the immunoproteasome with an alteration of the proteolytic specificity that is perhaps more appropriate for the immunological processing of endogenous antigens [25, 26]. It is likely that the acquisition of the immunoproteasome enabled the organism to produce MHC class I ligands more efficiently and thus combat pathogens more proficiently.

Sequence comparison of the 10 mammalian β-type subunit genes indicates that each proteasome subunit pair that undergoes exchanges upon γ-IFN stimulation emerged from the respective common ancestor by gene duplication [26]. Among the 10 β-type subunits, β1i is the most closely related to β1. The same is true with β2i and β2, and also with β5i and β5. The γ-IFN-inducible subunit genes appear to have been derived from the more ancient, constitutively expressed β1/*PSMB6*-, β2/*PSMB7*-, and β5/*PSMB5*-like genes (Table 7.1). The close evolutionary relationship of the exchangeable proteasome subunit pair is supported by the fact that yeast has the constitutively expressed β-type subunit genes β1/*PRE3*, β2/*PUP1*, and β5/*PRE2*, which resemble more closely mammalian β1/*PSMB6*, β2/*PSMB7*, and β5/*PSMB5* than their γ-IFN-inducible counterparts. For understanding the evolution of the immunoproteasome, we previously proposed a chromosomal duplication model explaining the emergence of the γ-IFN-regulated β-type subunits [26]. The basic assumption of this model is that all three sets of γ-IFN-regulated β-type subunits emerged simultaneously as a result of chromosomal duplication involving the MHC region. Many of the MHC-encoded genes including β1i/*PSMB9* and β5i/*PSMB8* appear to have emerged by an ancient chromosomal duplication that takes place as part of the genome-wide duplication, suggesting that modifications and renewal of preexisting non-immune genes were instrumental in the emergence of adaptive immunity.

7.3
Regulatory Factors

The crystal structure of 20S proteasomes reveals that the center of the α ring of the 20S proteasome is almost completely closed, preventing penetration of proteins into the inner surface of the β ring on which the proteolytically active sites are located. Thus, the 20S proteasome exists in a latent status in the cells. Accordingly, substrates gain access to the active sites only after passing through a narrow opening corresponding to the center of the α rings, and the amino termini of the α subunits form an additional physical barrier for substrates to reach the active sites. In certain cases, it is reported that unfolded proteins generated by stresses, e.g., due to oxidation, or naturally unfolded proteins without secondary structures, such as p21 and α-synuclein, are degraded directly by the 20S proteasome, but the mechanism that controls the gate opening of the closed α ring for interaction with these proteins remains a mystery [27, 28].

In most cases, however, additional protein factors that are associated with the 20S core particle are required to exert the proteolytic functions. In other words, the enzymatically active proteasome is generally capped on either and/or both ends of the central 20S proteasomal core by a regulator that can recognize target proteins and opens the α-ring channel for entry of the substrates for their ultimate breakdown. In turn, the proteasome hardly degrades substrates, because the active sites are usually masked by their location on the inside of the β-ring cavity, preventing free interaction with the substrate proteins. Due to this catalytic mechanism, the proteasome can be referred as a self-compartmentalizing protease [2]. So far, several factors have been identified that function as activators that presumably control proteolysis catalyzed by the proteasome. Below is a brief summary of these regulatory factors.

7.3.1 PA700

The regulatory complex PA700 (also termed 19S complex or regulatory particle [RP]), associates with the 20S proteasome in an ATP-dependent manner to form the proteasome with an apparent sedimentation coefficient of 26S and a molecular mass of ~2500 kDa (Figure 7.4). The 26S proteasome is mainly responsible for ATP-dependent selective degradation of polyubiquitylated substrates [1, 6, 29, 30]. This structure is a dumbbell-shaped particle, consisting of a centrally located, cylindrical 20S proteasome that functions as a catalytic machine and two large terminal PA700 modules attached to the 20S core particle in opposite orientations. PA700 contains approximately 20 heterogeneous subunits of 25–110 kDa, which can be classified into two subgroups: a subgroup of at least six ATPases, numbered from Rpt1 to Rpt6 (i.e., RP triple ATPases 1–6), that are structurally similar and have been highly conserved during evolution, and a subgroup of over 15 heterogeneous subunits, numbered from Rpn1 to Rpn15 (i.e., RP non-ATPases 1–15), that are structurally unrelated to the members of the ATPase family [31–33]. These subunits are listed in Table 7.1.

The PA700/RP structurally consists of two sub-complexes, known as "base" and "lid" [34], which, in the 26S proteasome, correspond to the portions of PA700 proximal and distal, respectively, to the 20S proteasome (Figure 7.4). The base is made up of six ATPases (Rpt1–Rpt6) and the two large regulatory components Rpn1 and Rpn2, while the lid contains multiple non-ATPase subunits (Rpn3–Rpn15 or over). The base complex is thought to bind in an ATP-dependent manner to the outer α ring of the central 20S proteasome and is considered to be involved in opening the gate of the α ring for entry of the protein substrate. On the other hand, the lid complex is thought to be involved in the recognition of target proteins (mostly polyubiquitylated proteins), indeubiquitylation for reutilization of ubiquitin, and in interactions with various other proteins (for details, see below).

These six ATPases are most similar in their central domains of approximately 200 amino acid residues, which contain a putative ATP-binding site. They are members of a large protein family termed AAA proteins (ATPases associated with a variety of cellular activities), characterized by the conserved 200-amino-acid domain containing a consensus sequence for an ATP-binding module [35]. These six ATPases are assembled into one ring complex. One role of the ATPase is to sup-

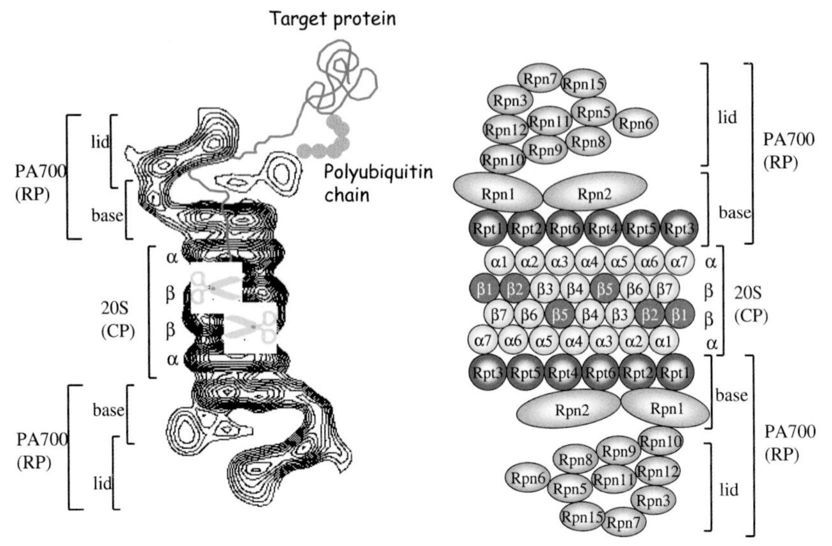

Fig. 7.4. Molecular organization of the 26S proteasome. Left panel: Averaged image of the 26S proteasome complex of rat based on electron micrographs. The α and β rings of the 20S proteasome are indicated. Photograph kindly provided by W. Baumeister. Right panel: Schematic drawing of the subunit structure. Ub: ubiquitin; CP: core particle (alias 20S proteasome); RP: 19S regulatory particle consisting of the base and lid sub-complexes; Rpn: RP non-ATPase; Rpt: RP triple ATPase.

ply energy continuously for the degradation of target proteins. In fact, the metabolic energy liberated by ATP consumption is probably utilized for assembly of the base complex with the 20S proteasome, although it may also be used for unfolding target proteins, gate opening of the 20S proteasome, and substrate translocation so that substrates can penetrate the channel of the α and β rings of the 20S proteasome [36, 37]. However, the exact reason for the presence of multiple homologous ATPases in the 26S proteasome complex remains largely unknown.

On the other hand, there are a number of Rpn subunits, but it is difficult to determine the number of bona fide stoichiometric subunits of the lid complex. As shown in Table 7.1, some subunits are not found in either mammal or yeast, but the details are largely unknown. Here we point out several mysterious aspects with a reference to subunit heterogeneity. Presumably, Rpn4 that binds to β6, Rpn13, S5b/p50.5, p27/NAS2, and p28/NAS6/gankyrin are not genuine subunits of the lid complex, which are transiently associated with the PA700, although they were initially identified by biochemical and genetic analyses as integral components of the proteasome. In addition, it is unknown whether or not Rpn14 is an actual subunit of PA700, because it was isolated by a comprehensive interaction analysis using yeast two-hybrid screening without biochemical evidence [38].

In budding yeast, all six Rpt subunits are essential, and most, if not all, Rpn subunits are also essential (Table 7.1). Nonessential Rpn subunits are of interest with respect to their roles, but their actual functions remain elusive. The 26S proteasome predominantly exists in the nucleus of the yeast [39], but the mechanism underlying the nuclear localization of PA700 has not yet been examined in detail. Indeed, the NLS motifs of PA700 subunits are largely unclear, although the sequences consisting of the basic amino acid cluster analogous to NLS are present in multiple Rpt and Rpn subunits. It has been reported that the NLS-like sequence of Rpn2 functions as the nuclear targeting signal in the budding yeast [40], but this does not seem to be the case in mammalian counterparts (our unpublished results). Indeed, when all human Rpt and Rpn subunits are fused with green fluorescence protein (GFP) and expressed in mammalian cells, their localizations showed three patterns, either in the nucleus, the cytosol, or both subcellular compartments (our unpublished results). To clarify this issue, further detailed biochemical and structural analyses of PA700 are required.

7.3.2
Rpn10

The Rpn10 subunit was identified as the first molecule capable of binding polyubiquitylated proteins *in vitro* [41]. Intriguingly, it has a unique sequence, referred to as the ubiquitin-interacting motif (UIM), that is identified as the minimal sequence bound with polyubiquitin. It is interesting to note that all subunits of the 26S proteasome conserved from yeast to mammal known so far have similar sizes, except for Rpn10. Yeast Rpn10 is approximately 30 kDa, which is 20 kDa smaller than that of other species, including human. Human Rpn10 has two UIM motifs, with cooperative roles to bind polyubiquitylated proteins [42]. In comparison, the yeast counterpart has a single UIM motif and lacks the C-terminal region containing the second UIM motif.

Importantly, Rpn10 also possesses acceptor sites for UBL domains of hHR23A/B, PLIC, and Parkin in higher eukaryotes [43]. Intriguingly, since Rad23 and Dsk2 have the ubiquitin-associated domain (UBA) that can bind polyubiquitylated proteins, beside the UBL domain, it is proposed that they may promote the targeting of substrates polyubiquitylated to the 26S proteasome [44]. Thus, there are multiple ways by which the 26S proteasome recognizes target substrates, but the pathway selected for each substrate is unknown [43, 45, 46]. Indeed, there is a genetic interaction between Rpn10 and Rad23 in yeast: the loss of both Rad23 and Rpn10 results in pleiotropic defects that are not observed in either single mutant, suggesting their functional redundancy and that Rad23 plays an overlapping role with Rpn10 [47, 48].

Surprisingly, we found that mouse Rpn10 mRNAs occur in at least five distinct forms, named Rpn10a to Rpn10e, and that they are generated from a single gene by developmentally regulated, alternative splicing [49]. Comparison of the genomic and cDNA sequences of Rpn10 revealed similar gene organizations in the medaka fish, *Oryzias latipes*, as an example of lower vertebrates, implying that the competence for all distinct forms of Rpn10 alternative splicing is widely retained in vertebrates [50]. In contrast, no Rpn10 isoforms have so far been found in EST databases of non-vertebrate species. Interestingly, the size of Rpn10e with a single UIM motif resembles that of yeast Rpn10, suggesting that Rpn10e is an ancient form and that other species may be evolutionarily generated from Rpn10e.

The multiplicity of Rpn10 indicates that the 26S proteasome exists in multiple functionally distinct forms with distinct Rpn10 isoforms. For example, the Rpn10a form (equivalent to that originally isolated as human S5a) is ubiquitously expressed, whereas Rpn10e is expressed only in embryos, with the highest levels of expression in the brain. While the former is thought to perform proteolysis constitutively in a wide variety of cells, the latter may play a specialized role in early development. In addition, we recently found that one of the alternative products of Rpn10c is specifically associated with an apoptotic factor, suggesting that it functions as an essential subunit linking the proteasome machinery to apoptotic regulation during *Xenopus* embryogenesis (Kawahara et al., submitted).

Deletion of the Rpn10 gene in the budding and fission yeasts *Physcometrilla patens*, and *Arabidopsis* has no effect on the cell proliferation, indicating that it is nonessential. However, the deletion caused larval-pupal lethality, but did not destabilize the regulatory complex of the 26S proteasome in *Drosophila* [51]. Likewise, knockout of the mouse Rpn10 gene was embryonically lethal (our unpublished results), but the reason this defect occurs remains elusive. It is interesting that Rpn10a knock-in mice lacking the Rpn10 gene are born normally without any apparent abnormalities, suggesting that Rpn10a can rescue the lethality caused by deletion of the gene and thus is an important Rpn10 family protein (our unpublished results). To date, Rpn10 is the only subunit of the 26S proteasome that displays a variety of subspecies in higher organisms.

7.3.3 Modulator

The modulator complex was isolated as a factor that stimulates PA700-dependent proteolysis in the presence of the 20S proteasome [52]. It promotes the transformation from the 20S proteasome and PA700 into the 26S proteasome without stable association with the resulting complex [53, 54]. Intriguingly, the modulator consists of three subunits: Rpt4, Rpt5, and one additional Rpt-unrelated p27 subunit. Whereas these two ATPases are essential for cell proliferation, the deletion of p27 had no effect on cell growth in yeast [55]. It is of interest that Rpt5 is capable of binding the polyubiquitylated proteins *in vitro*, functioning as a polyubiquitin receptor [56]. However, the straightforward role of the modulator remains elusive. It is possible that the modulator is an intermediate complex before maturation into the 26S proteasome. Alternatively, p27 could act as a mediator molecule that assists in the formation of the base complex containing the ATPase ring, but there is no experimental evidence in support of this function at present.

7.3.4 PA28

PA28 (or 11S regulator/REG) was discovered as an activator protein of the latent 20S proteasome when the peptide-hydrolyzing activity was assessed [30]. PA28 is composed of subunits of 28–32 kDa, but the native molecule has a molecular weight of about 170–180 kDa, leading to the assumption that it is an oligomeric complex, perhaps hexameric or heptameric [57]. Electron microscopic investigations indicate that PA28, free of the proteasome, is a ring-shaped particle and that it associates with the 20S proteasome by forming a conical structure on both ends of the complex [58], indicating that PA28 occupies the same site on the 20S core particle as the regulator complex does in the case of the 26S proteasome. This molecular figure is confirmed by the tertiary structural analysis [59, 60].

PA28 is composed of three related family proteins, named PA28α, PA28β, and PA28γ, with approximately 50% amino acid sequence, in which PA28α and PA28β form the heteropolymeric complex and PA28γ the homopolymeric complex [61]. Immunofluorescence analysis revealed that PA28α and PA28β are located mainly in the cytoplasm and present diffusely in the nucleus, whereas PA28γ is located predominantly in the nucleus without appreciable localization in the cytoplasm [62]. Therefore, the two types of PA28 complexes containing PA28$\alpha\beta$ and PA28γ are likely to function in different subcellular compartments.

Intriguingly, PA28α and PA28β are also markedly induced by γ-IFN in various types of cells, but no obvious influence of γ-IFN was found on PA28γ. Thus, γ-IFN can alter the subunit composition of PA28 in the cells, a process similar to the replacement of immunoproteasomal subunits observed upon γ-IFN treatment. These data suggest that PA28$\alpha\beta$ could play a role in the generation by the 20S proteasome of antigenic peptides that can be presented by MHC class I molecules [63]. Thus, these newly identified γ-IFN-regulated activator genes in combination with the three pairs of γ-IFN-regulated proteasome genes perhaps act synergistically to enhance antigen presentation.

Studies in mice deficient in both PA28α and PA28β genes show that the ATP-dependent proteolytic activities were decreased in PA28$\alpha^{-/-}/\beta^{-/-}$ cells, suggesting that PA28 is involved in protein degradation [64]. Splenocytes from PA28$\alpha^{-/-}/\beta^{-/-}$ mice displayed no apparent defects in processing of ovalbumin, and PA28$\alpha^{-/-}/\beta^{-/-}$ mice also showed apparently normal immune responses against infection with influenza A virus. However, they almost completely lost the ability to process a melanoma antigen TRP2-derived peptide. Hence, PA28$\alpha^{-/-}/\beta^{-/-}$ plays an essential role in the processing of certain antigens, but it is not prerequisite for antigen presentation in general [64]. Thus, the antigen-processing pathway is clearly separated into two routes, one dependent on PA28 and the other PA28-independent (Figure 7.5).

On the other hand, the function of PA28γ remains elusive. The PA28γ-deficient mice were born without apparent abnormalities in all tissues examined, but showed postnatal growth retardation compared to PA28$\gamma^{+/-}$ and PA28$\gamma^{+/+}$ mice

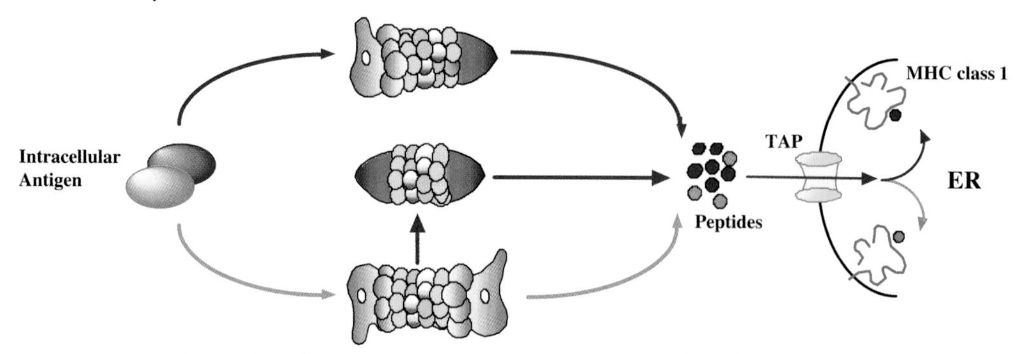

Fig. 7.5. Two distinct antigen-processing pathways mediated by the proteasome; one requires PA28, while the other is PA28-independent. Blue and red arrows represent the PA28-dependent and PA28-independent pathways, respectively. TAP: transporter associated with antigen processing; ER: endoplasmic reticulum. For details, see text.

[65], indicating that PA28γ functions as a regulator of cell proliferation and body growth in mice and suggesting that neither PA28α nor PA28β compensates for the PA28γ deficiency. In addition, PA28$\gamma^{-/-}$ mice display a slight reduction in $CD8^+$ T-cell numbers and do not effectively clear pulmonary fungal infections, indicating that that PA28$\gamma^{-/-}$ mice, like PA28$\alpha^{-/-}/\beta^{-/-}$ mice, are deficient in the processing of only specific intracellular antigens [66].

We argued earlier that the three γ-IFN-inducible subunits of the 20S proteasome most likely evolved for MHC class I–mediated antigen presentation. In this regard, it is likely that PA28α and PA28β might have coevolved with immunoproteasome subunits [26]. Because a PA28γ-like protein is found in invertebrates, the γ-IFN-inducible PA28α and PA28β subunits are probably derived from a PA28γ-like precursor [67]. This is analogous to the situation that the γ-IFN-inducible β-type subunits of the 20S proteasome emerged by gene duplication from the evolutionarily more ancient PSMB5-, PSMB6-, and PSMB7-like subunits.

7.3.5
Hybrid Proteasomes

The PA28 protein greatly stimulated multiple peptidase activities of the 20S proteasome without affecting destruction of large protein substrates, even though the proteins had already been polyubiquitylated. Thus, PA28 does not play a central role in the initial cleavage of protein substrates. It presumably has a stimulating effect on the degradation of polypeptides of intermediate size that are generated by the 26S proteasome, implying that the 26S proteasome and the PA28–proteasome complex may function sequentially or cooperatively [20, 26].

Recently, PA28 was found both in the previously described homo-PA28–proteasome complex and in the distinct proteasome complex that has one PA28 ring at one end of the 20S particle and a PA700 at the other [68]. We refer to the latter as "hybrid proteasome" [69], and its existence was directly demonstrated by electron microscopy [70, 71]. The formation of the hybrid proteasome proved to be ATP-dependent, like that of the 26S proteasome. The hybrid proteasome may contribute to more efficient proteolysis; perhaps intact substrate proteins are first recognized by PA700 and then fed into the cavity of the 20S proteasome, whose cleavage ability is greatly stimulated by the PA28$\alpha\beta$ complex. Indeed, it catalyzes ATP-dependent degradation of ornithine decarboxylase (ODC) without ubiquitylation but requires antizyme, an ODC inhibitory protein, as does the 26S proteasome. In contrast, the homo-PA28–proteasome complex cannot degrade ODC. Intriguingly, γ-IFN appreciably enhanced the ODC degradation through induction of the hybrid proteasome, which may also be responsible for the immunological processing of intracellular antigens (Figure 7.5). Indeed, the hybrid proteasome shows enhanced hydrolysis of small peptides and generates a pattern of peptides different from those generated by the 26S proteasome, without altering the mean product length [71]. Presumably, this change in the produced peptides accounts for the capacity of PA28 to enhance antigen presentation. Taken together, it is plausible that the two types of ATP-dependent proteases, the 26S and hybrid proteasomes, share the ATP-dependent proteolytic pathway in mammalian cells.

7.3.6 PA200

PA200 is a novel proteasome activator that stimulates 20S proteasomal hydrolysis of peptides, but not proteins, like PA28 [72]. Electron microscopy reveals that PA200 is attached to the 20S proteasome at both ends. It is a large protein of approximately 200 kDa with tandemly repeated HEAT-like motifs [73]. Homologues of PA200 are present in mammals, worms, plants, and budding yeast, but not in fruit fly and fission yeast. It is a nuclear protein, and the uniform nuclear distribution of PA200 changes to a strikingly punctate pattern in response to γ-irradiation, a behavior characteristic of many DNA-repair proteins. Indeed, mutation of the yeast *Bml3* gene-encoding mammalian PA200 ortholog results in hypersensitivity to bleomycin [74], and exposure to DNA-damaging agents induces the PA200 message [75]. Thus, it is plausible that PA200 operates in DNA repair, possibly by recruiting the proteasome to double-strand breaks. Interestingly, Blm3/PA200 was also identified as a new component of Ump1 (proteasome maturation factor)-associated precursor complexes (see Section 7.4.2). Lack of Blm3 resulted in an increased rate of precursor processing and an accelerated turnover of Ump1, suggesting that Blm3 prevents premature activation of proteasomal Cps [76]. Thus, Blm3 joins the core 20S proteasome inside the nucleus to coordinate late stages of nuclear proteasome assembly.

7.3.7 Ecm29

Ecm29 is identified as one of many proteins that are abundant in the affinity-purified proteasome, but it is absent from the proteasome, as defined previously, because elevated salt concentrations dissociate it during purification [77]. Ecm29 is a large protein of about 210 kDa with tandemly repeated HEAT-like motifs like PA200 [73]. The HEAT motif consists of two α helices and two turns; molecular modeling suggests that in the PA200 and Ecm29 repeats, the α helices may be slightly turned relative to their orientations in typical HEAT repeats. Both PA200 and Ecm29 are composed almost entirely of such repeats and therefore are likely to have α-helical solenoid structures. Based on the structural resemblance of PA200 and Ecm29, it is conceivable that they have overlapping roles in the cells.

Electron micrographs of free Ecm29 reveal a V-shaped morphology. Moreover, Ecm29 complexed with the core 20S proteasome displayed an open V-shaped morphology as well. The binding appeared to be the outer (α) ring of CP subunits. Ecm29 appears to bind the CP near the interface region, in which it contacts the RP/PA700 and CP, consistent with its function in stabilizing CP-RP association. Absence of Ecm29 leads to dissociation of the CP and RP when ATP is not provided, indicating that Ecm29 tethers the proteasome core particle to the regulatory particle. Ecm29 is conserved in various eukaryotes ranging from yeast to human.

7.3.8 PI31

PI31 was identified as a protein inhibitor of the 20S proteasome and has a molecular mass of approximately 30 kDa [78, 79]. PI31 is a proline-rich protein, particularly within its carboxyl-terminal half, where 26% of the amino acids are proline. Inhibition of the 20S proteasome by PI31 involved formation of the proteasome–PI31 complex. In addition to its direct inhibition of the 20S proteasome, PI31 inhibited the activation of the proteasome by each of two proteasome regulatory proteins, PA700 and PA28, suggesting that PI31 plays an important role in control of proteasome function, including that in ubiquitin-dependent pathways of protein degradation [79].

Previous studies reported that PI31 acts as a selective modulator of the proteasome-mediated steps in MHC class I antigen processing [20]. Consequently, overexpression of PI31 abrogates MHC class I presentation of an immunoproteasome-dependent CTL epitope and reduces the surface MHC class I levels on γ-IFN-treated mouse embryonic cells. Thus, PI31 represents a cellular regulator of proteasome formation and of proteasome-mediated antigen processing [80]. PI31 is localized at the nuclear envelope/endoplasmic reticulum membrane and selectively interferes with the maturation of immunoproteasome precursor complexes. Whereas homologues of PI31 are present in various higher organisms including mammal, *C. elegans* and budding and fission yeasts lack PI31.

7.4 Proteasome Assembly

While tremendous progress in uncovering the structure and functions of the proteasome system has been made, there is little information on the important issue of the regulatory mechanisms involved in the high-fidelity organization of the proteasome as a large multi-subunit complex. To understand this, it may be important to uncover the mechanism involved in the correct assembly of the proteasome. However, little is known about the assembly of the proteasome complex. Accumulating evidence suggests that assembly and maturation of the 20S proteasome is a precisely ordered multi-step event [20, 81]. That is, the α ring appears to assemble first, and the β subunits sequentially assemble onto the α ring, forming the 13–16S complex with an apparent size of 300 kDa that could be a pre-proteasome intermediate (alias half-proteasome), which contains one full α ring and one full β ring of unprocessed β-subunit precursors [81–84]. The processing of precursor β subunits takes place concomitantly with dimerization of half-proteasomes, forming enzymatically active mature 20S proteasomes. On the other hand, the biogenesis of 26S proteasomes remains largely elusive, especially in mammalian cells. In fact, there is no available information on how the base and lid complexes are assembled accurately. Moreover, how the 20S proteasome associates or dissociates with other multiple regulatory factors is also entirely unknown, although there is evidence that the formation of the 26S proteasome and the hybrid proteasome occurs in an energy-dependent fashion.

7.4.1 Roles of Propeptides

The three catalytic β-type proteasomal subunits $\beta1$, $\beta2$, and $\beta5$ are synthesized as proproteins (Table 7.1) and processed to their mature forms by removal of their N-terminal pro-sequences to become active assemblies, and this precursor processing occurs by an autocatalytic mechanism [85–88]. Intriguingly, precursor processing of $\beta5$ requires dimerization of the two halves of the proteasome particles (i.e., half-proteasomes) and prevents the formation of proteolytic sites until the central hydrolytic chamber is organized. Interestingly, propeptide processing itself is not required for proteasome assembly but is needed for maturation of a specific subset of active sites in yeast [86]. Unlike the propeptide of the $\beta5$ subunit, those of $\beta1$ and $\beta2$ are dispensable for cell viability and proteasome formation [89], although one study reported the importance of a propeptide of a $\beta2$ subunit whose deletion caused poor proliferation of yeast [90]. Thus, the propeptides of β subunits have unequal roles for efficient core particle maturation and a hierarchy of active-site formation [89]. In this regard, it is worth noting that another function of the propeptide is protection of the N-terminal catalytic threonine residue against $N\alpha$-acetylation [91, 92].

In addition, maturation of the catalytically inactive β-type subunits $\beta6$ and $\beta7$ appears to be exerted by active β-type subunits, forming the fully assembled 20S particle, but the role of propeptides of the non-catalytic subunits $\beta6$ and $\beta7$ has not been well documented so far.

7.4.2
Ump1

In considering the complex molecular architecture of the proteasome, a systematic pathway may be needed for the coordinated assembly of a large number of different subunits. We described above that PA200, Ecm29, and PI31 positively and negatively regulate the proteasome assembly, but there is a lack of definitive evidence in support of these actions. To date, only one molecule (termed yeast Ump1 and mammalian ortholog POMP [proteasome maturation protein], also known as proteassemblin) is known to play a crucial role in 20S proteasome assembly [93–96]. Ump1 exists in the 13–15/16S proteasome precursor complexes containing unprocessed β subunits but is not detected in the mature 20S proteasome. Upon the association of two half-proteasomes, Ump1 is rapidly degraded following the activation of proteolytic sites in the interior of the nascent proteasome, suggesting that it is a short-lived assembly chaperone. Yeast cells lacking Ump1 exhibit lack of coordination between the processing of β subunits and proteasome assembly, resulting in functionally impaired proteasomes [93]. The mammalian homologue hUmp1, POMP, or proteassemblin is a constituent of pre-proteasomes but is not a fully assembled 20S proteasome, as is Ump1 in yeast [94–96]. Moreover, it is also a constituent of the pre-immunoproteasome that contains the precursor of the γ-IFN-inducible subunit $\beta1i$ [96]. Intriguingly, POMP/proteassemblin is induced by γ-IFN [94, 96], although the effect is not great, indicating that it may be involved in the immunoproteasome assembly.

A central enigma about 20S proteasome assembly is the mechanism responsible for the correct positioning of the 14 different subunits. Apart from the known functions of Ump1, i.e., the linking of β subunits to the α ring of the 20S proteasome, the mechanism of assembly of the α ring is entirely unknown. We know that the assembly starts by the formation of the α ring, which is believed to be a spontaneous process, and then the α ring provides the docking sites for the β subunits. However, we recently identified a novel heterodimeric protein factor that specifically associates with the precursor forms of 20S proteasomes and facilitates the α-ring assembly and subsequent maturation of 20S proteasomes [116]. This factor is destroyed at a late maturation stage of the assembly pathway; perhaps its proteolysis is autocatalytic, like Ump1. Interestingly, this complex has no appreciable affinity to the β subunits. Based on these findings, we propose a multi-step-ordered mechanism for mammalian proteasome assembly.

7.4.3
Immunoproteasome Assembly

As described in the preceding section, γ-IFN induces a major structural reorganization of the standard proteasome, forming the "immunoproteasome." This alteration of the subunit pattern is presumably due to changes in the biosynthesis of immunoproteasome subunits, because γ-IFN has no effect on the levels of preexisting standard proteasome subunits synthesized before its addition. Thus, it is unlikely that exchange of post-translationally modified subunits with subunits of preexisting proteasomes is involved in the formation of γ-IFN-induced proteasomes [97]. Accordingly, the most probable explanation for the mechanism of subunit substitution is the preferential incorporation of γ-IFN-inducible subunits and the possibly rapid degradation of the unassembled standard proteasome subunits β1, β2, and β5.

Recent studies have provided insights into the molecular mechanisms underlying the assembly of immunoproteasomes [20]. Three sets of γ-IFN-regulated catalytic β subunits (β1i, β2i, and β5i) are synthesized as proproteins and processed to the mature forms by removal of their N-terminal pro-sequences, like β1, β2, and β5. Griffin et al. [98] showed that three γ-IFN-inducible subunits can replace constitutive catalytic 20S subunits during proteasome biogenesis. β2i requires β1i for its efficient incorporation into the pre-proteasome, and the pre-proteasome containing β1i and β2i requires β5i for efficient maturation. Thus, a mechanism exists that favors the assembly of the homogenous immunoproteasomes containing all three γ-IFN-inducible subunits. However, a recent study reported that β1i incorporation does not require β2i using $\beta 2i^{-/-}$ mice [90]. Indeed, there is clear evidence for the co-incorporation of β5i, rather than β5, with β1i and β2, whereas this specificity is reversed when the propeptides of β5i and β5 are switched [99]. Obviously, the β5i propeptide is responsible for the preferential incorporation, but not its catalytic activity. It is possible that β5/β5i propeptides play a critical role in preferential immunoproteasome assembly, suggesting that the differential interaction of Ump1 with β5 or β5i may play a role in the proteasome assembly [100].

7.4.4
Assembly of the 26S Proteasome

Emerging evidence indicates that mutation of certain lid subunits influences the integrity of the 26S proteasome in yeast. It was first reported that deletion of Rpn10 leads to separation of the lid–base sub-complex [31]. Thus, Rpn10 was thought to be present in the interface between the lid and base complexes and to stabilize the lid–base contacts, but it was later purified as an integral component in the lid complex [101]. Subsequently, some reports highlighted the role of certain Rpn subunits for 26S proteasome assembly [102]. Rpn9 is required for the incorporation of Rpn10 into the 26S proteasome, and it also participates in the efficient assembly and/or stability of the 26S proteasome [103]. Rpn5 plays a role in mediating correct proteasome localization and proper proteasome assembly [104]. On the other hand, Rpn6 is involved in maintaining the correct quaternary structure of the 26S proteasome, since depletion of Rpn6 affects both the structure and the peptidase activity of the 26S proteasome in the cell [105]. The loss of the temperature-sensitive mutant Rpn7-3 causes a defect in the lid complex, suggesting that Rpn7 is required for the integrity of the 26S complex by establishing a correct lid structure [106]. In evaluating these reports, it is rational to suspect that individual subunits of the 26S proteasome (if not all) require assembly of a regulatory complex containing both lid and base sub-complexes. Therefore, it is conceivable that deletion or malfunction of certain subunits causes disorganization of the complex composed of heterogeneous subunits.

Yeast two-hybrid analysis reveals a hierarchy of subunit interactions among the base and lid complexes [38, 107, 108]. Within the base, the Rpt4/5/3/6 subunits display their interaction cluster [107]. Within the lid, a structural cluster forms around Rpn5/8/9/11. Moreover, Rpn5/8/9/11 constitutes a sub-complex. However, under normal conditions, these sub-complexes are not evident, unlike half-proteasomes, suggesting that the assembly of the lid and base complexes is very rapid. Whether chaperone molecules assisting the assembly of the lid or base, or both, exist in the cell is an unresolved issue. Yeast Nob1 is a nuclear protein that forms a complex with PA700 of the 26S proteasome [109]. Nob1 serves as a chaperone-like factor to join the 20S proteasome with the 19S regulatory particle in the nucleus and facilitates the maturation of the 20S proteasome and degradation of Ump1p. Nob1 is then internalized into the 26S proteasome and degraded to complete the biogenesis of the 26S proteasome.

At present, the mechanism of 26S proteasome assembly is basically unknown, except that ATP energy is required for the association of the 20S proteasome and PA700 [110, 111]. Recently, we found a novel function for Hsp90 in the ATP-dependent assembly of the 26S proteasome [112]. Functional loss of Hsp90 using a temperature-sensitive mutant in yeast caused dissociation of the 26S proteasome. Conversely, these dissociated constituents reassembled in Hsp90-dependent fashion both *in vivo* and *in vitro*; the process required ATP hydrolysis and was suppressed by the Hsp90 inhibitor geldanamycin. We also found genetic interactions between Hsp90 and several proteasomal Rpn genes, emphasizing the importance of Hsp90 to maintain the integrity of the 26S proteasome. Thus, Hsp90 interacts with the 26S proteasome and plays a principal role in the assembly and maintenance of the 26S proteasome.

7.5
Perspectives

A recent comprehensive interactive study revealed the existence of miscellaneous molecules that could interact with the proteasome. For instance, several new proteins were identified by mild purification using affinity purification, which is coupled to high-throughput, sensitive, genome-wide proteomics analysis [77, 113]. In addition, the yeast two-hybrid analysis was introduced to define the interaction maps of multi-subunit complexes and to systematically identify new interacting proteins [38, 100, 107, 108, 114]. However, whether proteins identified by these methods are genuine subunits or transiently interacting proteins that are linked to proteasome functions await further studies.

Several factors could influence the functions of proteasomes. In this regard, many E3s and deubiquitylating enzymes (DUBs) are known to interact with the 26S proteasome. Indeed, the 26S proteasome is known to bind directly various E3 ubiquitin ligases. For example, certain E3s, such as SCF, APC, Ubr1, Ufd4, Hul5, Parkin, CHIP, and E6-AP, are reported to bind to the 26S proteasome. However, how these E3s interact with the 26S proteasome is largely unknown at present. Among these ligases, Parkin (the autosomal recessive juvenile parkinsonism–causing gene product) directly associates with Rpn10 via its UBL domain. As Hsp70 is associated with the 26S proteasome, possibly through Bag1, whose UBL motif interacts with Rpn1 and/or Rpn10, CHIP is also indirectly associated with the 26S proteasome, because CHIP binds to Hsp70 via its C-terminal EEVD sequence. The association of E3 with the 26S proteasome is functionally rational, considering the rapid destruction of substrate proteins.

In addition, a set of DUBs such as UBP6/Hsp14, UBP5, Doa4, and UCH37 are also capable of binding to the 26S proteasome. Previous studies reported that the C-terminal UBL sequence of UBP6 is responsible for the association with Rpn1. Moreover, $\beta 2$, Rpn5, and Rpn12 serve as acceptors for UBP5, Doa4, and UCH37, respectively, although the molecular basis of their interactions has not yet been defined [43, 115]. These DUBs collaborate with Rpn11 (as a genuine subunit of the lid complex) with a deubiquitylating metalloprotease activity, which allows ubiquitin peptide recycling before substrate degradation.

Recent studies have investigated the pathophysiological importance of the proteasome in the cells. For a full assessment of this issue, it is important to determine the biological significance of the diversity of the 26S proteasome system. Although we summarized our knowledge of this system in this chapter, the physiological roles of various interacting proteins are still largely unknown. Further studies should address the importance of the proteasome system in various cells and organs.

References

1 Coux, O., Tanaka, K. and Goldberg, A. L., Structure and functions of the 20S and 26S proteasomes. *Annu Rev Biochem* 1996. **65**: 801–847.

2 Baumeister, W., Walz, J., Zuhl, F. and Seemuller, E., The proteasome: paradigm of a self-compartmentalizing protease. *Cell* 1998. **92**: 367–380.

3 Hendil, K. B. and Hartmann-Petersen, R., Proteasomes: a complex story. *Curr Protein Pept Sci* 2004. **5**: 135–151.

4 Hershko, A., Ciechanover, A. and Varshavsky, A., Basic Medical Research Award. The ubiquitin system. *Nat Med* 2000. **6**: 1073–1081.

5 Hershko, A. and Ciechanover, A., The ubiquitin system. *Annu Rev Biochem* 1998. **67**: 425–479.

6 Voges, D., Zwickl, P. and Baumeister, W., The 26S proteasome: a molecular machine designed for controlled proteolysis. *Annu Rev Biochem* 1999. **68**: 1015–1068.

7 Pickart, C. M., Mechanisms underlying ubiquitination. *Annu Rev Biochem* 2001. **70**: 503–533.

8 Glickman, M. H. and Ciechanover,

A., The ubiquitin–proteasome proteolytic pathway: destruction for the sake of construction. *Physiol Rev* 2002. **82**: 373–428.

9 Schwartz, D. C. and Hochstrasser, M., A superfamily of protein tags: ubiquitin, SUMO and related modifiers. *Trends Biochem Sci* 2003. **28**: 321–328.

10 Pickart, C. M., Back to the future with ubiquitin. *Cell* 2004. **116**: 181–190.

11 Finley, D., Ciechanover, A. and Varshavsky, A., Ubiquitin as a central cellular regulator. *Cell* 2004. **116**: S29–32, 22 p following S32.

12 Zwickl, P., Seemuller, E., Kapelari, B. and Baumeister, W., The proteasome: a supramolecular assembly designed for controlled proteolysis. *Adv Protein Chem* 2001. **59**: 187–222.

13 Groll, M., Ditzel, L., Lowe, J., Stock, D., Bochtler, M., Bartunik, H. D. and Huber, R., Structure of 20S proteasome from yeast at 2.4 A resolution. *Nature* 1997. **386**: 463–471.

14 Unno, M., Mizushima, T., Morimoto, Y., Tomisugi, Y., Tanaka, K., Yasuoka, N. and Tsukihara, T., The structure of the mammalian 20S proteasome at 2.75 A resolution. *Structure (Camb)* 2002. **10**: 609–618.

15 Tanaka, K., Yoshimura, T., Tamura, T., Fujiwara, T., Kumatori, A. and Ichihara, A., Possible mechanism of nuclear translocation of proteasomes. *FEBS Lett* 1990. **271**: 41–46.

16 Nederlof, P. M., Wang, H. R. and Baumeister, W., Nuclear localization signals of human and Thermoplasma proteasomal alpha subunits are functional in vitro. *Proc Natl Acad Sci U S A* 1995. **92**: 12060–12064.

17 Knuehl, C., Seelig, A., Brecht, B., Henklein, P. and Kloetzel, P. M., Functional analysis of eukaryotic 20S proteasome nuclear localization signal. *Exp Cell Res* 1996. **225**: 67–74.

18 Velichutina, I., Connerly, P. L., Arendt, C. S., Li, X. and Hochstrasser, M., Plasticity in eucaryotic 20S proteasome ring assembly revealed by a subunit deletion in yeast. *Embo J* 2004. **23**: 500–510.

19 Tanaka, K., Tanahashi, N., Tsurumi, C., Yokota, K. Y. and Shimbara, N., Proteasomes and antigen processing. *Adv Immunol* 1997. **64**: 1–38.

20 Kloetzel, P. M., Antigen processing by the proteasome. *Nat Rev Mol Cell Biol* 2001. **2**: 179–187.

21 Rock, K. L., York, I. A., Saric, T. and Goldberg, A. L., Protein degradation and the generation of MHC class I-presented peptides. *Adv Immunol* 2002. **80**: 1–70.

22 Tanaka, K., Role of proteasomes modified by interferon-gamma in antigen processing. *J Leukoc Biol* 1994. **56**: 571–575.

23 Monaco, J. J. and Nandi, D., The genetics of proteasomes and antigen processing. *Annu Rev Genet* 1995. **29**: 729–754.

24 Rock, K. L., York, I. A. and Goldberg, A. L., Post-proteasomal antigen processing for major histocompatibility complex class I presentation. *Nat Immunol* 2004. **5**: 670–677.

25 Kloetzel, P. M., Generation of major histocompatibility complex class I antigens: functional interplay between proteasomes and TPPII. *Nat Immunol* 2004. **5**: 661–669.

26 Tanaka, K. and Kasahara, M., The MHC class I ligand-generating system: roles of immunoproteasomes and the interferon-gamma-inducible proteasome activator PA28. *Immunol Rev* 1998. **163**: 161–176.

27 Liu, C. W., Corboy, M. J., DeMartino, G. N. and Thomas, P. J., Endoproteolytic activity of the proteasome. *Science* 2003. **299**: 408–411.

28 Teoh, C. Y. and Davies, K. J., Potential roles of protein oxidation and the immunoproteasome in MHC class I antigen presentation: the 'PrOxI' hypothesis. *Arch Biochem Biophys* 2004. **423**: 88–96.

29 Rechsteiner, M., Hoffman, L. and Dubiel, W., The multicatalytic and 26 S proteases. *J Biol Chem* 1993. **268**: 6065–6068.

30 DeMartino, G. N. and Slaughter,

C. A., The proteasome, a novel protease regulated by multiple mechanisms. *J Biol Chem* 1999. **274**: 22123–22126.

31 Glickman, M. H., Rubin, D. M., Fried, V. A. and Finley, D., The regulatory particle of the Saccharomyces cerevisiae proteasome. *Mol Cell Biol* 1998. **18**: 3149–3162.

32 Finley, D., Tanaka, K., Mann, C., Feldmann, H., Hochstrasser, M., Vierstra, R., Johnston, S., Hampton, R., Haber, J., McCusker, J., Silver, P., Frontali, L., Thorsness, P., Varshavsky, A., Byers, B., Madura, K., Reed, S. I., Wolf, D., Jentsch, S., Sommer, T., Baumeister, W., Goldberg, A., Fried, V., Rubin, D. M., Toh-e, A. and et al., Unified nomenclature for subunits of the Saccharomyces cerevisiae proteasome regulatory particle. *Trends Biochem Sci* 1998. **23**: 244–245.

33 Tanaka, K., Molecular biology of the proteasome. *Biochem Biophys Res Commun* 1998. **247**: 537–541.

34 Glickman, M. H., Rubin, D. M., Coux, O., Wefes, I., Pfeifer, G., Cjeka, Z., Baumeister, W., Fried, V. A. and Finley, D., A subcomplex of the proteasome regulatory particle required for ubiquitin-conjugate degradation and related to the COP9-signalosome and eIF3. *Cell* 1998. **94**: 615–623.

35 Ogura, T. and Wilkinson, A. J., AAA+ superfamily ATPases: common structure–diverse function. *Genes Cells* 2001. **6**: 575–597.

36 Benaroudj, N., Zwickl, P., Seemuller, E., Baumeister, W. and Goldberg, A. L., ATP hydrolysis by the proteasome regulatory complex PAN serves multiple functions in protein degradation. *Mol Cell* 2003. **11**: 69–78.

37 Ogura, T. and Tanaka, K., Dissecting various ATP-dependent steps involved in proteasomal degradation. *Mol Cell* 2003. **11**: 3–5.

38 Bader, G. D. and Hogue, C. W., Analyzing yeast protein-protein interaction data obtained from different sources. *Nat Biotechnol* 2002. **20**: 991–997.

39 Wilkinson, C. R., Wallace, M., Morphew, M., Perry, P., Allshire, R., Javerzat, J. P., McIntosh, J. R. and Gordon, C., Localization of the 26S proteasome during mitosis and meiosis in fission yeast. *Embo J* 1998. **17**: 6465–6476.

40 Wendler, P., Lehmann, A., Janek, K., Baumgart, S. and Enenkel, C., The bipartite nuclear localization sequence of Rpn2 is required for nuclear import of proteasomal base complexes via karyopherin alphabeta and proteasome functions. *J Biol Chem* 2004. **279**: 37751–37762.

41 Deveraux, Q., Ustrell, V., Pickart, C. and Rechsteiner, M., A 26 S protease subunit that binds ubiquitin conjugates. *J Biol Chem* 1994. **269**: 7059–7061.

42 Young, P., Deveraux, Q., Beal, R. E., Pickart, C. M. and Rechsteiner, M., Characterization of two polyubiquitin binding sites in the 26 S protease subunit 5a. *J Biol Chem* 1998. **273**: 5461–5467.

43 Hartmann-Petersen, R., Seeger, M. and Gordon, C., Transferring substrates to the 26S proteasome. *Trends Biochem Sci* 2003. **28**: 26–31.

44 Madura, K., The ubiquitin-associated (UBA) domain: on the path from prudence to prurience. *Cell Cycle* 2002. **1**: 235–244.

45 Verma, R., Oania, R., Graumann, J. and Deshaies, R. J., Multiubiquitin chain receptors define a layer of substrate selectivity in the ubiquitin–proteasome system. *Cell* 2004. **118**: 99–110.

46 Elsasser, S., Chandler-Militello, D., Muller, B., Hanna, J. and Finley, D., Rad23 and Rpn10 serve as alternative ubiquitin receptors for the proteasome. *J Biol Chem* 2004. **279**: 26817–26822.

47 Lambertson, D., Chen, L. and Madura, K., Pleiotropic defects caused by loss of the proteasome-interacting factors Rad23 and Rpn10 of Saccharomyces cerevisiae. *Genetics* 1999. **153**: 69–79.

48 Hartmann-Petersen, R. and Gordon, C., Protein degradation: recognition of ubiquitinylated substrates. *Curr Biol* 2004. **14**: R754–756.

49 Kawahara, H., Kasahara, M., Nishiyama, A., Ohsumi, K., Goto, T., Kishimoto, T., Saeki, Y., Yokosawa, H., Shimbara, N., Murata, S., Chiba, T., Suzuki, K. and Tanaka, K., Developmentally regulated, alternative splicing of the Rpn10 gene generates multiple forms of 26S proteasomes. *Embo J* 2000. **19**: 4144–4153.

50 Kikukawa, Y., Shimada, M., Suzuki, N., Tanaka, K., Yokosawa, H. and Kawahara, H., The 26S proteasome Rpn10 gene encoding splicing isoforms: evolutional conservation of the genomic organization in vertebrates. *Biol Chem* 2002. **383**: 1257–1261.

51 Szlanka, T., Haracska, L., Kiss, I., Deak, P., Kurucz, E., Ando, I., Viragh, E. and Udvardy, A., Deletion of proteasomal subunit S5a/Rpn10/p54 causes lethality, multiple mitotic defects and overexpression of proteasomal genes in Drosophila melanogaster. *J Cell Sci* 2003. **116**: 1023–1033.

52 DeMartino, G. N., Proske, R. J., Moomaw, C. R., Strong, A. A., Song, X., Hisamatsu, H., Tanaka, K. and Slaughter, C. A., Identification, purification, and characterization of a PA700-dependent activator of the proteasome. *J Biol Chem* 1996. **271**: 3112–3118.

53 Adams, G. M., Falke, S., Goldberg, A. L., Slaughter, C. A., DeMartino, G. N. and Gogol, E. P., Structural and functional effects of PA700 and modulator protein on proteasomes. *J Mol Biol* 1997. **273**: 646–657.

54 Adams, G. M., Crotchett, B., Slaughter, C. A., DeMartino, G. N. and Gogol, E. P., Formation of proteasome-PA700 complexes directly correlates with activation of peptidase activity. *Biochemistry* 1998. **37**: 12927–12932.

55 Watanabe, T. K., Saito, A., Suzuki, M., Fujiwara, T., Takahashi, E., Slaughter, C. A., DeMartino, G. N., Hendil, K. B., Chung, C. H., Tanahashi, N. and Tanaka, K., cDNA cloning and characterization of a human proteasomal modulator subunit, p27 (PSMD9). *Genomics* 1998. **50**: 241–250.

56 Lam, Y. A., Lawson, T. G., Velayutham, M., Zweier, J. L. and Pickart, C. M., A proteasomal ATPase subunit recognizes the polyubiquitin degradation signal. *Nature* 2002. **416**: 763–767.

57 Zhang, Z., Krutchinsky, A., Endicott, S., Realini, C., Rechsteiner, M. and Standing, K. G., Proteasome activator 11S REG or PA28: recombinant REG alpha/REG beta hetero-oligomers are heptamers. *Biochemistry* 1999. **38**: 5651–5658.

58 Gray, C. W., Slaughter, C. A. and DeMartino, G. N., PA28 activator protein forms regulatory caps on proteasome stacked rings. *J Mol Biol* 1994. **236**: 7–15.

59 Knowlton, J. R., Johnston, S. C., Whitby, F. G., Realini, C., Zhang, Z., Rechsteiner, M. and Hill, C. P., Structure of the proteasome activator REGalpha (PA28alpha). *Nature* 1997. **390**: 639–643.

60 Whitby, F. G., Masters, E. I., Kramer, L., Knowlton, J. R., Yao, Y., Wang, C. C. and Hill, C. P., Structural basis for the activation of 20S proteasomes by 11S regulators. *Nature* 2000. **408**: 115–120.

61 Tanahashi, N., Yokota, K., Ahn, J. Y., Chung, C. H., Fujiwara, T., Takahashi, E., DeMartino, G. N., Slaughter, C. A., Toyonaga, T., Yamamura, K., Shimbara, N. and Tanaka, K., Molecular properties of the proteasome activator PA28 family proteins and gamma-interferon regulation. *Genes Cells* 1997. **2**: 195–211.

62 Wojcik, C., Tanaka, K., Paweletz, N., Naab, U. and Wilk, S., Proteasome activator (PA28) subunits, alpha, beta and gamma (Ki antigen) in NT2 neuronal precursor cells and HeLa S3 cells. *Eur J Cell Biol* 1998. **77**: 151–160.

63 Rechsteiner, M., Realini, C. and

Ustrell, V., The proteasome activator 11 S REG (PA28) and class I antigen presentation. *Biochem J* 2000. **345 Pt 1**: 1–15.

64 Murata, S., Udono, H., Tanahashi, N., Hamada, N., Watanabe, K., Adachi, K., Yamano, T., Yui, K., Kobayashi, N., Kasahara, M., Tanaka, K. and Chiba, T., Immunoproteasome assembly and antigen presentation in mice lacking both PA28alpha and PA28beta. *Embo J* 2001. **20**: 5898–5907.

65 Murata, S., Kawahara, H., Tohma, S., Yamamoto, K., Kasahara, M., Nabeshima, Y., Tanaka, K. and Chiba, T., Growth retardation in mice lacking the proteasome activator PA28gamma. *J Biol Chem* 1999. **274**: 38211–38215.

66 Barton, L. F., Runnels, H. A., Schell, T. D., Cho, Y., Gibbons, R., Tevethia, S. S., Deepe, G. S., Jr. and Monaco, J. J., Immune defects in 28-kDa proteasome activator gamma-deficient mice. *J Immunol* 2004. **172**: 3948–3954.

67 To, W. Y. and Wang, C. C., Identification and characterization of an activated 20S proteasome in *Trypanosoma brucei. FEBS Lett* 1997. **404**: 253–262.

68 Hendil, K. B., Khan, S. and Tanaka, K., Simultaneous binding of PA28 and PA700 activators to 20 S proteasomes. *Biochem J* 1998. **332 (Pt 3)**: 749–754.

69 Tanahashi, N., Murakami, Y., Minami, Y., Shimbara, N., Hendil, K. B. and Tanaka, K., Hybrid proteasomes. Induction by interferon-gamma and contribution to ATP-dependent proteolysis. *J Biol Chem* 2000. **275**: 14336–14345.

70 Kopp, F., Dahlmann, B. and Kuehn, L., Reconstitution of hybrid proteasomes from purified PA700-20 S complexes and PA28alphabeta activator: ultrastructure and peptidase activities. *J Mol Biol* 2001. **313**: 465–471.

71 Cascio, P., Call, M., Petre, B. M., Walz, T. and Goldberg, A. L., Properties of the hybrid form of the 26S proteasome containing both 19S and PA28 complexes. *Embo J* 2002. **21**: 2636–2645.

72 Ustrell, V., Hoffman, L., Pratt, G. and Rechsteiner, M., PA200, a nuclear proteasome activator involved in DNA repair. *Embo J* 2002. **21**: 3516–3525.

73 Kajava, A. V., Gorbea, C., Ortega, J., Rechsteiner, M. and Steven, A. C., New HEAT-like repeat motifs in proteins regulating proteasome structure and function. *J Struct Biol* 2004. **146**: 425–430.

74 Febres, D. E., Pramanik, A., Caton, M., Doherty, K., McKoy, J., Garcia, E., Alejo, W. and Moore, C. W., The novel BLM3 gene encodes a protein that protects against lethal effects of oxidative damage. *Cell Mol Biol (Noisy-le-grand)* 2001. **47**: 1149–1162.

75 Jelinsky, S. A., Estep, P., Church, G. M. and Samson, L. D., Regulatory networks revealed by transcriptional profiling of damaged Saccharomyces cerevisiae cells: Rpn4 links base excision repair with proteasomes. *Mol Cell Biol* 2000. **20**: 8157–8167.

76 Fehlker, M., Wendler, P., Lehmann, A. and Enenkel, C., Blm3 is part of nascent proteasomes and is involved in a late stage of nuclear proteasome assembly. *EMBO Rep* 2003. **4**: 959–963.

77 Leggett, D. S., Hanna, J., Borodovsky, A., Crosas, B., Schmidt, M., Baker, R. T., Walz, T., Ploegh, H. and Finley, D., Multiple associated proteins regulate proteasome structure and function. *Mol Cell* 2002. **10**: 495–507.

78 Zaiss, D. M., Standera, S., Holzhutter, H., Kloetzel, P. and Sijts, A. J., The proteasome inhibitor PI31 competes with PA28 for binding to 20S proteasomes. *FEBS Lett* 1999. **457**: 333–338.

79 McCutchen-Maloney, S. L., Matsuda, K., Shimbara, N., Binns, D. D., Tanaka, K., Slaughter, C. A. and DeMartino, G. N., cDNA cloning, expression, and functional characterization of PI31, a proline-rich inhibitor of the proteasome. *J Biol Chem* 2000. **275**: 18557–18565.

80 Zaiss, D. M., Standera, S., Kloetzel, P. M. and Sijts, A. J., PI31 is a modulator of proteasome formation and antigen processing. *Proc Natl Acad Sci U S A* 2002. **99**: 14344–14349.

81 Schmidtke, G., Schmidt, M. and Kloetzel, P. M., Maturation of mammalian 20 S proteasome: purification and characterization of 13 S and 16 S proteasome precursor complexes. *J Mol Biol* 1997. **268**: 95–106.

82 Frentzel, S., Pesold-Hurt, B., Seelig, A. and Kloetzel, P. M., 20 S proteasomes are assembled via distinct precursor complexes. Processing of LMP2 and LMP7 proproteins takes place in 13–16 S preproteasome complexes. *J Mol Biol* 1994. **236**: 975–981.

83 Nandi, D., Woodward, E., Ginsburg, D. B. and Monaco, J. J., Intermediates in the formation of mouse 20S proteasomes: implications for the assembly of precursor beta subunits. *Embo J* 1997. **16**: 5363–5375.

84 Mullapudi, S., Pullan, L., Khalil, H., Stoops, J. K., Tastan-Bishop, A. O., Beckmann, R., Kloetzel, P. M., Krueger, E. and Penczek, P. A., Rearrangement of the 16S precursor subunits is essential for the formation of the active 20S proteasome. *Biophys J* 2004.

85 Ditzel, L., Huber, R., Mann, K., Heinemeyer, W., Wolf, D. H. and Groll, M., Conformational constraints for protein self-cleavage in the proteasome. *J Mol Biol* 1998. **279**: 1187–1191.

86 Chen, P. and Hochstrasser, M., Autocatalytic subunit processing couples active site formation in the 20S proteasome to completion of assembly. *Cell* 1996. **86**: 961–972.

87 Seemuller, E., Lupas, A. and Baumeister, W., Autocatalytic processing of the 20S proteasome. *Nature* 1996. **382**: 468–471.

88 Zwickl, P., Kleinz, J. and Baumeister, W., Critical elements in proteasome assembly. *Nat Struct Biol* 1994. **1**: 765–770.

89 Jager, S., Groll, M., Huber, R., Wolf, D. H. and Heinemeyer, W., Proteasome beta-type subunits: unequal roles of propeptides in core particle maturation and a hierarchy of active site function. *J Mol Biol* 1999. **291**: 997–1013.

90 De, M., Jayarapu, K., Elenich, L., Monaco, J. J., Colbert, R. A. and Griffin, T. A., Beta 2 subunit propeptides influence cooperative proteasome assembly. *J Biol Chem* 2003. **278**: 6153–6159.

91 Arendt, C. S. and Hochstrasser, M., Eukaryotic 20S proteasome catalytic subunit propeptides prevent active site inactivation by N-terminal acetylation and promote particle assembly. *Embo J* 1999. **18**: 3575–3585.

92 Groll, M., Heinemeyer, W., Jager, S., Ullrich, T., Bochtler, M., Wolf, D. H. and Huber, R., The catalytic sites of 20S proteasomes and their role in subunit maturation: a mutational and crystallographic study. *Proc Natl Acad Sci U S A* 1999. **96**: 10976–10983.

93 Ramos, P. C., Hockendorff, J., Johnson, E. S., Varshavsky, A. and Dohmen, R. J., Ump1p is required for proper maturation of the 20S proteasome and becomes its substrate upon completion of the assembly. *Cell* 1998. **92**: 489–499.

94 Burri, L., Hockendorff, J., Boehm, U., Klamp, T., Dohmen, R. J. and Levy, F., Identification and characterization of a mammalian protein interacting with 20S proteasome precursors. *Proc Natl Acad Sci U S A* 2000. **97**: 10348–10353.

95 Witt, E., Zantopf, D., Schmidt, M., Kraft, R., Kloetzel, P. M. and Kruger, E., Characterisation of the newly identified human Ump1 homologue POMP and analysis of LMP7(beta 5i) incorporation into 20 S proteasomes. *J Mol Biol* 2000. **301**: 1–9.

96 Griffin, T. A., Slack, J. P., McCluskey, T. S., Monaco, J. J. and Colbert, R. A., Identification of

proteassemblin, a mammalian homologue of the yeast protein, Ump1p, that is required for normal proteasome assembly. *Mol Cell Biol Res Commun* 2000. **3**: 212–217.

97 Aki, M., Shimbara, N., Takashina, M., Akiyama, K., Kagawa, S., Tamura, T., Tanahashi, N., Yoshimura, T., Tanaka, K. and Ichihara, A., Interferon-gamma induces different subunit organizations and functional diversity of proteasomes. *J Biochem (Tokyo)* 1994. **115**: 257–269.

98 Griffin, T. A., Nandi, D., Cruz, M., Fehling, H. J., Kaer, L. V., Monaco, J. J. and Colbert, R. A., Immunoproteasome assembly: cooperative incorporation of interferon gamma (IFN-gamma)-inducible subunits. *J Exp Med* 1998. **187**: 97–104.

99 Kingsbury, D. J., Griffin, T. A. and Colbert, R. A., Novel propeptide function in 20 S proteasome assembly influences beta subunit composition. *J Biol Chem* 2000. **275**: 24156–24162.

100 Jayarapu, K. and Griffin, T. A., Protein-protein interactions among human 20S proteasome subunits and proteassemblin. *Biochem Biophys Res Commun* 2004. **314**: 523–528.

101 Saeki, Y., Toh-e, A. and Yokosawa, H., Rapid isolation and characterization of the yeast proteasome regulatory complex. *Biochem Biophys Res Commun* 2000. **273**: 509–515.

102 Ferrell, K., Wilkinson, C. R., Dubiel, W. and Gordon, C., Regulatory subunit interactions of the 26S proteasome, a complex problem. *Trends Biochem Sci* 2000. **25**: 83–88.

103 Takeuchi, J., Fujimuro, M., Yokosawa, H., Tanaka, K. and Toh-e, A., Rpn9 is required for efficient assembly of the yeast 26S proteasome. *Mol Cell Biol* 1999. **19**: 6575–6584.

104 Yen, H. C., Espiritu, C. and Chang, E. C., Rpn5 is a conserved proteasome subunit and required for proper proteasome localization and assembly. *J Biol Chem* 2003. **278**: 30669–30676.

105 Santamaria, P. G., Finley, D., Ballesta, J. P. and Remacha, M., Rpn6p, a proteasome subunit from Saccharomyces cerevisiae, is essential for the assembly and activity of the 26 S proteasome. *J Biol Chem* 2003. **278**: 6687–6695.

106 Isono, E., Saeki, Y., Yokosawa, H. and Toh-e, A., Rpn7 Is required for the structural integrity of the 26 S proteasome of Saccharomyces cerevisiae. *J Biol Chem* 2004. **279**: 27168–27176.

107 Fu, H., Reis, N., Lee, Y., Glickman, M. H. and Vierstra, R. D., Subunit interaction maps for the regulatory particle of the 26S proteasome and the COP9 signalosome. *Embo J* 2001. **20**: 7096–7107.

108 Davy, A., Bello, P., Thierry-Mieg, N., Vaglio, P., Hitti, J., Doucette-Stamm, L., Thierry-Mieg, D., Reboul, J., Boulton, S., Walhout, A. J., Coux, O. and Vidal, M., A protein-protein interaction map of the Caenorhabditis elegans 26S proteasome. *EMBO Rep* 2001. **2**: 821–828.

109 Tone, Y. and Toh, E. A., Nob1p is required for biogenesis of the 26S proteasome and degraded upon its maturation in Saccharomyces cerevisiae. *Genes Dev* 2002. **16**: 3142–3157.

110 Eytan, E., Ganoth, D., Armon, T. and Hershko, A., ATP-dependent incorporation of 20S protease into the 26S complex that degrades proteins conjugated to ubiquitin. *Proc Natl Acad Sci U S A* 1989. **86**: 7751–7755.

111 Hoffman, L. and Rechsteiner, M., Effects of nucleotides on assembly of the 26S proteasome and degradation of ubiquitin conjugates. *Mol Biol Rep* 1997. **24**: 13–16.

112 Imai, J., Maruya, M., Yashiroda, H., Yahara, I. and Tanaka, K., The molecular chaperone Hsp90 plays a role in the assembly and maintenance of the 26S proteasome. *Embo J* 2003. **22**: 3557–3567.

113 Verma, R., Chen, S., Feldman, R., Schieltz, D., Yates, J., Dohmen, J. and Deshaies, R. J., Proteasomal proteomics: identification of nucleotide-sensitive proteasome-interacting proteins by mass spectrometric analysis of affinity-purified proteasomes. *Mol Biol Cell* 2000. **11**: 3425–3439.

114 Cagney, G., Uetz, P. and Fields, S., Two-hybrid analysis of the *Saccharomyces cerevisiae* 26S proteasome. *Physiol Genomics* 2001. **7**: 27–34.

115 Hartmann-Petersen, R. and Gordon, C., Proteins interacting with the 26S proteasome. *Cell Mol Life Sci* 2004. **61**: 1589–1595.

116 Hirano, Y., Hendil, K. B., Yashiroda, H., Iemura, S., Nagane, R., Hioki, Y., Natsume, T., Tanaka, K., and Murata, M., A novel heterodimeric complex that promotes the assembly of mammalian 20S proteasomes, *Nature* in press.

117 Unno, M., Mizushima, T., Morimoto, Y., Tomisugi, Y., Tanaka, K., Yasuoka, N., and Tsukihara, T., The structure of the mammalian 20S proteasome at 2.75 Å resolution. *Structure* 2002. **10**: 609–618.

8
Proteasome-Interacting Proteins

Jean E. O'Donoghue and Colin Gordon

8.1
Introduction

8.1.1
The Proteasome

The proteasome is the key organelle within the cell responsible for the regulated degradation of intracellular proteins. It was originally found during the search for an ATP-requiring activity that was involved with protein degradation. It has since been discovered that the proteasome itself is made up of several subunits along with various interacting regulatory subunits. In addition, in order to carry out its function as a controlled method of degradation, it interacts with a wide array of proteins. It is these interacting proteins that confer subtlety of function upon what is, at its core, a protein-degrading machine. These interacting proteins serve to control, connect and activate proteolysis. In this chapter we will first introduce the proteasome itself and the system of ubiquitination which is used to target proteins for degradation before addressing the roles of its interactors. Then we will look at the regulators of the 20S core particle (CP) and those proteins involved in the assembly and stability of the proteasome. We will address proteins involved in the ubiquitination pathway that directly interact with the proteasome, namely E2 enzymes, E3 enzymes and deubiquitinating enzymes (DUBs). In addition we will deal with those proteins that function as the "go-betweens" for the ubiquitination system and proteasomal degradation, and finally we will address the growing evidence for proteasomal interaction with proteins involved with transcription, translation, and DNA repair. Many of the proteins discussed in this chapter have different names in different species. In general we have used the orthologue name relevant to the species in which the work was conducted – with other orthologue names in parenthesis where possible. For a complete list of orthologue names and their appropriate species see table 8.1.

Protein Degradation, Vol. 2: The Ubiquitin-Proteasome System.
Edited by R. J. Mayer, A. Ciechanover, M. Rechsteiner

ISBN: 3-527-31130-0

Table 8.1. This table shows the various names assigned to the proteins discussed in this chapter. Some proteins have multiple names in one species, others have different names for different orthologues. Here *D.m.* = *Drosophila melanogaster*

	S. pombe	*S. cerevisiae*	*H. sapiens*	*Other (Species)*
Proteasome subunits	Mts1	Rpn9	S11	
	Mts2	Rpt2	S4	
	Mts3	Rpn12	S14	
	Mts4	Rpn1	S2	
	Mts8	Pre6	β1	
	Pus1	Rpn10	S5a	
	Pad1	Rpn11	POH1/S13	
UBL-UBA proteins	Rhp23	Rad23	hHR23a/b/c	
	Dph1	Dsk2	hPLIC1/2	
DUB		Doa4/Ubp4		
	Uch2		UCH37	p37A (*D.m.*)
	Ubp6	Ubp6	USP14	
Others		Blm10	PA200	
		Hul5	KIAA10	
	Cdc48	Cdc48	VCP/p97	
	Ubx3	Shp1	p47	
	Sum1	TIF34	eIF3i	
	Int6/Yin6		eIF3e/Int6	

8.1.2
Structure of the 26S Proteasome

The 26S proteasome is made up of the 20S core particle (CP) and the 19S regulatory particle (RP). The core particle contains the proteases that can degrade proteins to small peptides. It consists of 28 subunits – 14 α and 14 β proteins which form 4 stacked rings of 7 subunits each – 2 α rings and 2 β rings. The α rings sandwich the β rings to form a cylindrical structure. In this way a central channel is formed with three chambers: two antechambers on either side of a central chamber. This central chamber is lined with at least three active sites whose combined specificities can act to hydrolyse almost all peptide bonds. Access to these active sites is controlled by the α subunits which form the antechambers and can exhibit closed or open conformations. The protein to be degraded passes through this pore and the proteases degrade it to 6–9 amino acid peptide products which are released and recycled.

Access to these catalytic sites is controlled by the regulatory particle. This particle is made up of a "base" and a "lid" structure which attach to either end of the cylindrical CP (see figure 8.1). The RP functions to recognise ubiquitinated substrates and unfold proteins thus controlling access to the potent proteases contained within the 20S CP. The RP is made up of approximately 20 different protein subunits. A subset of these are the AAA ATPases which are required for the unfolding of proteins to be degraded.

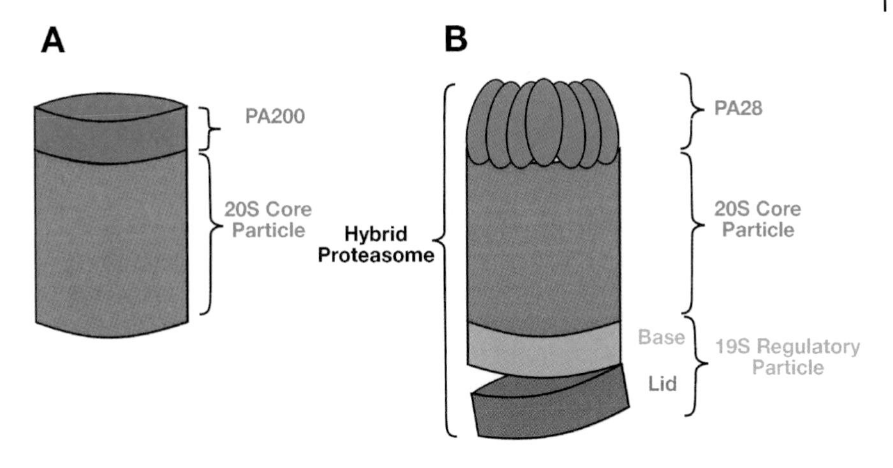

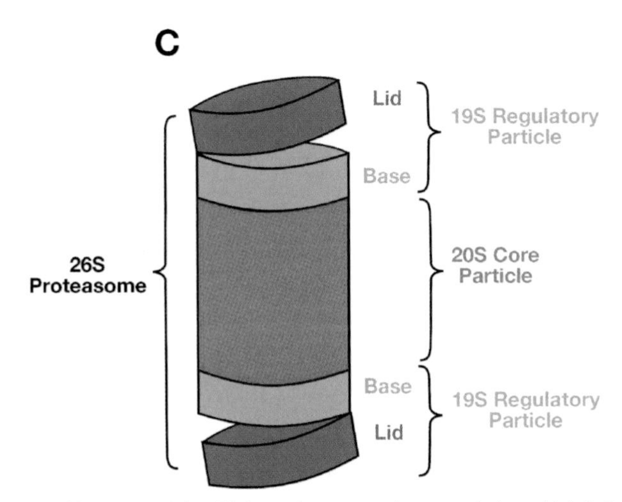

Fig. 8.1. 20S core particle with its various regulators.
A. 20S core particle with PA200. This may function in DNA repair.
B. Hybrid proteasome with one 19S particle and one PA28 particle. PA28 is a heptamericring which is interferon-inducible.
C. The classic 26S proteasome with two 19S regulatory particles for every one 20S core particle. This is the main proteasome species seen in the cell and is responsible for ubiquitin-mediated degradation.

8.1.3
Marking Proteins for Proteasomal Degradation – the Ubiquitin System

Ubiquitin is a 76 amino acid protein. It can modify proteins covalently by conjugating them through lysine linkages with ubiquitin chains forming through similar lysine linkages. This "ubiquitination" of proteins is carried out by a cascade of enzymes – E1, E2 and E3. E1 uses ATP to activate the ubiquitin moiety, generating a high-energy thiolester intermediate in the process. The activated ubiquitin is transferred from the E1 to a cysteine residue of an E2 enzyme, thus generating another thiolester intermediate. An E3 enzyme is then required to catalyse the transfer of the ubiquitin moiety from a thiolester intermediate on the E2 to an amide linkage on the target protein or another ubiquitin moiety to create a chain. As the E3s interact directly or indirectly with targeted protein, these enzymes confer specificity on the system of ubiquitination. A protein can be monoubiquitinated, or multiubiquitinated and there is some evidence to suggest that these are not equivalent with regard to the fate of the substrate.

E4 enzymes have also been described. These enzymes target substrates that are already ubiquitinated but only have chains one or two molecules long. E4 enzymes thus serve to increase the ubiquitin chain length.

8.2 Regulators of the Holoenzyme and Chaperones Involved in Assembly of the Proteasome

8.2.1 Proteasome Assembly and Integrity

Some proteasome interactors are involved in the initial assembly of the proteasome such as the *S. cerevisiae* protein Ump1p. This was first described in 1998 as a short-lived chaperone required for the correct maturation of the 20S proteasome [1]. Ump1p was discovered in a screen for mutants defective for the degradation of test substrates. *ump1* null mutants were found to exhibit classic phenotypes of disrupted proteasome function; that is, they were hypersensitive to cadmium, canavanine and thermal stress and they showed an accumulation of ubiquitin-protein conjugates. Upon further experimentation, Ump1 was found to be a component of proteasome precursor complexes that was degraded upon the formation of the 20S proteasome and was, in fact, required to co-ordinate the proteasome's assembly and activation.

Another protein vital to the proteasome is Ecm29. This protein has been proposed to tether the 19S RP to the 20S CP to form the 26S proteasome [2].

Recently it has been proposed that Ecm29 and PA200 (see section 8.2.2) are composed almost entirely of HEAT-like repeat motifs [3] suggesting they have α-helical solenoid structures – similar to those proposed for Rpn1 and 2 [4]. The functional significance of these helical structures, however, is unclear.

8.2.2 Regulators of the Holoenzyme

While the 19S regulatory particle is the usual companion to the 20S holoenzyme, other complexes can also bind the core particle to modify its function (see figure 8.1).

The 11S REGs/PA28 proteins form a heptameric ring structure that can bind and activate the 20S proteasome. They were originally identified and characterised as molecules that could strongly activate the 20S proteasome to degrade small fluorogenic peptides [5, 6]. This heptameric structure is made up of REGα, REGβ and REGγ proteins which form homo or hetero-oligomers that bind the proteasome with differing affinities: REGα/β > REGγ > REGα > REGβ [7]. These PA28 rings can activate the proteasome without affecting the active sites of the 20S CP. This is thought to be achieved through the facilitation of entry or exit of the substrates to/from the 20S core [8]. Hybrid proteasomes, that is 20S CP attached to both 19S and PA28, have been found, and can make up a quarter of the proteasome population in mammalian cell line extracts [9]. As for the role of PA28 and its activation of the 20S CP, it is known that PA28 is inducible by interferon-γ. It has been known for some time that the proteasome has a role to play in the production of antigens for MHC class I presentation. It now seems that PA28 is a major player in this role of the proteasome. Recently hybrid proteasomes have also been shown to be induced by interferon-γ. It is worth noting that PA28 is not present in either budding or fission yeast and so it is possible that the genes that encode an interferon-γ-inducible regulator for the 20S CP evolved along with other genes responsible for adaptive immunity [10].

Another complex that can interact with the 20S proteasome has been studied in mammals and is called the PA200. This 200kD complex has been shown to exist in a monomer-dimer equilibrium and is found in the nucleus of mammalian cells [11]. There appear to be orthologues of PA200 in *C. elegans*, *Arabidopsis* and *S. cerevisiae*, but not in *S. pombe* or *Drosophila*. In mammals (mice), there are slightly differing forms of PA200. Here, while the 200kD species is abundant in testis, a 160kD form is more reactive elsewhere and a 60kD form is seen in the liver, lung and brain [11]. It is also worth noting that while the 200kD and 160kD forms are nuclear, the 60kD species is cytoplasmic. The authors suggest these different forms could arise from splicing variants – but their physiological relevance is unknown. PA200, like PA28 promotes proteasomal hydrolysis of peptides but not that of large folded proteins. As for the physiological role of PA200, or PA200-20S proteasomes, there are several pieces of evidence to support its having a role in DNA repair. The yeast orthologue of PA200, Blm10 (originally termed Blm3) was discovered in a screen for mutants sensitive to the DNA damaging agent bleomycin [12] and has also been shown to complex with Sir4p [13]. This is a chromatin component that leaves the telomeres and relocates to DSBs (Double Strand Breaks) where it binds Ku70 [14]. PA200 in mammals is most abundant in testis where double strand breaks in DNA occur during meiotic recombination and it forms intracellular foci upon γ-irradiation similar to a number of DNA repair factors [11]. In addition to its putative role in DNA repair, recent work has also shown a three- to four-fold upregulation of PA200 in four different models of muscle wasting [15]. The true physiological function of PA200, as such, remains to be determined. There has been some speculation whether Blm10 in yeast was actually a functional homologue of PA200 when evidence was found to suggest that Blm10 functions to suppress premature activity of newly-formed 20S core particles and regulate the maturation of the 20S CP [16]. Recently, however, evidence has been found that

Blm10 can associate with the mature, active proteasome, with Blm10 docking onto the end of the CP cylinder and strongly activating its peptidase activity [17]. Blm10 is usually found in a hybrid Blm10-CP-RP complex. Why these complexes were not seen in the earlier study is unclear.

8.3
Enzymes Controlling Ubiquitination and Deubiquitination

8.3.1
E2 Ubiquitin-Conjugating Enzymes

As described in section 8.1.3 the enzymes that allow for the addition of ubiquitin to its substrates are classified as the ubiquitin-activating enzymes (E1s), ubiquitin-conjugating enzymes (E2s) and substrate-recognition enzymes (E3s). We now know that E2s have a role to play at the proteasome itself.

In *S. cerevisiae*, Tongaonkar *et al.* explored the possibility that E2s may interact with the proteasome [18]. This, it could be reasoned, could lend greater efficiency to the ubiquitin-proteasome system, that is, if the machinery that constructed the ubiquitin-tagged proteins were linked to the mechanism of their degradation. Initially the intra-cellular location of Ubc4, a highly abundant E2 in *S. cerevisiae* was addressed. Orthologues of Ubc4 are seen in mammals, plants and humans and it contains a conserved catalytic domain present in all E2s.

It was shown that when the proteasome was isolated by immunoprecipitation, Ubc4 could be seen to be associated with it [18]. This was done using two different tags to pull out the proteasome, and both times Ubc4 was present in the proteasome fractions. Also, a catalytically inactive Ubc4 could still interact with the proteasome. Indeed it was seen that purified proteasomes could act as an E2 enzyme *in vitro*, that is, a test substrate could be ubiquitinated when provided with the appropriate E1 ligase and purified proteasomes [18]. Using the same strategy, Ubc1, 2 and 5 were also found to be associated with the proteasome. It appears that the E2s do not compete with each other for proteasome-binding sites, as overexpression of one Ubc does not lead to the reduction in binding of another. The authors offered two possible explanations for this. Firstly it could be possible that only a small number of proteasomes are bound to E2s at any one time – such that the increase in the number bound to one E2 is not enough to compete out the binding of another. Secondly, if the E2s each have different binding sites on the proteasome, or if they bind different E3s associated with the proteasome (see section 8.3.2), then the overexpression of one and the increase in occupation of its binding site will not necessarily impinge on another's binding.

It had been known previously that *ubc4Δubc5Δ* double mutants are susceptible to heat stress [19]. Upon examination of the association of Ubc4 with the proteasome under heat stress, Ubc4 levels in the proteasome fraction were found to increase dramatically (approx. 25-fold).

It appears therefore that E2 enzymes can, at least in part, mediate the close interaction between the ubiquitin-tagging system and the proteasome. In the case of Ubc4 this interaction was most important in the heat-stressed cell. Presumably the close interaction between an E2 enzyme and the proteasome would facilitate the timely removal of misfolded proteins as a result of the increase in temperature. Both the proximity of the ubiquitin-tagging system and, one can imagine, the lack of access of cellular deubiquitinating enzymes to the tagged substrate would allow for an highly efficient quality control mechanism when the cell is under stress.

8.3.2
E3 Ubiquitin Ligases

E2–E3 complexes allow for the construction of multiubiquitin chains bound to a specific substrate with the E3 conferring substrate-specificity upon this activity. Given that E2s appear to associate with the proteasome, it is not surprising that some E3s possess this property as well.

Xie *et al.* describe the proteasome-binding properties of Ubr1p and Ufd4p, the E3 components of two independent ubiquitin-related proteolytic pathways in *S. cerevisiae* [20]. This would seem to suggest that association with the proteasome has some advantage for E3 enzymes given that two separate pathways have adopted this strategy.

Ubr1p is the E3 for the N-end rule pathway whereby proteins with destabilising N-termini can be marked for degradation. Both *in vitro* and *in vivo* work proved that Ubr1 can interact with the 19S subunits Rpt1, Rpt6 and Rpn2 [20]. In addition, the *in vivo* experiments revealed that Ubr1 can also interact with Pre6 – a protein of the 20S core of the proteasome. It therefore appears that Ubr1p can bind multiple members of the 19S proteasome and potentially one of the 20S core proteases also.

The UFD (Ubiquitin Fusion Degradation) pathway provides the means to remove those proteins that have a "non-removable" ubiquitin moiety. By "non-removable" it is meant that the ubiquitin moiety shows resistance to the deubiquitinating process. This can be caused by two things, a change in the last residue of the ubiquitin moiety or, the existence, in the substrate, of a proline residue immediately C-terminal to the lysine to which the ubiquitin is attached.

Ufd4p is the E3 for this particular pathway in *S. cerevisiae*. It has been found that this E3, despite being from a different pathway and having no significant sequence similarity to Ubr1, also binds Rpt6 (but not any other proteasome subunit examined *i.e.* 9 other subunits of the 19S proteasome) both *in vitro* and *in vivo*. In the *in vivo* experiment it was shown that Ufd4 could co-immunoprecipitate Rpt6 (bearing in mind that both of these fusion proteins were overexpressed). Interestingly it was also shown that it could co-immunoprecipitate Rpn1. This meant that while Ufd4 does not interact directly with Rpn1 (GST-tagged Rpn1 did not pull down Ufd4); Ufd4, through its interaction with Rpt6 is associated with the mature proteasome *in vivo*.

In further studies on the role of Ufd4 and its association with the proteasome it was found that Ufd4 also directly interacts with Rpt4 and that the binding of both

Rpt4 and Rpt6 is dependent on the presence of the 201-residue N-terminal region of Ufd4 [21]. The N-terminal 201 amino acid residues were found to be important such that when they were deleted, Ufd4 (Ufd4$^{\Delta N}$) could no longer bind GST-Rpt4 or GST-Rpt6. This was repeated *in vivo* for the Ufd4 – Rpt6 interaction. In addition, it was shown *in vivo* that its N-terminal region was required for Ufd4 to interact with the 26S proteasome; that is, for Ufd4 to co-immunoprecipitate Pre6.

To examine the effect of this N-terminal region on the function of Ufd4, a series of experiments were performed using the β-galactosidase-based substrate Ub^{V76}-V-βgal [21]. In *ufd4Δ* cells, Ub^{V76}-V-βgal is long-lived and when ubiquitinated, displays only one ubiquitin moiety. When functional Ufd4 is put back in the system both of these phenotypes are rescued; that is, the half-life of Ub^{V76}-V-βgal decreases, and it could be multiubiquitinated. If instead an *ufd4*$^{\Delta N}$ expressing plasmid is transfected into the *ufd4Δ* cells, the degradation of Ub^{V76}-V-βgal remains slow while the ubiquitination effect is rescued. This implies that the loss of the 201 N-terminal residues does not affect the ability of Ufd4 to ubiquitinate its substrates but rather the speed of their degradation. This would seem to imply that the delivery of the substrate to the proteasome is important in the rate of degradation of the substrate and can be carried out by the E3 of that substrate. Therefore one would expect that if the delivery of the ubiquitinated substrate to the proteasome were impaired in some way, overexpression of a protein like Ufd4 could compensate by delivering the substrate to the proteasome. This was found to be true in *cdc48-1* and *rpn10Δ* cells [21]. In both these strains the degradation of Ub^{V76}-V-βgal is impaired due to the loss of the ubiquitin-binding properties of either Cdc48 or Rpn10. However when Ufd4 is overexpressed in these strains, the rate of degradation of Ub^{V76}-V-βgal is strongly increased while overexpression of Ufd4$^{\Delta N}$ does not change the degradation kinetics in these mutant strains.

An E3 enzyme in mammalian cells – KIAA10 – also appears to interact with the proteasome. When KIAA10 was being purified as an E3 specific to erythroid cells, contaminating proteins such as S1 (Rpn2) and S2 (Rpn1) were also co-purified [22]. Given that Ufd4 interacts with the proteasome – it was investigated if this was also true for KIAA10. This was shown to be the case *in vitro* using a GST-binding assay. Additionally it was found that GST-tagged KIAA10 could interact with the intact 19S RP by western blotting for S8/p48/Rpt6. These results were confirmed *in vivo* where KIAA10 was co-immunoprecipitated with proteasomes, using both anti-S8 and anti-S10a antibodies. As regards the region of KIAA10 that facilitates this interaction, the results are unclear. While the loss of the first 132 amino acids reduces KIAA10 – S2 binding, it does not ablate it and the *in vivo* results for KIAA10 without 132 residues of its N-terminus were inconclusive. This suggests that while there is an S2-binding site within the first 132 amino acids of KIAA10, it may not be the only point of contact between KIAA10 and the proteasome. The budding yeast orthologue of KIAA10 is Hul5, and there is also evidence that Hul5 interacts with the proteasome in *S. cerevisiae* [2].

Loss-of-function mutations in the gene encoding the Parkin protein, are implicated in causing a form of autosomal recessive juvenile parkinsonism in humans [23]. It was subsequently found that Parkin was an E3 ubiquitin ligase [24] and that this E3 ligase contains a UBL domain which binds the 26S proteasome subunit Rpn10 in mammals [25]. There are now a growing number of Parkin substrates including α-synuclein [26] and poly-glutamine proteins [27] that suggest that Parkin is required to target potentially toxic proteins for degradation. Its location at the proteasome therefore, would appear to facilitate the efficient disposal of these potentially harmful proteins.

Other E3 ligases shown to be associated with the proteasome include SCF (Skp1/Cullin/F-box) and APC (Amphase Promoting Complex) [28]. Here, the Cdc4 subunit of SCF, tagged with a polyoma epitope, was able to interact with purified 26S proteasomes (in the presence of ATP). This was shown by the presence of Rpt1 and Rpt6 on western blots. Similarly it was shown that epitope-tagged APC could be co-immunoprecipitated with Rpt1.

8.3.3
Deubiquitinating Enzymes (DUBs)

One of the most important classes of proteins that associate with the proteasome is the deubiquitinating enzymes (DUBs). Important, because their function is integral to that of the proteasome. In order to degrade substrates efficiently the ubiquitin chain must be removed from that substrate. There are two good reasons for this to occur. Firstly, ubiquitin chains are highly thermodynamically stable and so their unfolding and degradation along with the protein to which they are attached takes a large effort. Secondly, it is more efficient for cells to recycle the ubiquitin moieties rather than to constantly translate and degrade them. In order to accomplish this, it makes sense for the deubiquitinating activity to be situated at the proteasome so that ubiquitin is released by that activity, while ensuring that the substrate gets degraded.

There are four DUBs known to interact with the proteasome; Pad1/Rpn11, Ubp6, Uch2/UCH7 and Ubp4/Doa4.

Work on DUBs revealed a "cryptic" deubiquitinating activity was associated with the proteasome in *S. cerevisiae* [29] and mammals [30]. The importance of removal of ubiquitin before degradation was illustrated though the *in vitro* use of ubiquitin, mutated to be irremovable from an ovomucoid moiety (Ub^{m}-OM) by bovine 26S proteasomes. When the degradation of this construct and that of removable Ub-OM was compared, it was found that the rate of degradation was reduced when ubiquitin was mutated and that the non-removable ubiquitin was degraded along with the substrate. Removal of ubiquitin was therefore important to allow efficient degradation of Ub-OM. It is worth noting that similar results were obtained using pentaubiquitin chains attached to OM. This implied the presence of a DUB that removed ubiquitin to promote efficient degradation. This DUB was also unusual in that it was resistant to Ub-aldehyde, a chemical that inhibits the majority of DUBs which are cysteine proteases. Therefore this DUB was not a cysteine protease. In addition this deubiquitinating activity actually promoted degradation rather than inhibiting it, a surprising result since one would imagine that the removal of the ubiquitin moiety/chain from a substrate would actually stabilise a substrate, as it would no longer be targeted for degradation. However if the deubiquitinating activity takes place at the proteasome, then the removal of ubiquitin can promote degradation, as the ubiquitin chain is highly thermodynamically stable and difficult to unfold and degrade with the substrate.

In addition to its Ub-aldehyde resistance, the proteasome-associated deubiquitinating activity was also dependent on ATP. However this dependence was only seen when the ubiquitinated substrate was incubated with 26S proteasomes. When the substrate was incubated with isolated 19S complex, deubiquitination occurred in the absence of ATP. This implied the ATP was not required for the removal of Ub but potentially for coupling the deubiquitination to downstream degradation by the ATP-dependent proteolysis of the 20S CP. The best candidate for this activity was POH1 (Rpn11 in budding yeast, Pad1 in fission yeast). POH1 is the most highly conserved 19S subunit potentially due to the presence of a catalytic domain.

In addressing the role of Rpn11 in budding yeast – the active site residues were identified and mutated [30]. This resulted in a lethal phenotype and when expression of Rpn11 was decreased there was an increase in ubiquitin conjugates. The identity of the active site residues suggested Rpn11 could be a zinc metalloprotease and when a zinc chelator was incubated with the bovine 19S proteasome – there was no deubiquitination of Ub-OM. However Rpn11 alone *in vitro* could not be shown to have deubiquitinating activity suggesting that perhaps only in the context of the 19S RP does Rpn11 have its deubiquitinating activity [30].

In *S. cerevisiae*, similar experiments were performed using Sic1-Ub as a test substrate [29]. Again it was shown that the proteasome-associated deubiquitinating activity was insensitive to Ub-aldehyde and required ATP. They also independently identified Rpn11 as the best candidate subunit and they characterised the active site of Rpn11 as a JAMM domain (Jab1/Pad1/MPN).

The JAMM domain of Rpn11 has also been examined in humans and *Drosophila* where its orthologue is S13 [31]. It was found that the *Drosophila* and human S13s are functional homologues, and also that if the JAMM domain is mutated, there is a loss of deubiquitinating activity. Another motif, similar to the cysteine box of other ubiquitin hydrolases, was also described in the JAMM domain-containing S13 and Csn5.

Ubp6 was first purified and characterised in 1997 by Park *et al.* [32] as a 58kD protein. Ubp6 proved to be sensitive to ubiquitin aldehyde and iodoacetamide suggesting it too is a cysteine protease. Its ability to hydrolyse Ub-αNH extensions and release free ubiquitin from poly-Ub-εNH protein conjugates confirmed its role as a deubiquitinating enzyme (DUB) [32].

It was initially suspected that Ubp6 could bind the proteasome given that it contained a UBL (UBiquitin Like) domain. This domain was known to be involved in binding Rad23 to the proteasome [33] and so when it was discovered in Ubp6 [34] a similar location was envisaged. It was also shown that the UBL domain was not required for Ubp6's deubiquitinating activity *in vitro*, implying that this domain has no effect on the catalytic site of the enzyme.

Association of Ubp6 with the proteasome was proven through the use of mass spectrometry to analyse affinity purified proteasomes [28]. This technique had been used successfully in the past to identify the protein subunits of the ribosome and here it was employed to examine some of the proteins that associate most closely with the proteasome. One of these proteins was Ubp6. To confirm this finding, epitope-tagged Ubp6 was subsequently shown to co-immunoprecipitate subunits of the 19S proteasome such as Rpt1, Rpt6 and Rpn10.

A more in-depth study of this relationship between Ubp6 and the proteasome was subsequently carried out in mammalian cells (where the Ubp6 orthologue is USP14). A C-terminally modified ubiquitin derivative, ubiquitin vinyl sulphone (UbVS) was used to irreversibly label those DUBs that are cysteine proteases and in doing so, block the active site and repress the activity of these proteases [35]. This allowed an examination of activity of USP14 at the proteasome as well as confirming its association with it. [I^{125}] UbVS-labelled USP14 was detected in immunoprecipitated samples of the 26S proteasome, but not in fractions containing the 20S core particle alone. This suggested that USP14 associates only with mature 26S proteasomes, potentially via the 19S RP. In addressing some of the characteristics of USP14's deubiquitinating activity and how this is influenced by its association with the proteasome, it was found that upon inhibition of the proteasome by NLVS there was a 15-fold increase in USP14 active site labelling. Because UbVS labels the active site of cysteine proteases, an increase in labelling corresponds to an increase in activity and so this implies that when the proteasome is prevented from degrading proteins there is an increase in the deubiquitinating activity of USP14. This is not due to the increase in *de novo* synthesis of USP14 as a similar increase in activity was seen in cells where translation was inhibited by puromycin. It could be due to an increase in the recruitment of USP14 to the proteasome upon inhibition of proteolysis but for the fact that only a small amount of USP14 appears to be soluble – not enough for the associated increase in activity. It therefore seems that the activity of the USP14 present at the proteasome is enhanced upon inhibition of the proteasome. How and why this occurs is unclear. Does the inhibition of the proteasome change the physical conformation of the proteasome in the vicinity of USP14, thus allowing substrates increased access to its active site? Does this allow for the "unclogging" of proteasomes – that is releasing potential proteasome substrates when the CP pore is blocked? It is worth noting here that this active site labelling also picks up another proteasome-associated DUB – UCH37, but that the activity of this enzyme is not affected by proteasome inhibition. This seems to suggest that this effect is inextricably bound up with the function of USP14 at the proteasome – a function not provided by other DUBs. It is known, however that the Ubp6 null mutant in *S. cerevisiae* is viable [34] (but sensitive to canavanine) implying that the budding yeast cell can compensate for the loss of function of USP14/Ubp6.

Regarding the Ubp6 null mutant's sensitivity to canavanine, it was found that this sensitivity was rescued by the overexpression of free ubiquitin [2]. This would seem to imply that the toxicity of canavanine is due to the depletion of free pools of ubiquitin – that is not enough ubiquitin is released from substrates by Ubp6 at the proteasome to replenish free pools of ubiquitin. This was corroborated by an experiment that showed if cycloheximide was added to cells to prevent synthesis of new ubiquitin, ubiquitin in Ubp6 null mutants was unstable over time, that is the lack of a DUB to remove ubiquitin from proteasome substrates resulted in the degradation of ubiquitin as well as their substrates [36].

Further analysis showed that Ubp6 bound the proteasome preferentially at the base of the 19S proteasome rather than the lid or 20S CP and that this binding was indeed mediated by its UBL domain [2]. Upon examination of the binding of Ubp6 to the proteasome it was found that while the UBL domain was necessary and sufficient for binding to the 19S base, binding to the lid required the presence of the catalytic site. While binding to the 19S RP stimulated the activity of Ubp6, binding to the base alone did not stimulate Ubp6 to the same extent. This suggests that although it binds the base via its UBL domain, the presence of the lid of the 19S RP is required for maximal activity. Upon closer examination it was found that the UBL domain bound Rpn1 [2]. That Ubp6's activity is important to the proteasome *in vivo* was seen when the hydrolysing activity of Ub-AMC was examined. While wild type proteasomes exhibit high levels of this activity, those from Ubp6 null cells do not, suggesting that most of this activity is attributable to the presence of Ubp6. It is also worth noting that levels of Ubp6 are similar to that of integral 19S subunits, again highlighting its importance *in vivo*.

However, while this work on Ubp6 in *S. cerevisiae* suggested that it was the main DUB at the proteasome [2], it is important to note that budding yeast do not have an orthologue of Uch2/UCH37, another proteasome-associated DUB which is found in *S. pombe* and mammals. Therefore a study of the relative importance of Ubp6 and Uch2 in *S. pombe* may be a more valid model for what occurs in mammals. A study was carried out to examine this by looking at *ubp6* and *uch2* null mutants and by using the same ubiquitin-AMC assay employed in budding yeast [37].

Neither *ubp6*Δ nor *uch2*Δ mutants were lethal, but the *ubp6*Δ mutant did exhibit synthetic lethality with *mts1* (*rpn9*), *mts2* (*rpt2*) or *mts3* (*rpn12*) temperature sensitive mutants, but not with *mts4* (*rpn1*), *mts8* (*β1*), *pad1* (*rpn11*) or *pus1* (*rpn10*) mutants. Further studies showed that this synthetic lethality was not in fact due to Ubp6's role as a DUB, but rather to its role in binding the 19S RP. Ubp6 without a UBL domain showed the synthetic lethality, while Ubp6 without a catalytic site did not. It is worth noting that the $rpn11^{D122A}$ and *ubp6*Δ are synthetically lethal in *S. cerevisiae* [38] suggesting that Ubp6 plays a different role in budding yeast as compared to fission yeast.

This difference was seen again when the deubiquitinating activity associated with the 26S proteasome was examined. Here, using the same assay that showed the importance of Ubp6 at the budding yeast proteasome [2], it was found that the main deubiquitinating activity at the fission yeast proteasome was Uch2 rather than Ubp6. It would appear therefore that with the presence of Uch2 in *S. pombe* cells, and by extension, UCH37 in mammalian cells, the importance of Ubp6 is far less than that seen in the Uch2-less *S. cerevisiae*. This has implications in the use of the budding yeast as a model for deubiquitination at the mammalian proteasome.

UCH37 (Uch2) was first recognised as an isopeptidase activity associated with the 19S of bovine proteasomes that disassembled polyubiquitin by "chewing" off the ubiquitins one by one [39]. This activity was identified and explored further when it was described as Uch2 (Ubiquitin Carboxy-terminal Hydrolase) in fission yeast [40]. This was the *S. pombe* orthologue of mammalian UCH-L5 (mouse) / UCH37 (human). Both of these UCH DUBs had the UCH domain along with a C-terminal extension unlike other UCH proteins. The sub-cellular localisation of this protein was examined and it exhibited a perinuclear localisation during interphase and mitosis. However if the C-terminal extension was removed, Uch2 exhibited a more diffuse cell-wide location. From this evidence the authors decided to examine whether Uch2 was associated with the 26S proteasome, and if this was accomplished via the C-terminal domain. This proved to be true as Uch2 was co-immunoprecipitated with the 26S proteasome, while Uch2 lacking the C-terminal domain was not. Also upon the analysis of fractions of glycerol gradient centrifugation, Uch2 appeared to be closely associated with the proteasome and perhaps a 19S subunit [40]. Subsequent work in *Drosophila* identified the Uch2 orthologue, p37A as a subunit of the 19S RP by 2D gel electrophoresis and protein sequencing [41].

Another DUB thought to associate, albeit weakly, with the 26S proteasome is Doa4/Ubp4 [42]. Here the particular domain interacting with the proteasome is suggested to be the N-terminal 310 residues of the protein. Interestingly, *S. pombe* does not have an obvious orthologue of Doa4.

8.4 Shuttling Proteins: Rpn10/Pus1 and UBA-UBL Proteins

One of the major fields of interest regarding proteasome-interacting proteins is how the ubiquitination system and proteasome degradation system are connected or how are ubiquitinated proteins delivered to and recognised by the proteasome. One of the ways this has been studied is by looking for ubiquitin-chain interacting proteins and determining if these interact with the proteasome also.

One of the first proteins to be found to bind ubiquitin chains was, unsurprisingly, a proteasome subunit itself, the 19S RP's Rpn10 (*S. cerevisiae*) / Pus1 (*S. pombe*) / S5a (mammals). It was found that Rpn10 in *S. cerevisiae* bound ubiquitin with a preference for longer chains but that *rpn10*Δ mutants were viable [43]. It was known that when the ubiquitin-proteasome system is not functioning, for example in the case of other proteasome subunit mutants, cells are no longer viable [44]. Therefore the fact that the *rpn10*Δ mutant was viable implied that if Rpn10's role was to recognise ubiquitinated substrates, it was not the only protein to carry out that function. It was also found that while Rpn10 was a proteasome subunit, it could also exist as a slower sedimenting species indicating that a fraction of Rpn10 exists free of the proteasome [43].

Later a motif essential for multiubiquitin binding was found in Rpn10. This was a stretch of conserved hydrophobic amino acids in the C-terminal half of the protein – LAM/LALRL/V [45] – later described as a UIM (Ubiquitin Interacting Motif) domain [46]. However this motif was not required for Ub-Pro-β-gal degradation nor did its loss affect sensitivity to canavanine. A similar result was found for the *S. pombe* orthologue, Pus1 [47]. Genetic interactions between *pus1*$^{+}$ and other 19S subunits, in particular *mts3*$^{+}$ (*RPN12* in *S. cerevisiae*), were also found. Overexpression of Pus1 could rescue the temperature sensitive mutant *mts3-1* at 32 °C, while *pus1Δmts3-1* was synthetically lethal at the permissive temperature. However Pus1 could not rescue the *mts3* null mutant, which is lethal. This suggests that while Pus1 and Mts3 may interact *in vivo*, their functions were not identical. These two proteins were also shown to interact *in vitro*. However when the ubiquitin-binding motif was altered, then Pus1 could not rescue *pus1Δmts3-1* or *mts3-1*, despite its still being able to bind Mts3. Interactions were also described between *pus1* and *mts4-1* (*rpn1*) and *pus1* and *pad1-1* (*rpn11*) although in this case synthetic lethality was rescued equally well by Pus1 with or without its LAMAL motif.

It was subsequently found that the DNA repair protein Rad23 (*S. cerevisiae*) could bind the proteasome through another domain – the UBL domain (Ubiquitin-Like) [33]. This domain is similar to the amino acid sequence of ubiquitin itself and had been known for sometime, as had its presence in Rad23. However its function up until then was unclear. Initially it was thought that given its similarity to ubiquitin, it targeted the protein containing it for rapid degradation [48]. However Rad23 had been found to be a highly stable protein, despite its possession of a UBL domain [49]. The significance of the UBL domain was not known therefore until it was shown that it was the means by which Rad23 could bind the proteasome [33]. There was also evidence to suggest that Rpn10 (Pus1) and Rad23 (Rhp23) played overlapping roles as the double null in budding [50] and fission [51] yeast exhibited cold sensitivity, canavanine sensitivity, slow growth and a G2/M phase delay – a more severe phenotype that either mutation alone. The increased sensitivity to the arginine analogue canavanine of *rad23Δrpn10Δ* mutants suggested a defect in the ubiquitin-proteasome pathway. The same phenotype was seen in cells lacking Rpn10 and the UBL domain of Rad23. In addition, accumulation of multiubiquitinated substrates in the double mutant indicated a proteasome deficiency [50].

Other UBL-containing proteins were found to bind proteasomes in human cell lines [52]. These proteins, hPLIC1 and 2, are homologous to the *S. cerevisiae* Dsk2 and *S. pombe* Dph1. Both could immunoprecipitate elements of the proteasome although it appeared hPLIC2 associated with a subset of proteasomes bound to the cytoskeleton as well as some free proteasomes [52]. In *S. cerevisiae* a double knockout of *dsk2* and *rad23* display a G2/M cell cycle arrest [53]. This suggested that the two UBL domain-containing proteins encoded by these genes have overlapping functions.

Rad23 and Dsk2 have another domain in common – the UBA domain. The first description of a UBA (UBiquitin Associated) domain was as a theoretical motif found in several known E2s, E3s and DUBs. The function of the UBA domain was first suggested by Berolaet *et al.* when they provided evidence that Rad23 and Ddi1 in *S. cerevisiae* interacted with ubiquitin *in vitro* and *in vivo*, and that this interaction required the presence of their UBA domains [54].

The function of Rad23, Dsk2 and how these proteins overlap with Rpn10's role in proteasome-mediated degradation was illustrated conclusively in *S. pombe*. It was shown that Rhp23 (Rad23) and Dph1 (Dsk2) could bind multiubiquitin chains (with far greater affinity than monoubiquitin) through their UBA domains and the 26S proteasome through their UBL domain *in vivo*. It was also found that the two Rhp23 UBA domains were not equivalent in their capacity to bind multiubiquitin, with the central UBA domain more potent than the C-terminal one. Given that these two proteins, Rhp23 and Dph1, were now known to bind ubiquitin chains and the proteasome, similar to Pus1 (Rpn10) – the triple deletion was constructed and found to be lethal. It should be noted here that the triple deletion in *S. cerevisiae* is viable but exhibited a severe slowed growth phenotype and a large accumulation of polyubiquitinated proteins [55]. This is probably due to the presence in *S. cerevisiae* of Ddi1, a UBL-UBA domain protein not seen in *S. pombe*. This would suggest that these three proteins perform overlapping functions that are vital to fission yeast cells. The next most severe phenotype was found in *pus1Δrhp23Δ* cells which had a severe growth defect at 25 °C and were not viable at 36 °C. It was found that, while Rhp23 could rescue this phenotype, Rhp23 without either its UBA domain or its UBL domain was unable to do so. (A similar result was obtained later in *S. cerevisiae* [56].) It could therefore be suggested that multiubiquitinated proteins can be delivered to the proteasome by these "shuttling proteins" – Pus1 (Rpn10), Rhp23 (Rad23) or Dph1 (Dsk2), and that the loss of this delivery system is lethal to the cell.

This model of substrate delivery has been confirmed and expanded in *S. cerevisiae* where it was found that cells lacking both Rad23 and Dsk2 were deficient in protein degradation mediated by the UFD pathway and that mutation of the UBA domain of Dsk2 resulted in cells deficient in proteolysis [57]. Also the two UBA domains of Rad23 appear to act synergistically [56]. Later Ddi1 was added to the growing list of UBA-UBL proteins that could interact with ubiquitin chains and the proteasome [55].

To elucidate the pathway further, experiments were then undertaken in *S. cerevisiae* to find where exactly on the proteasome the UBL domain of Rad23 binds. The answer appears to be Rpn1, a subunit of the base complex of the 19S RP. It was found that Rpn1 specifically recognised the UBL domain through an N-terminal region of Leucine Rich Repeat-like or LRR-like repeats with a short adjacent sequence, and that Rad23 and Dsk2 competed with each other to bind Rpn1 at the same site [58]. In *S. pombe*, it was shown that Rhp23 (Rad23) could bind Mts4 (Rpn1) with its UBL domain binding a region of Mts4 between amino acid 181 and 407 [59]. It was also shown that this domain of Mts4 could only be found in Mts4 orthologues and that another UBL-containing protein Udp7 (SPCC1442.07c) could bind this domain via its UBL domain. It therefore appeared that this region is a UBL-binding domain in the proteasome. This work also showed that Pus1 (Rpn10) could also bind Mts4 (Rpn1) but not in the same region as the UBL proteins. Pus1 bound a region of Mts4 between amino acid 408 and 582 termed the PC-repeat domain [59].

Subsequent work in human cell lines included the discovery of a hHR23 (Rad23) "client" protein *in vivo* – p53 [60]. Here evidence showed that hHR23 binds a polyubiquitinated p53 via its UBA domain protecting it from deubiquitination *in vitro* and *in vivo*, with downregulation of hHR23 resulting in accumulation of p53.

More recently a cell-free system in *S. cerevisiae* has been used to examine the role of UBA-UBL proteins in substrate collection [61]. The biochemical evidence obtained here supports the genetic evidence in yeast for the shuttle protein hypothesis [51]. Wild type proteasomes (affinity purified) degraded the test-substrate (ubiquitinated-Sic1) efficiently, while those from *rpn10Δ* and *rad23Δ* did not. The proteasomes from the mutant strains were also deficient in deubiquitination. However if recombinant Rad23 were added to the Rad23-deficient proteasomes, efficient degradation and deubiquitination were restored. This effect was dosage dependent; at low concentrations recombinant Rad23 restored wild type degradation and deubiquitination, while at higher concentrations it inhibited these processes.

Further work has shown that the UBL domain of Rad23 can also bind Ufd2, an E4 enzyme and that mutation of the UBL domain alters this interaction and impairs the UFD proteolytic pathway [62]. This is also true for Dsk2, but not Ddi1, suggesting that Rpn1 and Ufd2 compete for the binding of Rad23 via its UBL domain. These findings raise the possibility that the UBL is not strictly a proteasome-binding domain but serves other functions in Rad23 and other UBL-carrying proteins. Also it is possible that the binding of an E4 like Ufd2 could enhance Rad23's shuttling function by placing it in the vicinity of the protein being ubiquitinated through binding the E4 via its UBL domain.

There is also evidence to suggest that Cdc48 along with its co-factors Npl4 and Ufd1 (see section 8.6) plays a role in recruiting Ufd2 to the oligoubiquitinated substrate allowing the polyubiquitination which will in turn recruit shuttle proteins like Rad23 and Dsk2 [63].

8.5 Other UBL-Containing Proteins

There are other UBL-containing proteins without UBA domains, therefore not shuttling proteins, that have been shown to interact with the proteasome. Bag1 (Bcl2-Associated athanoGene) is well conserved in eukaryotes except for the budding yeast orthologue which has no UBL domain. In mammals it had been known for some time that BAG1 could act as a cofactor of HSC70 (constitutive) and HSP70 (heat-induced) chaperone proteins [64], when it was found that its UBL domain did indeed allow it to interact with the 26S proteasome. It was also found that this association occurred in an ATP-dependent manner, and that it promoted the binding of Hsc70 and Hsp70 to the proteasome [65]. Bag1 has also been shown to stimulate the release of substrates from Hsp70 suggesting that perhaps if a protein cannot be correctly refolded by Hsp70, Bag1 functions to promote that protein's degradation.

Bag1 has also been shown to associate with an E3 ubiquitin ligase CHIP [66] (Carboxyl terminus of Hsc70-Interacting Protein) along with Hsc70 and Hsp70 and in fact that it is ubiquitinated by CHIP itself. However this ubiquitin chain does not target Bag1 for degradation, but rather promotes the association of Bag1 with the proteasome [67, 68]. Therefore, both ubiquitinated and non-ubiquitinated Bag1 can interact with the proteasome.

Nub1 (Nedd8 Ultimate Buster) was found originally as an inhibitor of Nedd8 expression at the translation level [69] before subsequently being found to be a UBL domain-containing protein that interacts with Rpn10 in the proteasome [70]. It was also shown that Nub1 interacts with Nedd8-conjugated proteins and promotes their proteolysis by the 26 proteasome.

8.6 VCP/p97/cdc48

VCP (Valosin-Containing Protein)/p97/Cdc48 is an abundant ATPase of the AAA family of proteins. Since its initial discovery in budding yeast as a cell cycle mutant [71], it has been shown to play a role in a wide variety of cellular processes. Here we are concerned with its function as a proteasome interactor and the effect this has on the cell. In budding yeast it was found that Cdc48p is necessary for the degradation of a ubiquitin-fusion reporter protein [72] and this was shortly followed by a study in mammalian cells that described how it associates with ubiquitinated IκBα and the proteasome thereby targeting IκBα for degradation [73]. It was found that VCP interacts preferentially with polyubiquitinated IκBα, and that this interaction was necessary for the degradation of IκBα. Ubiquitinated forms of IκBα were stabilised in the absence of VCP. In addition, VCP could co-immunoprecipitate subunits of the 26S proteasome in an ATP-dependent manner.

It later became clear that Cdc48 seems to work with various adaptor proteins that aid in "targeting" its ATPase activity toward different cellular functions. It binds to p47 (Shp1/Ubx3) to carry out a Golgi membrane fusion role [74], and it forms a complex with Ufd1 and Npl4, which can both bind ubiquitin chains. This Cdc48-Ufd1-Npl4 complex performs a role in ubiquitin-mediated proteolysis at the endoplasmic reticulum, a process known as ERAD (Endoplasmic Reticulum Associated Degradation) [75]. Cdc48 uses its ATPase activity to physically remove ubiquitinated proteins from complexes for transport to the proteasome [76]. Specifically it has been shown that Cdc48-Ufd1-Npl4 is required to relocate ubiquitinated substrates (preferentially polyubiquitinated chains) from the ER into the cytosol for ubiquitin-mediated degradation [77].

Subsequently however, it became clear that p47/Shp1/Ubx3 can also bind ubiquitin, especially monoubiquitin, via a UBA domain and that this domain is essential for the role of Cdc48-p47 in membrane fusion [78]. It was therefore suggested that perhaps while Ufd1-Npl4 bound Cdc48 performs a role in ERAD by binding multiubiquitinated proteins and targeting them for degradation, p47-Cdc48 perhaps binds monoubiquitin to some other end. p47 (Shp1 in *S. cerevisiae* Ubx3 in *S. pombe*) contains a UBX domain (Ubiquitin regulatory X) a well-conserved domain believed to mediate Cdc48 binding [79]. It was established that there are seven family members and that all seven can bind Cdc48. In addition, *shp1* and *ubx2* null strains show defects in the degradation of a test substrate [80], although not in that of an ERAD substrate [81]. It has also been shown that *ubx3* null mutants in *S. pombe* are heat and canavanine sensitive and display synthetic lethality with *pus1* (*rpn10*) null mutants [81]. This suggests that p47/Shp1/Ubx3 function in ubiquitin-mediated proteolysis as well as membrane fusion, and that it serves an overlapping function with Pus1 (Rpn10).

Recently, new evidence has been found to support a novel model of Cdc48's role in ubiquitin-mediated proteolysis [63]. Richly and co-workers found that Cdc48 is required for the E4 enzyme Ufd2 to bind ubiquitin chains, and in mutants lacking Cdc48 or its co-factors, Rad23 (or Dsk2) is no longer able to bind Ufd2. This suggests that not only does Cdc48 facilitate Ufd2 binding to oligoubiquitinated substrates but it is also involved in recruiting Rad23 to substrates for degradation. It had been found previously, in a screen for ERAD-defective mutants, that Rad23 and Dsk2 were important in this process [82]. This implies a model whereby a substrate has a chain of one or two ubiquitin molecules added by its E1, E2 and E3 enzymes which in turn allows the binding of Cdc48 via its co-factors Npl4 and Ufd2. This recruits the E4 (such as Ufd2) to increase the length of the multiubiquitin chain which in turn allows the binding of UBL-UBA shuttle proteins (such as Rad23) which transfer the ubiquitinated substrate to the proteasome [63].

8.7
Proteasome Interactions with Transcription, Translation and DNA Repair

We have seen that the proteasome associates closely with various components of the ubiquitin system in order to make its recognition of proteins ready for degradation both efficient and accurate. But what of the role it plays in other cellular processes? Does the proteasome interact with proteins in order to connect itself to those processes requiring efficient degradation? Here we look at the proteasome's role in transcription and translation through the interactors that connect these vital cellular processes.

There is some evidence to suggest that the proteasome has a part to play in transcription, in particular that the proteasome can interact with RNA polymerase II. It was shown that some subunits of the 19S regulatory particle could be recruited to the Gal1-10 promoter upon induction of transcription [83]. These subunits included Rpt1-6 which are found in the base of the 19S RP. No other subunits from the lid or the 20S proteasome were found to associate with activated promoters and so the authors termed the subset of 19S subunits that were present at active promoters, the APIS complex – AAA Proteins Independent of the 20S proteasome. However there is no evidence to confirm the existence of such a complex in the cell. This reflected earlier work by the same group, where evidence was presented that the 19S complex is required for efficient elongation of RNA polymerase II [84]. Recent work by the same authors suggests that the 26S proteasome is associated physically with regions of induced genes that correlate with a build-up of RNA polymerase II [85]. These regions include the 3′ end of genes, sites of UV damage and other locations that could represent pauses in elongation. The authors suggest therefore, that the 26S proteasome may be involved in some aspect of transcription termination. However the ubiquitin-proteasome system has been shown to be involved with transcription in more ways than one. These include regulation of chromatin structure and the controlled degradation of transcription activators. (See review [86]). Therefore the relative importance of the observed proteasome–RNA polymerase II interactions are unknown.

The proteasome also has a role to play at the level of translation. eIF3 is a translation initiation factor made up of many subunits. One of these, Sum1 (eIF3i) was found to change its intracellular localisation in *S. pombe* upon stresses such as osmotic and heat shock [87]. Following heat shock, Sum1 relocated to the 26S proteasome at the nuclear rim; this localisation of Sum1 was also shown to be dependent upon Cut8, the protein believed to localise the proteasome to the nuclear periphery. In temperature sensitive mutants of proteasomal subunits such as *mts4-1*, *mts2-1* or *pad1-1*, Sum1 no longer goes to the nuclear periphery upon heat shock. This implies that the relocation observed depends on a fully-functioning proteasome. In addition when Sum1 is over-expressed in *mts4-1* cells, they exhibit an elongated phenotype consistent with cell cycle arrest. In fact Mts4 and Sum1 physically interact *in vivo* [87].

In a similar vein, the mammalian eIF3 subunit eIF3e or Int6 was found to interact with Rpt4 in the yeast two hybrid system and *in vivo* by co-immunoprecipitation [88]. Int6 also co-immunoprecipitated the 20S proteasome subunit HC3 under conditions that allowed for an intact 26S proteasome suggesting that Int6 associates with the mature 26S proteasome.

This relationship between eIF3e/Int6 and the 26S proteasome was confirmed in fission yeast [89]. Here it was noticed that mutants of Yin6, the *S. pombe* orthologue of eIF3e/Int6, exhibit a similar phenotype to proteasome subunit mutants. They are sensitive to canavanine, they accumulate ubiquitinated proteins and they can increase the severity of proteasome subunit mutants. This showed that not only does Yin6 associate with the proteasome, but that it has an important role to play in the degradation of ubiquitinated protein, given that its loss had an effect on proteasome function. In addition *yin6* null mutants exhibited a cell cycle defect in that they had abnormally long mitosis and inefficient chromosome segregation.

Closer examination of the *yin6* null mutant uncovered a particular effect upon the 19S lid subunit Rpn5. It was found that while the *yin6* null mutant exacerbated proteasome subunit mutants, it did not affect *rpn5* mutants, suggesting their presence in the same pathway. The localisation of Rpn5 was also affected in the *yin6* null mutant in that it localised to the cytoplasm rather than at the nuclear periphery while Yin6 location was unaffected in the *rpn5* mutant. The overexpression of Rpn5 also partially rescued the *yin6* null phenotype but not vice versa. This seems to suggest that Yin6 functions upstream of Rpn5, and also that it may be required to target Rpn5 to the 19S lid. It is worth noting that Rpn5 lacks a nuclear localisation signal and so it is likely that it enters the nucleus as part of a complex. Yin6 has a nuclear export signal and a nuclear localisation signal. In fact, the whole proteasome appears misassembled in *yin6* null cells and the authors report they have found some evidence that this is the case in *rpn5* null cells too. Whether the proteasome needs all its subunits to be correctly assembled or Rpn5 plays a particularly important role in proteasome assembly is unclear.

It is also worth noting that a *ras1* null mutant exacerbates the phenotypes of the *yin6* null mutant, while overexpression of Ras1 can rescue these phenotypes. This suggests a pathway whereby Ras1 can affect the proteasome via the interaction of Yin6 and Rpn5.

One of the more interesting aspects of this interaction between Yin6/Int6/eIF3e and the proteasome is the possibility of the crosstalk between the translation apparatus and proteolytic degradation apparatus via this interaction. Conceivably, translation and degradation need to be linked as a proofreading method of degrading incorrectly folded newly-translated proteins. This interaction could also be required as a means of co-ordinating the levels of certain critical proteins by the translation machinery receiving information from the degradation machinery and/or vice versa.

This relationship could also include crosstalk with the COP9 signalosome as there is evidence that these three complexes – eIF3, COP9 and the lid component of the 26S proteasome have similar structures and may have evolved in a similar fashion [90].

While we have discussed a shuttling role for Rad23 with regard to the ubiquitin-proteasome system (see section 8.4), it should be noted that Rad23 also has a vital role to play in DNA repair where it interacts with the nucleotide excision-repair (NER) factor Rad4 forming a dimer that can bind damaged DNA. What impact therefore does Rad23's interaction with the proteasome make on its role in DNA repair? It has been found that when Rad23 lacks a functional UBL domain *S. cerevisiae* cells are more susceptible to UV damage, and that this domain is required for optimal levels of NER [33]. Also Rad23 possesses a Rad4 binding domain that binds and stabilises Rad4 and in itself can rescue NER in *rad23Δ* cells. This action of Rad23 appears to occur entirely separately from its UBL-proteasome mediated function in NER [91]. There are also data to suggest that the Rad23-proteasome role in NER is independent of the 20S proteasome [92], but the mechanism of this role in NER is unknown.

8.8 Concluding Remarks

The proteasome is a simple machine – it degrades proteins to peptides. We have shown here, however that that the role the proteasome plays in the cell is far from simple. The first layer of control of degradation is via the 19S regulatory particle which restricts access to the protease activity of the 20S core particle. The ubiquitin signal introduces another level of regulation upon degradation. In this review we have discussed how the proteins which interact with the proteasome affect its function and direct its activity such that it becomes a finely controlled tool regulating most cell processes. For a summary of these interactions see Figure 8.2.

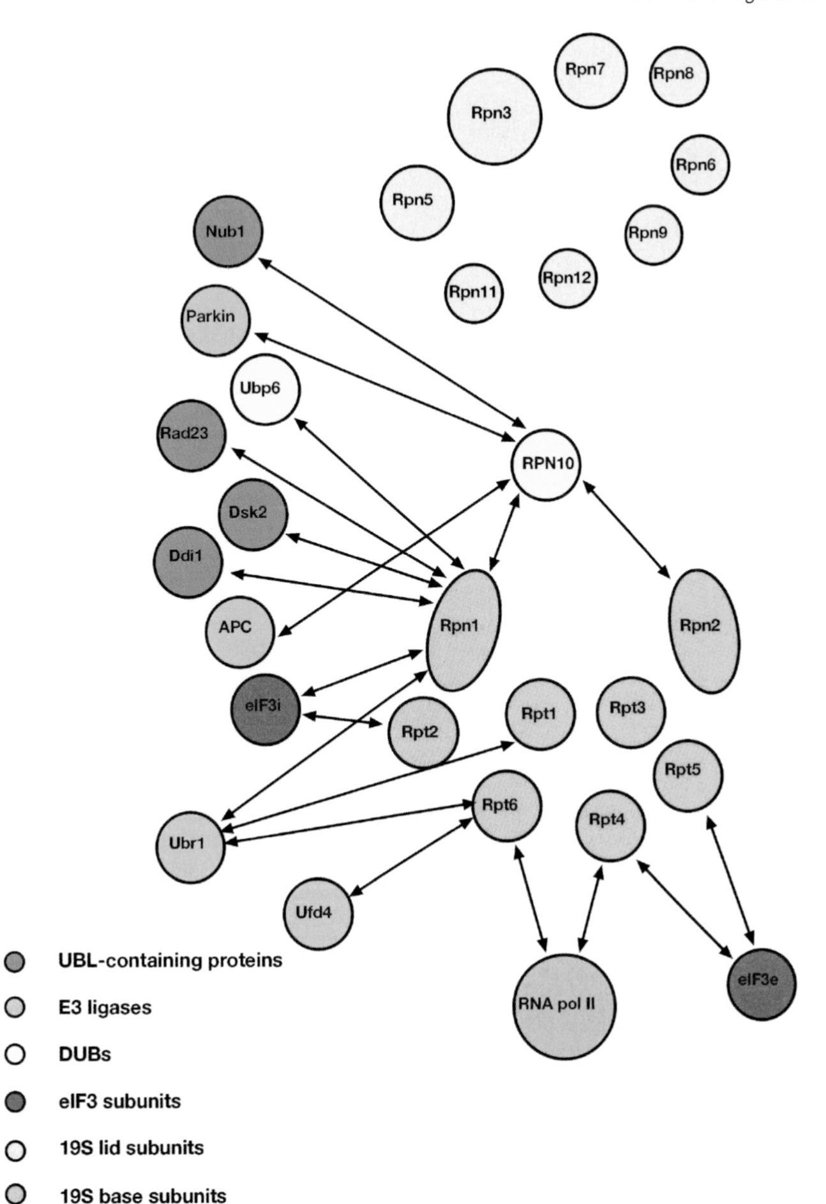

Fig. 8.2. Proteasome-interacting proteins that have been shown to interact with specific subunits of the 19S regulatory particle. Note: There is evidence that a pool of Rpn10 exists free in the cell as well as being a 19S subunit. It also functions as a shuttling protein in a similar fashion to UBA-UBL proteins. Therefore the interactions between it and subunits of the 19S RP are also shown here.

References

1 Ramos, P.C., et al., Ump1p is required for the proper maturation of the 20S proteasome and becomes its substrate upon completion of the assembly. Cell, 1998. 92: p. 489–499.

2 Leggett, D.S., et al., Multiple associated proteins regulate proteasome structure and function. Mol Cell, 2002. 10: p. 495–507.

3 Kajava, A.V., et al., New HEAT-like repeat motifs in proteins regulating proteasome structure and function. J Struct Biol, 2004. 146: p. 425–430.

4 Kajava, A.V., What curves alpha-solenoids? Evidence for an alpha-helical toroid structure of Rpn1 and Rpn2 proteins of the 26S proteasome. J Biol Chem, 2002. 277: p. 49791–49798.

5 Dubiel, W., et al., Purification of an 11S regulator of the multicatalytic protease. J Biol Chem, 1992. 267: p. 22369–22377.

6 Chu-Ping, M., C.A. Slaughter, and G.N. DeMartino, Identification, Purification, and Characterization of a Protein Activator (PA28) of the 20S Proteasome (Macropain). J Biol Chem, 1992. 267(15): p. 10515–10523.

7 Realini, C.A., et al., Characterisation of recombinant REGα, REGβ, and REGγ proteasome activators. J Biol Chem, 1997. 272: p. 25483–25492.

8 Stohwasser, R., et al., Kinetic evidences for facilitation of peptide channelling by the proteasome activator PA28. Eur J Biochem, 2000. 276: p. 6221–6229.

9 Tanahashi, N., et al., Hybrid proteasomes: Induction by interferon-γ and contribution to ATP-dependent proteolysis. J Biol Chem, 2000. 275: p. 14336–14345.

10 Tanaka, K. and M. Kasahara, The MHC class I ligand-generating system: roles of immunoproteasomes and the interferon-gamma-inducible proteasome activator PA28. Immunol Rev, 1998. 163: p. 161–76.

11 Ustrell, V., et al., PA200, a nuclear proteasome activator involved in DNA repair. EMBO J, 2002. 21(13): p. 3516–3525.

12 Moore, C.W., Further characterisation of bleomycin-sensitive (*blm*) mutants of *Saccharomyces cerevisiae* with implications for a radiomimetic model. J Bacteriol, 1991. 173: p. 3605–3608.

13 Ho, Y., et al., Systematic identification of protein complexes in *Saccharomyces cerevisiae* by mass spectrometry. Nature, 2002. 415: p. 180–183.

14 Martin, S.G., et al., Relocalisation of telomeric Ku and SIR proteins in response to DNA strand breaks in yeast. Cell, 1999. 97: p. 621–633.

15 Lecker, S.H., et al., Multiple types of skeletal muscle atrophy involve a common programme of changes in gene expression. FASEB J, 2004. 18: p. 39–51.

16 Fehlker, M., et al., Blm3 is part of nascent proteasomes and is involved in a late stage of nuclear proteasome assembly. EMBO Rep, 2003. 4(10): p. 959–963.

17 Schmidt, M., et al., The HEAT repeat protein Blm10 regulates the yeast proteasome by capping the core particle. Nat Struct Mol Biol, 2005. 12(4): p. 294–303.

18 Tongaonkar, P., et al., Evidence for an interaction between ubiquitin-conjugating enzymes and the 26S proteasome. Mol Cell Biol, 2000. 20(13): p. 4691–4698.

19 Seufert, W. and S. Jentsch, Ubiquitin-conjugating enzymes *UBC4* and *UBC5* mediate selective degradation of short-lived and abnormal proteins. EMBO J, 1990. 9: p. 543–550.

20 Xie, Y. and A. Varshavsky, Physical association of ubiquitin ligases and the 26S proteasome. Proc. Natl. Acad. Sci., 2000. 97(6): p. 2497–2502.

21 Xie, Y. and A. Varshavsky, UFD4 lacking the proteasome-binding region catalyses ubiquitination but is impaired in proteolysis. Nature Cell Biology, 2002. 4: p. 1003–1007.

22 You, J. and C.M. Pickart, A HECT domain E3 enzyme assembles novel polyubiquitin chains. J Biol Chem, 2001. 276(23): p. 19871–19878.

23 Kitada, T., et al., Mutations in the parkin gene cause autosomal recessive juvenile parkinsonism. Nature, 1998. 392: p. 605–608.

24 Shimura, H., et al., Familial Parkinson disease gene product, parkin, is a ubiquitin-protein ligase. Nat Genet, 2000. 25(3): p. 302–305.

25 Sakata, E., et al., Parkin binds the Rpn10 subunit of 26S proteasomes through its ubiquitin-like domain. EMBO Rep, 2003. 4(3): p. 301–306.

26 Shimura, H., et al., Ubiquitination of a new form of alpha-synuclein by parkin from human brain: implications for Parkinson's disease. Science, 2001. 293(5528): p. 263–269.

27 Tsai, Y.C., et al., Parkin facilitates the elimination of expanded polyglutamine proteins and leads to preservation of proteasome function. J Biol Chem, 2003. 278: p. 22044–22055.

28 Verma, R., et al., Proteasomal Proteomics: Identification of Nucleotide Sensitive Proteasome-interacting Proteins by Mass Spectrometric Analysis of Affinity-purified Proteasomes. Mol Cell Biol, 2000. 11: p. 3425–3429.

29 Verma, R., et al., Role of Rpn11 metalloprotease in deubiquitination and degradation by the 26S proteasome. Science, 2002. 298: p. 611–615.

30 Yao, T. and R.E. Cohen, A cryptic protease couple deubiquitination and degradation by the proteasome. Nature, 2002. 419: p. 403–407.

31 Lundgren, J., et al., Use of RNA interference and complementation to study the function of the *Drosophila* and human proteasome subunit S13. Mol Cell Biol, 2003. 23(15): p. 5320–5330.

32 Park, K.C., et al., Purification and characterisation of UBP6, a new ubiquitin-specific protease in *Saccharomyces cerevisiae*. Archives of Biochemistry and Biophysics, 1997. 347(1): p. 78–84.

33 Schauber, C., et al., Rad23 links DNA repair to the ubiquitin/proteasome pathway. Nature, 1998. 391: p. 715–718.

34 Wyndham, A.M., R.T. Baker, and G. Chelvanayagam, The Ubp6 family of deubiquitinating enzymes contains a ubiquitin-like domain: SUb. Protein Sci, 1999. 8: p. 1268–1275.

35 Borodovsky, A., et al., A novel active site-directed probe specific for deubiquitylating enzymes reveals proteasome association of USP14. EMBO J, 2001. 20(18): p. 5187–5196.

36 Hanna, J., D.S. Leggett, and D. Finley, Ubiquitin depletion as a key mediator of toxicity by translational inhibitors. Mol Cell Biol, 2003. 23(24): p. 9251–9261.

37 Stone, M., et al., Uch2/Uch37 is the major deubiquitinating enzyme associated with the 26S proteasome in fission yeast. J Mol Biol, 2004. 344: p. 697–706.

38 Guterman, A. and M.H. Glickmann, Complementary roles for Rpn11 and Ubp6 in deubiquitination and proteolysis by the proteasome. J Biol Chem, 2004. 279(3): p. 1729–1738.

39 Lam, Y.A., et al., Specificity of the ubiquitin isopeptidase in the PA700 regulatory complex of 26S proteasomes. J Biol Chem, 1997. 272(45): p. 28438–28446.

40 Li, T., et al., Identification of a 26S proteasome-associated UCH in fission yeast. Biochem Biophys Res Commun, 2000. 272: p. 270–275.

41 Holzl, H., et al., The regulatory complex of *Drosophila melanogaster* 26S proteasomes: subunit composition and localisation of a deubiquitylating enzyme. J Cell Biol, 2000. 150(1): p. 119–129.

42 Papa, F.R., A.Y. Amerik, and M. Hochstrasser, Interaction of the Doa4 deubiquitinating enzyme with the yeast 26S proteasome. Mol Cell Biol, 1999. 10: p. 741–756.

43 van Nocker, S., et al., The multiubiquitin-chain-binding protein Mcb1 is a component of the 26S proteasome in *Saccharomyces cerevisiae* and plays a nonessential, substrate-

specific role in protein turnover. Mol Cell Biol, 1996. 16(11): p. 6020–6028.
44 Gordon, C., et al., Defective mitosis due to a mutation in the gene for a fission yeast 26S protease subunit. Nature, 1993. 366: p. 355–357.
45 Fu, H., et al., Multiubiquitin chain binding and protein degradation are mediated by distinct domains within the 26S proteasome subunit Mcb1. J Biol Chem, 1998. 273(4): p. 1970–1981.
46 Hofmann, K. and L. Falquet, A ubiquitin-interacting motif conserved in components of the proteasomal and lysosomal protein degradation systems. Trends Biochem Sci, 2001. 26(6): p. 347–350.
47 Wilkinson, C.R.M., et al., Analysis of a Gene Encoding Rpn10 of the Fission Yeast Proteasome Reveals That the Polyubiquitin-binding Site of This Subunit Is Essential When Rpn12/Mts3 Activity Is Compromised. J Biol Chem, 2000. 275(20): p. 15182–15192.
48 Johnson, E.S., et al., Ubiquitin as a degradation signal. EMBO J, 1992. 11(2): p. 497–505.
49 Watkins, J.F., et al., The *Saccharomyces cerevisiae* DNA repair gene RAD23 encodes a nuclear protein containing a ubiquitin-like domain required for biological function. Mol Cell Biol, 1993. 13: p. 7757–7765.
50 Lambertson, D., L. Chen, and K. Madura, Pleiotropic defects caused by loss of the proteasome-interacting factors Rad23 and Rpn10 of Saccharomyces cerevisiae. Genetics, 1999. 153(1): p. 69–79.
51 Wilkinson, C., et al., Proteins containing the UBA domain are able to bind to multi-ubiquitin chains. Nat Cell Biol, 2001. 3: p. 939–943.
52 Kleijnen, M.F., et al., The hPLIC proteins may provide a link between the ubiquitination machinery and the proteasome. Mol Cell, 2000. 6: p. 409–419.
53 Biggins, S., I. Ivanovska, and M.D. Rose, Yeast ubiquitin-like genes are involved in duplication of the microtubule organising centre. J Cell Biol, 1996. 133: p. 1331–1346.
54 Bertolaet, B.L., et al., UBA domains of DNA damage-inducible proteins interact with ubiquitin. Nat Struct Biol, 2001. 8(5): p. 417–422.
55 Saeki, Y., et al., Ubiquitin-like proteins and Rpn10 play co-operative roles in ubiquitin-dependent proteolysis. Biochem Biophys Res Commun, 2002. 293: p. 986–992.
56 Chen, L. and K. Madura, Rad23 promotes the targeting of proteolytic substrates to the proteasome. Mol Cell Biol, 2002. 22(13): p. 4902–4913.
57 Rao, H. and A. Sastry, Recognition of specific ubiquitin conjugates is important for the proteolytic functions of the ubiquitin-associated domain proteins Dsk2 and Rad23. J Biol Chem, 2002. 277(14): p. 11691–11695.
58 Elsasser, S., et al., Proteasome subunit Rpn1 binds ubiquitin-like protein domains. Nat Cell Biol, 2002. 4: p. 725–730.
59 Seeger, M., et al., Interaction of the anaphase-promoting complex/cyclosome and proteasome protein complexes with multiubiquitin chain binding proteins. J Biol Chem, 2003. 278(19): p. 16791–16796.
60 Glockzin, S., et al., Involvement of the DNA repair protein hHR23 in p53 degradation. Mol Cell Biol, 2003. 23(24): p. 8960–8969.
61 Verma, R., et al., Multiubiquitin chain receptors define a layer of substrate selectivity in the ubiquitin-proteasome system. Cell, 2004. 118: p. 99–110.
62 Kim, I., K. Mi, and H. Rao, Multiple interactions of Rad23 suggest a mechanism for ubiquitylated substrate delivery important in proteolysis. Mol Biol Cell, 2004. 15: p. 3357–3365.
63 Richly, H., et al., A series of ubiquitin binding factors connects CDC48/p97 to substrate multi-ubiquitination and proteasomal targeting. Cell, 2005. 120: p. 73–84.
64 Takayama, S., et al., BAG-1 modulates the chaperone activity of Hsp70/Hsc70. EMBO J, 1997. 16(16): p. 4887–96.
65 Luders, J., J. Demand, and J.

Hohfeld, The ubiquitin-related BAG-1 provides a link between the molecular chaperones Hsc70/Hsp70 and the proteasome. J Biol Chem, 2000. 275(7): p. 4613–4617.

66 Demand, J., et al., Cooperation of a ubiquitin domain protein and an E3 ubiquitin ligase during chaperone/proteasome coupling. Curr Biol, 2001. 11: p. 1569–1577.

67 Alberti, S., et al., Ubiquitylation of BAG-1 suggest a novel regulatory mechanism during the sorting of chaperone substrates. J Biol Chem, 2002. 277(48): p. 45920–45927.

68 Alberti, S., et al., Correction to "Ubiquitylation of BAG-1 suggests a novel regulatory mechanism during the sorting of chaperone substrates to the proteasome". J Biol Chem, 2003. 278(20): p. 18702–18703.

69 Kito, K., E.T.H. Yeh, and T. Kamitani, NUB1, a NEDD8-interacting Protein, Is Induced by Interferon and Down-regulates the NEDD8 Expression. J Biol Chem, 2001. 276(23): p. 20603–20609.

70 Kamitani, T., et al., Targeting of NEDD8 and Its Conjugates for Proteasomal Degradation by NUB1. J Biol Chem, 2001. 276(49): p. 46655–46660.

71 Moir, D., et al., Cold-sensitive cell-division-cycle mutants of yeast: isolation, properties and pseudoreversion studies. Genetics, 1982. 100: p. 547–563.

72 Ghislain, M., et al., Cdc48p interacts with Ufd3p, a WD repeat protein required for ubiquitin-mediated proteolysis in *Saccharomyces cerevisiae.* EMBO J, 1996. 15: p. 4884–4899.

73 Dai, R.M., et al., Involvement of valosin-containing protein, an ATPase co-purified with IκBα and 26S proteasome, in ubiquitin-proteasome-mediated degradation of IκBα. J Biol Chem, 1998. 273: p. 3562–3573.

74 Rabouille, C., et al., Syntaxin 5 is a common component of the NSF- and p97-mediated reassembly pathways of the Golgi cisternae from mitotic Golgi fragments in vitro. Cell, 1998. 92: p. 603–610.

75 Bays, N.W., et al., HRD4/NPL4 is required for the proteasomal processing of ubiquitinated ER proteins. Mol Biol Cell, 2001. 12: p. 4114–4128.

76 Rape, M., et al., Mobilization of processed, membrane-tethered SPT23 transcription factor by $CDC48^{UFD1/NPL4}$, a ubiquitin-selective chaperone. Cell, 2001. 107: p. 667–677.

77 Jarosch, E., et al., Protein dislocation from the ER requires polyubiquitination and the AAA-ATPase Cdc48. Nat Cell Biol, 2002. 4: p. 134–139.

78 Meyer, H.H., M. Wang, and G. Warren, Direct binding of ubiquitin conjugates by the mammalian p97 adaptor complexes, p47 and Ufd1-Npl4. EMBO J, 2002. 21(21): p. 5645–5652.

79 Buchberger, A., et al., The UBX domain: a widespread ubiquitin-like module. J Mol Biol, 2001. 307: p. 17–24.

80 Schuberth, C., et al., Shp1 and Ubx2 are adaptors of Cdc48 involved in ubiquitin-dependent protein degradation. EMBO Rep, 2004. 5(8): p. 818–824.

81 Hartmann-Petersen, R., et al., The Ubx2 and Ubx3 cofactors direct Cdc48 activity to proteolytic and nonproteolytic ubiquitin-dependent processes. Curr Biol, 2004. 14: p. 824–828.

82 Medicherla, B., et al., A genomic screen identifies Dsk2p and Rad23p as essential components of ER-associated degradation. EMBO Rep, 2004. 5(7): p. 692–697.

83 Gonzalez, F., et al., Recruitment of a 19S proteasome subcomplex to an activated promoter. Science, 2002. 296: p. 548–550.

84 Ferdous, A., et al., The 19S regulatory particle of the proteasome is required for efficient translation elongation by RNA polymerase II. Mol Cell, 2001. 7: p. 981–991.

85 Gillettte, T.G., et al., Physical and functional association of RNA polymerase II and the proteasome.

Proc Natl Acad Sci, 2004. 101(16): p. 5904–5909.

86 Muratani, M. and W.P. Tansey, How the ubiquitin-proteasome system controls transcription. Nat Rev Mol Cell Biol, 2003. 4: p. 1–10.

87 Dunand-Sauthier, I., et al., Sum1, a component of the fission yeast eIF3 translation initiation complex, is rapidly relocalised during environmental stress and interacts with components of the 26S proteasome. Mol Biol Cell, 2002. 13: p. 1626–1640.

88 Hoareau Alves, K., et al., Association of the mammalian proto-oncoprotein Int-6 with the three protein complexes eIF3, COP9 signalosome and 26S proteasome. FEBS Lett, 2002. 527: p. 15–21.

89 Yen, H.-C., C. Gordon, and E.C. Chang, *Schizosaccharomyces pombe* Int6 and Ras homologues regulate cell division and mitotic fidelity via the proteasome. Cell, 2003. 112: p. 207–217.

90 Glickman, M.H., et al., A subcomplex of the proteasome regulatory particle required for ubiquitin-conjugate degradation and related to the COP9-signalosome and eIF3. Cell, 1998. 94(5): p. 615–623.

91 Ortolan, T.G., et al., Rad23 stabilizes Rad4 from degradation by the Ub/proteasome pathway. Nucl Acids Res, 2004. 32(22): p. 6490–6500.

92 Russell, S.J., et al., The 19S regulatory complex of the proteasome functions independently of proteolysis in nucleotide excision repair. Mol Cell, 1999. 3(6): p. 687–95.

9
Structural Studies of Large, Self-compartmentalizing Proteases

Beate Rockel, Jürgen Bosch, and Wolfgang Baumeister

9.1
Self-compartmentalization: An Effective Way to Control Proteolysis

Within the metabolic pathways of a cell, proteolysis plays a key role at different levels. The basic or "housekeeping" function is the degradation of proteins that are nonfunctional or misfolded due to mutations or as a result of stresses such as heat or oxidation. Such proteins are prone to aggregation and therefore should be removed. Regulatory proteins such as transcription factors or components of signal transduction chains need to be degraded at specific times of their life span. In the immune system, the activity of proteases ensures the availability of immunocompetent peptides that are produced via degradation of foreign proteins.

Intracellular proteolysis, however, is a hazard, and the destruction of proteins not destined for degradation must be prevented. An effective strategy for this purpose is to confine proteolysis to secluded compartments, where access is limited to proteins exhibiting degradation signals. Such a compartment can be a membrane-delimited organelle – such as the lysosome – or the proteolytic chamber of a self-compartmentalizing protease, a structural design that has evolved in prokaryotic cells, which are devoid of membrane-bound compartments [1]. This principle of self- or auto-compartmentalization has been implemented successfully in several unrelated proteases, the proteolytic subunits of which self-assemble into barrel-shaped complexes. The active sites of these protease complexes are sequestered physically in internal chambers and thus are accessible only for unfolded polypeptides. Accessory proteins – either transiently or continuously associated with the protease – recognize their target proteins, unfold them in an energy-dependent manner, and finally aid in translocating them into the interior. The translocation occurs through narrow orifices, which are likely to prevent immediate discharge and enforce a retention period, eventually leading to a minimum product size.

This concept has been realized successfully in all kingdoms of life, and the ATP-dependent proteolytic systems typically are linear assemblies, where the accessory proteins flank the protease unit. Examples for such adaptor-protease complexes are the 26S proteasome in eukaryotes; ClpX, ClpA, and ClpY associated with the

Protein Degradation, Vol. 2: The Ubiquitin-Proteasome System.
Edited by R. J. Mayer, A. Ciechanover, M. Rechsteiner

ISBN: 3-527-31130-0

Clp proteases ClpP or ClpQ in bacteria [2]; and the proteasome-activating nucleotidase PAN, which prepares proteins for degradation by the 20S proteasome in archaea.

In the successive degradation of the resulting, relatively small, products into amino acids, large complexes are also involved. Despite the relatively small size of their substrates, some of them, such as the tricorn protease in archaea and the eukaryotic tripeptidyl peptidase II, have masses of several megadaltons [3, 4]. In addition, their molecular architecture differs considerably from the linear barrel-shaped assemblies, and the route along which substrates enter and exit is less obvious.

In this review, we describe the structural and functional organization of key proteolytic complexes that are found in eukarya, eubacteria, and archaea, with the main emphasis on the giant proteases that have thus far been visualized in their fully assembled and fully functional oligomeric form only via electron microscopy (Figure 9.1).

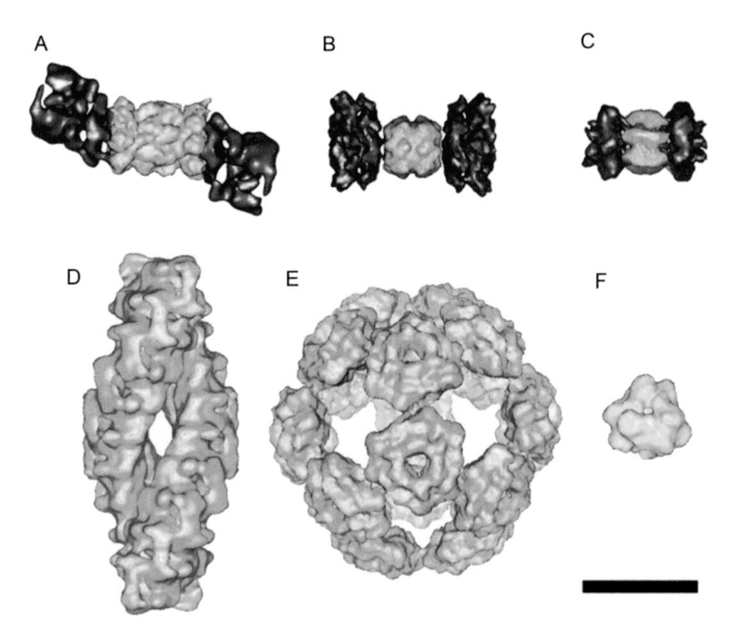

Fig. 9.1. Surface representations of large cytosolic proteolytic complexes drawn to scale. (A) 26S proteasome (20S: yellow; 19S caps: blue); (B) ClpAP complex (ClpP: yellow; ClpA: blue); (C) HslUV (HslV: yellow; HslU: blue); (D) tripeptidyl peptidase II; (E) tricorn protease capsid; (F) tetrahedral aminopeptidase. The structures of the 19S caps, tripeptidyl peptidase II, and tricorn were obtained from electron microscopy; for all other structures, the respective crystal structure was low-pass filtered to a resolution of 1.5 nm.

9.2
ATP-dependent Proteases: The Initial Steps in the Proteolytic Pathway

9.2.1
The Proteasome

The proteasome has frequently been described as "the paradigm" for a self-compartmentalizing protease. It is an ancestral particle that is ubiquitous and essential in eukarya and ubiquitous but not essential in archaea [5]. In eubacteria, where the proteolytic systems are redundant and the proteases Lon, Clp, HslV, and FtsH coexist, proteasomes are rare and not essential; genuine proteasomes have hitherto been found only in actinomycetes [6].

9.2.1.1 The 20S Proteasome

Architecture The quaternary structure of 20S proteasomes is the same in all kingdoms: the barrel-shaped 700-kDa complex consists of 28 subunits (termed α and β) that are arranged into four seven-membered rings. The four rings enclose three large cavities that are separated by narrow constrictions. One α ring and one β ring jointly form the outer (ante-) chambers, whereas two β rings enclose the central, proteolytic chamber. Crystal structures of 20S proteasomes from eukaryotes, archaea, and bacteria illustrate that α and β subunits share the same basic fold. This fold, which is typical for the superfamily of Ntn (N-terminal nucleophile) hydrolases [7, 8], consists of a pair of five-stranded β sheets flanked on both sides by α helices [9–13].

The 20S proteasome of the archaeon *Thermoplasma acidophilum*, like most prokaryotic proteasomes, exhibits $\alpha_7\beta_7\beta_7\alpha_7$-stoichiometry. The 20S proteasomes of the actinomycete *Rhodococcus erythropolis* contain two different α and β subunits. In eukaryotes, α- and β-type subunits have each diverged into seven distinct subunits; thus, eukaryotic proteasomes have a $\alpha_{1-7}\beta_{1-7}\beta_{1-7}\alpha_{1-7}$ stoichiometry and show pseudo-sevenfold symmetry.

The β subunits, which enclose the proteolytic chamber, are N-terminal hydrolases; the N-terminal threonine acts as both the catalytic nucleophile and the primary proton acceptor [9, 14]. In contrast to archaea, where all 14 β subunits are active, there are only six proteolytically active sites in a fully assembled eukaryotic proteasome, since four out of seven β subunits lack the N-terminal threonine residue. However, this discrepancy in the number of active sites in eukaryotic and prokaryotic proteasomes is not reflected in the size of the degradation products. Their average length is 7–8 residues and appears to be independent of number, specificity, and spatial arrangement of the active sites [15–17].

Assembly Unassembled proteolytically active β subunits carry propeptides, i.e., N-terminal extensions of variable lengths, which require post-translational removal for the formation of the active sites. Removal of these propeptides occurs autocatalytically and is delayed until the 20S complex is fully assembled, thereby ensuring that the active sites reside in a secluded, proteolytic chamber. The α subunits from *T. acidophilum* form seven-membered rings in the absence of β subunits [18], whereas the α subunits of the *Rhodococcus* proteasome do not form rings on their own, probably due to the size of their contact region, which is considerably smaller than corresponding α-subunit contact regions in *Thermoplasma*, yeast, and mammalian 20S proteasomes [11]. Instead, in the presence of β subunits, they assemble into $\alpha\beta$ heterodimers and subsequently into half-proteasomes, which dimerize and eventually are activated by propeptide cleavage.

While the assembly of prokaryotic proteasomes proceeds independent of cofactors, assembly of eukaryotic proteasomes requires extrinsic maturation factors and must be carefully orchestrated to ensure the correct positioning of each of the 14 different subunits [19]. Some of the eukaryotic α subunits also assemble into ring structures that serve as a scaffold for subsequent beta-subunit assembly [20, 21].

Substrate access The three inner compartments of the *Thermoplasma* 20S proteasome are accessible only through the narrow entry ports at both ends of the particle. This was corroborated by electron microscopic studies of *Thermoplasma* 20S proteasomes where Nanogold-labeled insulin was used as substrate [22]. In the crystal structure of the *Thermoplasma* 20S proteasome, the entrance ports are 1.3 nm wide and are constricted by an annulus built from turn-forming segments of the seven α subunits. Thus, they appear to be open and accessible for unfolded proteins [9]. In contrast, entrance of substrate into yeast 20S proteasomes appeared to be blocked by a plug formed by the interdigitating N-terminal tails of the α subunits [23]. These N-terminal tails were disordered in *Thermoplasma* 20S proteasomes and thus not visible in the crystal structure. Deletion of the nine N-terminal residues of $\alpha 3$ of the yeast 20S proteasome led to an opening similar to that seen in *Thermoplasma* and to a much-enhanced peptidase activity compared to eukaryotic, wild-type 20S proteasomes, which have a low basal activity [23]. In fact, a similar activation also occurs in *Thermoplasma* proteasomes when the N-termini of the α subunits are deleted [24].

In vivo, eukaryotic proteasomes associate with regulatory complexes such as 19S caps, PA28, or PA26, which are known to activate the 20S complex. All of these adaptors interact with the terminal α rings of the proteasome and are likely to function by mechanisms that open the gate for substrate uptake (for reviews, see Refs. [25–29]). The same mechanism might be valid for archaeal proteasomes, which assemble at least transiently with hexameric ATPase complexes [30–33].

9.2.1.2 **The PA28 Activator**

The PA28 activator of the eukaryotic 20S proteasome is restricted to organisms with an adaptive immune system [34]. It is ATP-independent and stimulates the hydrolysis of small peptides, albeit not of denatured or ubiquitinylated proteins [35, 36]. PA28 induces dual substrate cleavages by the 20S proteasome and thus can enhance the generation of antigenic peptides [37, 38]. The expression of PA28 and that of the immunoproteasome subunits β1i, β2i, and β5i is induced by γ-interferon [39].

PA28 is a predominantly cytosolic complex of 200 kDa consisting of two related 28-kDa subunits, PA28α and PA28β, which assemble into a heteroheptamer [40, 41]. The PA28α heptamer, which is able to stimulate 20S proteasomes similarly to native PA28α/β, is composed of a bundle of α helices forming a cone-shaped structure traversed by a central channel [42, 43]. PA28 complexes can bind to both ends of 20S proteasomes, as has been shown by electron microscopy [44, 45].

PA26, a PA28-related protein in *Trypanosoma brucei* that stimulates the peptidolytic activity of *Trypanosoma*, rat, and yeast 20S proteasomes [46, 47], has been crystallized in complex with the yeast 20S proteasome. The structure reveals that the C-terminal regions of the PA26 subunits insert into pockets formed by the α subunits of the proteasome. This in turn induces conformational changes in the 20S α subunits, resulting in the proposed gate opening: the α tails are straightened out and pushed away from the entrance gates (Figure 9.2) [28]. Interestingly, the residues that stabilize the open conformation in eukaryotic proteasomes (i.e., Y8, D9, P17, and Y26) are also conserved and are important for proteolysis in archaeal proteasomes [33].

As shown by electron microscopy, opening of the axial channel also occurs when PA200, a large, 200-kDa nuclear protein with a dome-like structure, binds to the alpha-ring of 20S proteasomes [48].

9.2.1.3 The 19S Cap Complex

The major player in intracellular proteolysis in eukaryotes is the 26S proteasome, a 20S proteasome flanked by one or two 19S regulatory complexes that associate with the 20S core in an ATP-dependent manner [49–52]. The 26S proteasome links the ubiquitin system for targeting substrates for degradation with the machinery executing their degradation. The 19S regulatory complexes recognize ubiquitinylated proteins and prepare them for degradation via the 20S complex. These preparatory steps involve the binding of ubiquitinylated substrates, their deubiquitinylation, unfolding, and subsequent translocation into the 20S complex. The 19S complex is composed of two sub-complexes, the lid and the base, which are located distally and proximally in relation to the 20S core, respectively (Figure 9.3). 19S caps of *Drosophila melanogaster* have a mass of approximately 890 kDa and consist of 18 subunits, which were identified by two-dimensional gel electrophoresis [53]. Recognition and binding of ubiquitin-tagged substrates appear to be mediated by the eight subunits of the lid complex, since the 26S holoenzyme but not the 20S-base complex is capable of degrading ubiquitinylated proteins [54, 55]. 19S particles from mammals, *Drosophila*, and fission yeast, but not budding yeast, contain a deubiquitinylating subunit [53, 56, 57], which, using gold-labeled ubiquitin aldehyde [53], was mapped to the lid–base interface close to the proposed location of Rpn10, where multi-ubiquitinylated chains have been proposed to bind *in vitro* [58].

The base of the 19S cap contains the two largest subunits of the 26S proteasome S1/Rpn2 and S2/Rpn1 as well as six paralogous AAA ATPases. Attachment of the

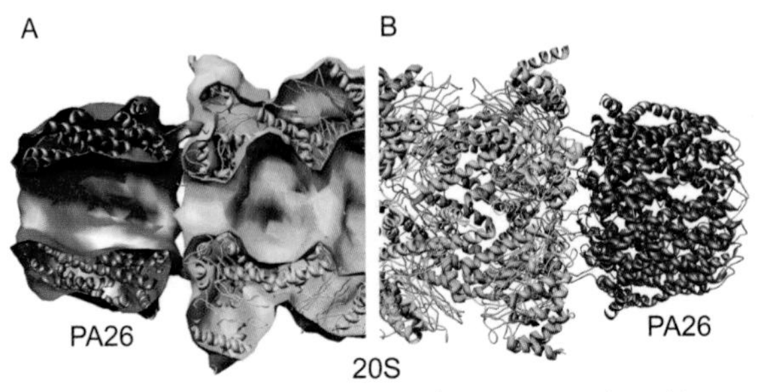

Fig. 9.2. The PA26–20S–PA26 complex. Structure of the hybrid complex between *T. brucei* PA26 (blue) and yeast 20S (yellow). (A) Cut-open view of the crystal structure low-pass filtered to a resolution of 1.5 nm combined with a ribbon representation showing the orifice of the proteasome. (B) Crystal structure of the PA26–20S–PA26 complex (PDB entry 1FNT [28]).

base complex only is sufficient to activate the 20S proteasome; thus, the ATPases must be involved in the gating of the α-ring channel and in controlling access to the proteolytic core. However, the symmetry mismatch between the 19S base and the 20S α ring suggests a gate-opening mechanism differing from the one found in PA26–20S complexes. Since both the base complex and its evolutionary ancestor PAN exhibit chaperone activity *in vitro* [32, 59], the ATP-dependent unfolding of substrates can be attributed to the ATPases acting in a "reverse chaperone" or unfoldase mode [31, 60].

The six paralogous ATPases of the base contain one copy of the AAA module [61–63] and an N-terminal coiled-coil region, which might mediate the binding of substrate proteins [64] and promote interactions between individual ATPases [65], hence playing a role in the assembly of the heterohexameric ring. All 19S ATPases are essential, as was shown by deletion analysis in fission and budding yeast [66–68]. Site-directed mutations in the Walker A motif of individual yeast ATPase sub-

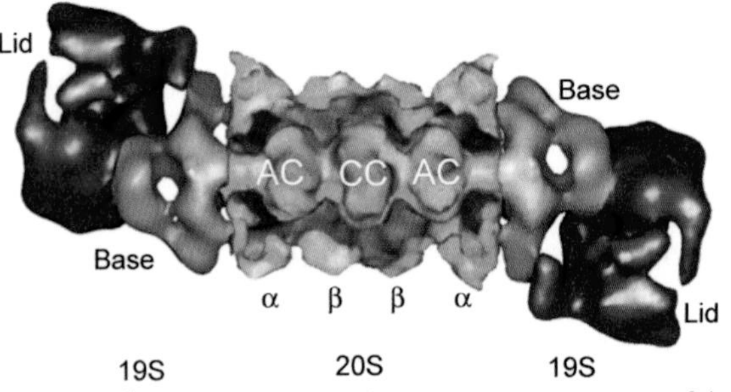

Fig. 9.3. The 26S proteasome. Composite image of the 3D structure of the 19S caps (lid: blue; base: gray) from *Drosophila melanogaster* [78] with a cut-away view of the crystal structure of the activated yeast proteasome (PDB entry: 1FNT [28]) low-passed filtered to 1.5 nm. AC: antechamber; CC: catalytic chamber.

units resulted in different phenotypes, indicating that the six ATPases of the base complex are not functionally redundant [69–71].

When observed in the electron microscope, the fully assembled 26S proteasome appears as an elongated dumbbell-shaped particle. 26S proteasomes have been isolated from different organisms; all preparations show a mixture of 20S particles associated with either one or two 19S units, resulting in total lengths of 30 nm and 45 nm, respectively [72–74]. Immunoprecipitation experiments suggest that 19S–20S–PA28 heterocomplexes also occur *in vivo* [75]; however, these appear to be very labile and have thus far been visualized only in the electron microscope after *in vitro* reconstitution [76, 77]. 2D averages of 26S proteasomes feature the characteristic "dragon head," where the 19S complexes in the double-capped particles face in opposite directions, reflecting the C2-symmetry of the eukaryotic 20S particle. In general, structural studies with 26S proteasomes are hampered by their low stability and tendency to dissociate into various sub-complexes. It is notoriously difficult to obtain a sufficiently homogenous and stable 26S preparation. Therefore, the only 3D structure of fully assembled 26S complexes available so far has been obtained from negatively stained 26S particles adsorbed on carbon film. Here, the 19S caps appear not to be in a fixed position with respect to the core but rather undergo an up-and-down "wagging-type" movement with a maximum amplitude of 2°, the functional relevance of which is not yet clear [78]. An even larger variety of states are observed when 26S proteasomes are frozen in a thin self-supporting layer of vitreous ice, where nearly mirror, symmetric, double-capped complexes also occur. This might hint at a possible rotary movement between the α rings of the 20S complex and the ATPase heterohexamer of the 19S base complex (Kapelari et al., unpublished results, [49]). In contrast, a recent immunoelectron microscopic study employing monoclonal antibodies against subunits $\alpha 4$ and $\alpha 6$ of the 20S complex outer rings suggests that the 19S cap complexes are attached to the 20S core complex in a defined orientation [79].

The 19S complex is an intricate, spongy structure from which it is not obvious what path a substrate protein will take before it is translocated to the 20S core (Figure 9.3) [78]. The observed flexible linkage between the caps and the core complicates structural analysis of the 26S complex, and a detailed analysis of a large set of ice-embedded 26S particles will be required to monitor the proposed rotational freedom of the 19S caps in the holocomplex and to confirm the proposed gate opening by the base complex.

9.2.1.4 **Archaeal and Bacterial AAA ATPases Activating the 20S Proteasome**

While eukaryotic 20S proteasomes can be isolated in association with a regulatory complex, interactions of activators with the 20S complex in bacteria and archaea seem to be rather transient. Some archaea contain the AAA ATPase proteasome-activating nuclease (PAN), which was first discovered in *Methanococcus jannaschii* as a homologue to the ATPases of the eukaryotic 19S base complex [80]. When mixed with 20S proteasomes from *Thermoplasma*, *Methanococcus* PAN stimulates the degradation of substrate proteins [81], and 20S–PAN complexes have been visualized whereby PAN apparently associates with the ends of the 20S proteasome cylinder [82]. Heterologously expressed PAN assembles into a 650-kDa complex, probably representing a dodecamer [81]. It recognizes ssrA-tagged green fluorescent protein as a substrate and mediates its energy-dependent unfolding and subsequent translocation into the 20S proteasome for degradation [24, 32, 83]. No homologues of PAN exist in *T. acidophilum* and its close relatives [84, 85]. Thus, in these organisms other complexes must substitute for the missing PAN function. In *T. acidophilum* the role of PAN is likely to be fulfilled by VAT, an archaeal AAA ATPase for which chaperone-like activity was demonstrated [86, 87]. Its eukaryotic homologues p97 and Cdc48 have been implicated in the degradation of substrate proteins via the ubiquitin proteasome pathway, thus making a role for VAT in protein degradation plausible [88, 89]. However, the putative substrate-binding domain of VAT and p97 is not a coiled coil, as characteristic for the proteasomal AAA ATPases, but rather a two-domain structure consisting of a double ψ barrel and a six-stranded β-clam fold [90, 91]. Both VAT and p97 form homohexameric 500-kDa toroids that have been characterized by electron microscopy [91–95]. The detailed intersubunit contacts have been revealed recently by crystal structures of the p97 complex [96–98].

A candidate for interaction with the bacterial proteasome is the ARC complex, a more distant member of the AAA family found in most bacteria possessing genuine 20S proteasomes [2, 99, 100]. ARC from *Rhodococcus erythropolis* is composed of two hexameric rings and, like the proteasomal ATPases, it contains an N-terminal coiled-coil domain [101, 102]. However, a functional interaction between the ARC complex and the *Rhodococcus* proteasome has not been demonstrated yet.

All of the hexameric complexes described above contain a central channel, which might serve to translocate the unfolded substrate into the 20S complex. However, the route taken by substrates during binding, unfolding, and translocation has not been visualized for the eukaryotic 26S proteasome, where the molecular understanding of the steps involved in transferring ubiquitinylated subunits from the lid to the base and eventually into the 20S core is substantially incomplete, nor has it been visualized for the corresponding 20S-activator complexes in bacteria and archaea. Although a pore-threading mechanism is conceivable, substrate unfolding via PAN occurs at the surface of the AAA ATPase ring [83]. Furthermore, in one of the recently solved crystal structures of the VAT homologue p97, a zinc ion occludes the central pore of the hexamer, which might prevent threading of the substrate [96, 97].

9.2.2
The Clp Proteases

Clp proteases are proteolytic complexes found in bacteria as well as in mitochondria and chloroplasts of eukaryotic cells [103]. Like the 26S proteasome, they function in ridding the cell of abnormal proteins as well as in regulatory circuits. Aside from Clp proteases, two other protease families are found in bacteria, Lon and

FtsH, in which both ATPase activity and proteolytic activity are joined in a single polypeptide chain.

As in the 26S proteasome, ATPase activity and proteolytic activity reside in different sub-complexes within ClpAP: ClpP is the proteolytic component, and ClpA is the ATPase [103, 104]. A second ATPase, ClpX, also assembles with ClpP to form ClpXP complexes [105, 106]. ClpA and ClpX recognize different signals for degradation and fulfill different regulatory functions within the cell [107].

The 21.5-kDa subunits of the serine protease ClpP assemble into two heptameric rings, isologously bonded, that enclose the proteolytic chamber lined with 14 active sites. The chamber is large enough to accommodate a protein of about 51 kDa [104, 108]; however, as in the 20S proteasome, the entry port is very narrow (10 Å) and thus not wide enough to admit most folded proteins [109].

The ATPase complexes ClpA and ClpX belong to the family of AAA^+ ATPases. ClpA consists of three structural domains: an N-terminal domain, representing a helical pseudo-dimer, followed by two AAA^+ modules [110]. In contrast, the N-domain of ClpX is shorter, is unrelated to that of ClpA, and is followed by a single AAA^+ module only. As was revealed by its solution structure, the ClpX N-domain contains a zinc-binding domain that forms a stable dimer [111, 112]. Both ClpA and ClpX are homohexameric in the presence of ATP or non-hydrolyzable ATP analogues [113, 114] and both have been crystallized [110, 115], but co-crystals with ClpP are not available as yet. The fully functional assembly has thus far been visualized only by electron microscopy, where heterocomplexes of the ClpAXP type also have been observed [116, 117].

In the assembled ClpAP and ClpXP complexes, the ATPase rings are positioned over the entry ports to the proteolytic chamber, analogous to the base complexes in the 26S proteasome. Substrates interact with ClpA and ClpX on the surface distal to ClpP, where the N-domains are likely to provide additional interaction sites [117–119]. Before being translocated through the narrow entrance site of ClpP, target proteins must be unfolded. Both ClpA and ClpX can act as chaperones and unfold proteins even in the absence of proteolytic activity [120–124].

In the fully assembled proteasome and Clp complexes, the protease and the ATPases form linear assemblies, an arrangement reflecting the sequence of steps necessary for substrate degradation. Time-resolved electron microscopic studies of ClpAP and ClpXP incubated with their respective model substrates RepA (ClpA) and λO (ClpX) allowed the visualization of the substrate pathway through a proteolytic machine "par excellence" [125–127]. The interaction of ClpAP and ClpXP with protein substrates involves several steps. Firstly, substrate binds to specific sites on the distal surface of the ATPase. Subsequently, the substrate is unfolded and fed into the digestion chamber of ClpP. Translocation – at least in ClpAP – appears to be a stepwise process whereby portions of substrate accumulate at the inner surface of the ATPase [125, 126] (Figure 9.4).

As is the case for the 26S proteasome, there is a symmetry mismatch between the (sixfold) ATPase unit and the (sevenfold) proteolytic unit in the ClpAP and ClpXP complexes [113]. For the ClpAP system, small rotational increments of

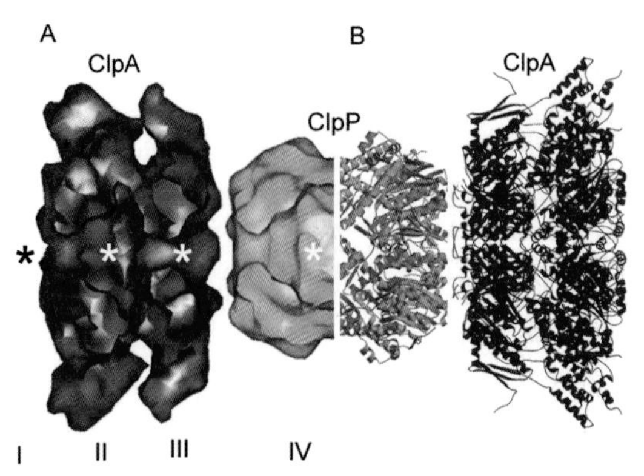

Fig. 9.4. The ClpAP complex. (A) Composite model of ClpAP as cut-away view. Crystal structures of ClpA (blue, PDB entry 1KSF [110]) and ClpP (yellow, PDB entry 1TYF [109]) have been low-pass filtered to 1.5 nm. The numbers describe the route of a protein substrate through ClpAP: (I) Substrates bind to specific sites on the distal side of ClpA. (II) Substrates are unfolded and translocated into the proteolytic chamber (IV). (III) Translocation in ClpAP appears to be a stepwise process during which portions of substrate accumulate at the inner surface of ClpA [125, 126]. (B) Composite crystal structure of ClpAP.

8.6° are sufficient to bring six- and sevenfold rings into (pseudo-)equivalent positions, a movement which might occur during the processive translocation of substrate into ClpP [30, 113]. Presumably, no symmetry mismatch occurs in HslUV (also called ClpYQ). Here, unlike the heptameric ClpP, the protease HslV (ClpQ) is a dimer of hexamers [128], whereas the ATPase HslU (ClpY) is dimorphic and assembles *in vitro* into hexameric and heptameric rings [129]. However, in the crystal structure of HslUV, both the ATPase and the protease complexes are hexameric [130–132]. The relative orientation of the ATPase HslU to the protease HslV has been a controversial issue: whereas in the crystal structure of HslUV obtained by Bochtler et al. [131] the "intermediate" (I) domains of HslU are pointing towards HslV, in the crystal structure of Sousa et al. [130] they extend outwards from the complex and the hexameric ATP-binding rings of HslU bind intimately to the HslV protease, an arrangement that is also consistent with the appearance of the HslUV complex in the electron microscope [129, 133].

While in proteasome and in Clp complexes the protease unit is flanked by one or two ATPase complexes, recent electron microscopic studies of CodWX, an N-terminal serine protease complex from *Bacillus subtilis*, reveal a strikingly different picture: the protease CodW, consisting of two stacked hexameric rings, binds to either one or both ends of a "spool-like" CodX-ATPase double ring, and even more elongated particles are found, in which an additional CodX double ring is bound to CodW [134, 135]. The physiological relevance of these linear assemblies with alternating ATPase and protease units is currently unclear.

9.3
Beyond the Proteasome: ATP-independent Processing of Oligopeptides Released by the Proteasome

The length of the fragments generated by the proteasome, Clp, and HslV varies from eight to 15 amino acids [15–17], but peptides of this length cannot be recycled by the cell and require further degradation down to the level of single amino acids. This task is performed by a number of ATP-independent proteases. In eukaryotes, proteasome degradation products are cleaved further by prolyl oligopeptidase (POP), thimet oligopeptidase (TOP), and tripeptidyl peptidase II. In archaea, this downstream processing is performed by the tricorn protease and its cofactors F1, F2, and F3 or by the tetrahedral aminopeptidase TET, which is found in archaea not containing any tricorn homologues (see Ref. [136] for a review).

9.3.1
Tripeptidyl Peptidase II

Among the proteases involved in downstream processing of oligopeptides released by the 26S proteasome, the tripeptidyl peptidase II (TPP II) complex has attracted attention owing to its extraordinary size, its versatility, and its apparent potential to substitute for some of the proteasome's functions.

TPP II was discovered in 1983 in the extralysosomal fraction of rat liver during the search for peptidases specific for proteins phosphorylated by cAMP-dependent protein [137]. It is a serine peptidase of the subtilisin type, which has broad substrate specificity, except that proline is not accepted in the P1 or P1′ positions [138]. The basic activity of TPP II is the removal of tripeptides from the free N-terminus of oligopeptides [137], but in addition to this exopeptidase activity, a much lower endopeptidase activity of the trypsin type was demonstrated [139]. Aside from soluble TPP II, a membrane-bound TPP II form exists in brain and in liver, which cleaves and inactivates the neuropeptide cholecystokinin as well as a number of different neuropeptides *in vitro* [140, 141].

Mammalian TPP II has a molecular subunit mass of 138 kDa, whereas the insect, worm, plant, and fungal forms carry an additional insert in the C-terminal region and have a molecular mass of 150 kDa [136, 142, 143]. The cDNA for TPP II does not contain any obvious signal peptide or membrane-spanning domain; thus, the membrane-bound TPP II is believed to bind to the membrane via a glycosyl phosphatidyl inositol anchor [141]. Since a number of variants of the TPP II–encoding mRNA exist (i.e., mRNA with long and short untranslated 3′ ends [144, 145]), one of them might encode this membrane-bound form. Pyrolysin, a membrane-bound serine endopeptidase from the archaeon *Pyrococcus furiosus* is the closest structural, albeit not functional, homologue [146].

Thus far, TPP II has been found only in eukaryotes and has been purified to apparent homogeneity from a variety of sources [4, 139, 142, 147, 148]. Like the proteasome, it occurs in a variety of tissue types, and its broad substrate specificity indicates its participation in general intracellular protein turnover.

In addition to its proposed housekeeping function, cytosolic TPP II participates in the trimming of antigenic peptides to be presented by the MHC class I complex. Whereas the proteasome releases some antigenic peptides directly in their final form, others are produced as precursor peptides, possessing the correct C-terminus of the final antigenic peptide, but with an extended N-terminus that requires additional trimming [149]. Although this additional trimming activity occurs in the endoplasmic reticulum, where the interferon gamma–inducible aminopeptidase ERAP1 is involved [150–153], a screen using antigenic peptide precursors released from the proteasome revealed two cytoplasmic peptidases that are involved in this process: puromycin-sensitive aminopeptidase (PSA, a widely expressed monomeric cytosolic amino peptidase of 100 kDa) and TPP II [154]. Since TPP II never cleaved within the antigenic peptide sequence in this screen, it is speculated that it might play a role in protecting antigenic peptides from their complete hydrolysis in the cytosol [155].

Some studies suggest that TPP II can in part compensate for the loss of proteasomal function. Thus, cells treated with a high dose of proteasome inhibitor normally undergo apoptosis, but EL-4 cells adapted to this treatment responded with an increased TPP II activity [156]. When TPP II was overexpressed in EL-4 cells, they resisted otherwise-lethal proteasome inhibitor concentrations and did not accumulate polyubiquitinylated proteins [157]. Such an upregulation of TPP II activity also occurred in apoptosis-resistant cells derived from large *in vivo* tumors exhibiting decreased proteasome activity [158]. The same was observed in Burkitt's lymphoma cells, which are likewise apoptosis-resistant and did not accumulate polyubiquitinylated proteins in response to normally lethal doses of proteasome inhibitors. When TPP II activity was inhibited, this apoptosis resistance was abolished [159].

Whether TPP II is *de facto* able to substitute for the proteasome is still discussed. Cells with low proteasome activity apparently are still able to process ubiquitinylated proteins; however it is not clear whether this is due to residual proteasomal activity or to increased TPP II activity. Princiotta et al. [160] found that EL-4 cells adapted to NLVS (4-hydroxy-5-iodo-3-nitrophenylacetyl-Leu-Leu-leucinal-vinyl sulfone) still require proteasome function for the degradation of polyubiquitinylated proteins as well as antigen processing, and they concluded that the adaptation is not due to an induced alternative protease. They suggest that a shared substrate pool between proteasomes and TPP II might explain the observed survival of NLVS-treated EL-4 cells in which TPP II was overexpressed [157].

Many functional data are controversial, and our understanding of the structure is incomplete. As was shown by electron microscopy, the 138-kDa subunits of mammalian TPP II occur as discrete double-bow structures (ca. 50×20 nm), as single-bow structures, and as dissociation products of lower mass [161]. However, the predominant species is a double-bow oligomer composed of two single bows twisted together to a short double helix. Dissociation experiments to produce homogeneous breakdown products led to an accumulation of 8×9-nm structures, which

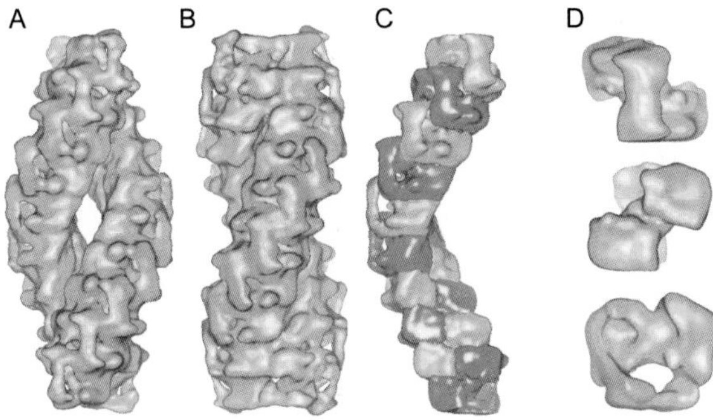

Fig. 9.5. Surface representations of TPP II as obtained by single-particle electron microscopy and 3D reconstruction. (A, B) Two perpendicular views of the TPP II complex. (C) Single TPP II strand with alternating coloring of the dimers in order to stress their interdigitation. (D) Different views of a computationally excised dimer.

in the presence of polyethylene glycol assembled into linear paracrystalline arrays. It is likely that these 8×9-nm structures represent the dimeric or tetrameric state of the enzyme with intersubunit contacts different from those in the bow structures [162].

TPP II complexes isolated from *Drosophila melanogaster* embryos are spindle-shaped 28×60-nm particles consisting of two segmented and twisted strands [4]. In the 3D reconstruction of the *Drosophila* TPP II complex at 2.2-nm resolution, each of the two strands is composed of a linear assembly of 10 interdigitated dimers (Figure 9.5) [163]. Intact TPP II complexes isolated from mammals and *Drosophila* are of defined length [4, 139], while TPP II particles heterologously expressed in *E. coli* often possess extensions beyond the spindle poles or occur as single strands of variable lengths, presumably as a consequence of the comparatively high TPP II concentration in those preparations. Treatment of such extended particles and single strands with destabilizing agents led to trimming of extensions and disassembly of single strands and demonstrated that the spindles observed in native preparations are the thermodynamically favored conformation. This stabilization of the spindles probably results from a double-clamp structure at the spindle poles in which the terminal dimer of one strand "locks" the two terminal dimers of its neighboring strand and vice versa [163].

TPP II exhibits the highest activity only when assembled into strands. Dissociation of the complex (e.g., upon dialysis) results in loss of activity. The minimal, active unit has been described as a dimer [143, 164]. Residue G252, which is conserved in all homologues, is apparently involved in complex formation since in mammalian TPP II the mutation G252R led to impaired complex formation and loss of activity. This effect is possibly a consequence of the location of G252 within the catalytic domain of TPP II (Asp-44, His-264, Ser-449) and its proximity to the catalytic His-264 [165, 166]. In wild-type TPP II, association and dissociation appear to be reversible provided the protein concentration is sufficiently high [164] and the equilibrium between both states is considered as a means of regulating the enzyme's activity; however, the underlying mechanism of this regulation remains unknown at present. Some clues might be obtained by investigating conditions where TPP II activity is upregulated. Is the proposed equilibrium between associated and dissociated forms shifted to the fully assembled complex, resulting in higher TPP II activity, or is the increase in activity always accompanied by an increase in TPP II protein and mRNA, as was observed for septic muscles [167]?

Up to now, no substrate-localization studies on the TPP II complex have been performed. Thus, the questions of whether the active sites are buried in a channel, how many peptides are processed simultaneously, and where they enter and leave the TPP II complex remain unsolved. From electron microscopic studies of TPP II constructs containing a bulky tag at their N-terminus, we can conclude that the N-terminal domains including the catalytic residues are located at the inner backbone of the strands. Indeed, the linear stacking of the dimers into strands leads to the formation of a channel or an arcade with lateral openings through which substrate flow might occur [163].

9.3.2
Tricorn Protease

In the course of searching for regulatory components of the proteasome from *T. acidophilum*, a new, high-molecular-weight protease was found [168]. The purified protein eluted with a molecular mass of 720 kDa from size-exclusion columns; when subjected to SDS-polyacrylamide gel electrophoresis, only a single polypeptide chain of 121 kDa was observed. At the C-terminal end (residues 878–1036), this polypeptide showed significant homology to the *E. coli* tail-specific protease (Tsp) and to the mammalian interphotoreceptor retinol–binding protein (IRBP). Subsequent biochemical experiments indicated a preference for trypsin-like substrates, although one chymotrypsin-like substrate (alanyl-alanyl-phenylalanyl-7-amino-4-methylcoumarin, H-AAF-AMC) was also cleaved. This homohexameric protease was named tricorn due to its triangular shape in electron micrographs. Simultaneously, a much larger assembly of tricorn, approximately 50 nm in diameter and exhibiting icosahedral symmetry, was observed. These capsid-like structures were found in *Thermoplasma* cells as well as in the void volume of Superose 6-fractionated *Thermoplasma* lysate; recombinantly overexpressed His6-tagged tricorn failed to produce such capsid structures. Three-dimensional reconstructions of ice-embedded capsids led to a significantly improved representation of the tricorn hexamers within the tricorn capsids, with a nominal resolution of 1.3 nm [3, 169]. Such a tricorn capsid is assembled from 20 homohexameric toroids, resulting in a combined molecular mass of 14.6 MDa (Figure 9.5B). This form of supramolecular organization has been proposed to make substrate channeling more efficient by positioning the tricorn protease into close spatial relationship with potentially interacting aminopeptidases. Hitherto, a physiological function of these icosahedral complexes in the *Thermoplasma* cell could not be demonstrated experimentally; however, Tamura and colleagues characterized three aminopeptidases, F1, F2, and F3, the tricorn-interacting factors, that enable tricorn to accept a broader substrate range when mixed together [170, 171]. Like the eukaryotic, functionally homologous TPP II, the tricorn protease seems to act downstream of the proteasome, as polypeptides that are degraded by the proteasome to 6–12 mers are further degraded to di-, tri-, and tetrapeptides by tricorn.

These oligopeptides have to be processed further in order to recycle the amino acids within the cell, a task accomplished by the aminopeptidases F1, F2, and F3. F1 is a proline iminopeptidase (PIP) with 14% sequence identity to the catalytic domain of prolyl oligopeptidase (POP) [172]. A number of homologues of the F1 peptidase are known in bacteria and eukaryotes, all of them belonging to the superfamily of α/β hydrolases [136]. As judged by gel filtration studies, F1 migrates as a monomeric enzyme with a molecular mass of 33.5 kDa. The active site residues S105, H271, and D244 were identified by sequence alignments and verified by mutational studies. While functioning as a peptidase, releasing proline residues from short oligomers, F1 also enhances the cleavage activity of tricorn and, moreover, generates novel peptidase activities when assayed together with the tricorn peptidase [170].

The aminopeptidases F2 and F3 are closely related to one another, with an overall sequence identity of 56.3%, but are unrelated to F1. They harbor zinc finger motifs in the N-terminal half of their respective polypeptide sequences, and homologues are known in yeast and bacteria (e.g., PepN [173]). The enzymatic activities of F2 and F3 are inhibited completely by removing the coordinated zinc by metal chelators. Both enzymes migrate as monomers with a molecular mass of 89 kDa in gel filtration studies. They have overlapping substrate spectra, but each of them hydrolyzes specific substrates as well. All three interacting factors F1, F2, and F3 are efficient in cleaving only very short peptides of two to four residues; while F2 is mainly responsible for the release of basic residues, F3 releases acidic residues.

In vitro incubation studies revealed a sequential manner of peptide degradation in *T. acidophilum*: peptides released by the proteasome are further cleaved by the tricorn protease to di- and tetrapeptides, which eventually are degraded to free amino acids by either the tricorn-interacting factors alone or in their interaction with tricorn. Changing the incubation order of tricorn, F1, F2, and F3 with substrates released from the proteasome failed to produce free amino acids in an efficient way [171].

Formerly, the occurrence of tricorn protease appeared to be restricted to thermophilic archaeal genera such as *Thermoplasma* and *Sulfolobus*, but meanwhile, orthologous tricorn genes have also been found in several bacterial genomes [136]. For instance, the structural and functional characteristics of the tricorn-like enzyme from *Streptomyces coelicolor*, which was studied in some detail by Tamura and coworkers, are very similar to the respective complex from *T. acidophilum*. Interestingly, the genome of *S. coelicolor* harbors two genes with significant similarity to tricorn, one with a calculated molecular mass of 115 kDa and another of 125 kDa [174]. In *Streptomyces* cells, however, tricorn was expressed as a homohexamer selectively composed of the 115-kDa polypeptide type and not as a hetero-oligomer from both types. It is presently unclear whether the second tricorn gene is expressed in *S. coelicolor* at all or only under special environmental conditions. Judg-

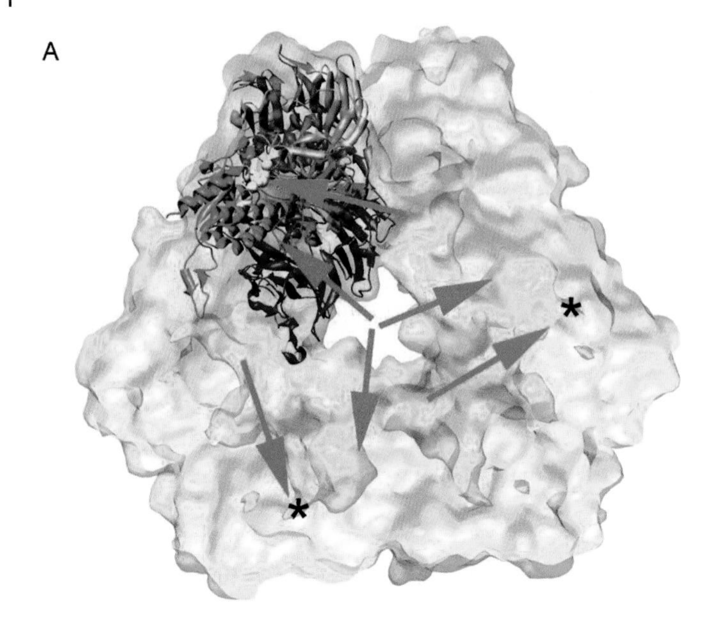
A
*
*

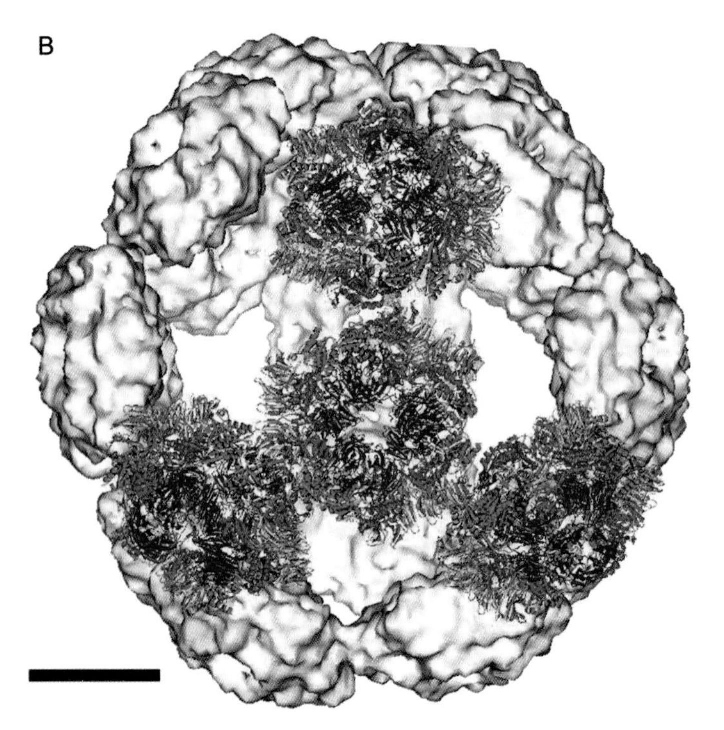
B

ing from the scattered distribution of tricorn genes across species, it seems plausible that different protein-degradation pathways exist in different species whereby functional homologues could replace tricorn's peptidase function (see Section 9.3.3).

In order to reveal the atomic structure of the tricorn protease and, by that means, to obtain more detailed functional and mechanistic insights, several groups have performed crystallization experiments. Two different crystal forms (C2, $P2_1$) of the *T. acidophilum* tricorn protease were reported [175, 176]. In both cases, the crystal lattice was built from hexameric toroids of 720 kDa. The D3-symmetric tricorn particle is assembled from two staggered and interdigitating trimeric rings. Due to the different contact areas within a tricorn hexamer, its assemblage can be described as a trimer of dimers. The 2.0-Å structure solved by Brandstetter et al. [176] revealed the subunit structure (Figure 9.6A). Thus, a single subunit is divided into five subdomains: a six-bladed β-propeller (residues M39–D310) followed by a seven-bladed β-propeller (A326–K675), and a PDZ-like domain (R761–D855) integrated between two mixed α-β domains (S681–G752 and R856–N1061). Both β6- and β7-propellers are open, Velcro-like structures, enabling a certain structural plasticity, possibly even a widening of the domain during substrate uptake [177, 178]. Such open, Velcro-like structures have so far been observed in POP, dipeptidyl peptidase IV (DPP IV/CD26), and alpha-L-arabinanase [172, 179, 180].

Co-crystallization experiments with specific tricorn inhibitors identified the residues S965 (nucleophile), H746 (proton donor), and D966 together with G918 (oxyanion hole) as the peptidase active site [176]. This active-site arrangement suggests that hydrolysis follows the classical mode of action of trypsin-like serine proteases (catalytic triad). After covalently binding the substrate C1 position to the active-site serine, a negatively charged tetrahedral transition state is formed. The other residues of the peptide are kept in place by hydrogen bonding, usually along a β strand or extended loop. During this step, the peptide bond is cleaved, one peptide product is attached to the enzyme in the acyl-enzyme intermediate, and the other peptide product diffuses away rapidly. In a second step of the reaction, the acyl-enzyme intermediate is hydrolyzed by a water molecule to release the second peptide product and to restore the active-site serine.

The dissociation of the product (di- or tripeptide) from the active-site residues generates the space necessary for the unprimed product to move forward for further processing [181]. A prominent cluster of basic residues (R131, R132) delineates the binding site of the substrate carboxy terminus. These basic residues together with the primed site topology clearly identify tricorn as a carboxypeptidase.

Fig. 9.6. Tricorn protease. (A) Cut-away representation of a tricorn hexamer with an overlay of one chain modeled as ribbons (PDB entry 1K32 [176]). The active-site residues are depicted as white spheres in the ribbon display and as asterisks on the cut-away representation. Blue: β7-propeller domain; yellow: β6-propeller domain; red: PDZ domain; purple and green: mixed alpha-beta domains. The main substrate pathway through the β7-propeller domain is marked by red arrows, whereas the alternative pathway through the inner cavities is highlighted by green arrows (see Section 9.3.2). (B) Reconstructed, three-dimensional density map of the icosahedral capsid at 1.3-nm resolution with an insert of four tricorn hexamers represented as ribbons.

The geometric dimensions explain tricorn's preferential di- and tripeptidase activity. In contrast to PDZ domains in other structures, the tricorn PDZ domain is not involved in substrate recognition; instead, it mainly serves to scaffold the subdomains [182].

Neighboring subunits have an effect on substrate recognition, especially the mobile side chain of residue D936, which is provided by a symmetry-related subunit and serves as a substrate-specificity switch accommodating both hydrophobic and basic P1 residues (preceding the cleavable bond). The P1 residue is held in place by main-chain interactions with the oxyanion hole (G918, D966). P2–P4 residues are bound through unsaturated main-chain hydrogen bonds at the strand I994-P996.

Based on the domain topology, the β7-propellers were suggested to serve as substrate filters for the active site in analogy to POP, while the β6-propellers might release the hydrolyzed peptides from the active site. A double cysteine mutant located at the entrance to the β7 channel (R414C and A643C) resulted in significant decreases in activity towards fluorogenic substrate and insulin B-chain, of 20% and 40%, respectively, compared to the wild-type activity. Co-crystallization experiments with long peptides with a C-terminal, inhibitory chloromethyl ketone group revealed a trapped peptide density stretching from the active site towards the β7 tunnel. Therefore, it has been suggested that the β7 domains might be suitable as major channels for substrate access to the active site, although the entrance of the β7 pores is obstructed by basic side chains, which form a lid (R369, R414, R645 and K646). An alternative but longer substrate pathway also seems plausible (Figure 9.6A). Following this scenario, the substrate peptide would reach the active site directly through funnel-like cavities emerging from the central channel of the tricorn hexamer. Convincing evidence for the role of the β6-propeller domain was obtained with the L184C mutant, which restricts the channel through the β6 domain when alkylated, thus leading to a reduced enzyme activity due to product accumulation [181]. Accordingly, F1 might bind in close proximity to the exit pore of the β6 domain, thus being well positioned for further degrading the released di- and tripeptides to free amino acids. This configuration would also be consistent with an arrangement in the whole icosahedral capsid, where only three F1 molecules could bind to a tricorn hexamer β6 domain, while the other three β6 domains would serve as anchoring points for neighboring tricorn hexamers [183] (Figure 9.6B).

Earlier biochemical studies indicated the presence of certain elements of the modular tricorn protease in humans, including their cofactor proteins [184]; however, other functional tricorn homologues might be difficult to detect in eukaryotes. Tricorn is not built of a single domain, but of five folding domains. Therefore, tricorn homologues might assemble noncovalently from different gene products.

9.3.3
Tetrahedral Aminopeptidase

Recently, a role analogous to that of tricorn in *Thermoplasma* has been ascribed to tetrahedral aminopeptidase (TET) from the halophilic archaeon *Haloarcula maris-*

mortui [185], a 500-kDa complex that degrades peptides of different length in organisms where tricorn is not present. TET is an aminopeptidase with broad substrate specificity that has a preference for neutral and basic residues and can progressively degrade peptides of up to 30–35 amino acids in length. This property assigns TET a role in processing peptides released by the proteasome (6–12 mers) or Lon (3–24 mers) [15]. Its 42-kDa subunits assemble into a dodecamer with a tetrahedral shape (edge length: 15 nm). In the 3D reconstruction obtained from negatively stained TET particles, four 2.1-nm-wide central channels emanate from the middle of each facet and converge into a central cavity. Additionally, a smaller channel of 1.7 nm emanates from the apices. Their asymmetry is discussed as reflecting the route of substrates through the complex, and it is suggested that the wider channels serve as entry ports for the peptide chain and the narrower channels as exit sites for the released peptides [185]. In the recently determined crystal structure of the TET homologue FrvX from *Pyrococcus horikoshii* (Figure 9.7), the smaller channel is nearly completely blocked by a phenylalanine residue [186]. Based on the crystal structure of TET from *Pyrococcus horikoshii*, a unique mechanism of substrate attraction and orientation is discussed. Here, the interior of the four central openings leading to the central, proteolytic chamber is negatively charged, forcing the peptide substrates to enter the proteolytic chamber with their N-termini. While substrate access is proposed to occur through the 1.8-nm-wide central openings, the proposed exit channels are not located at the vertices of the tetrahedron but are arranged in close proximity to the central openings at the facets [187] (Figure 9.7C).

9.4
Conclusions

Large, self-compartmentalizing proteases occur in all three kingdoms of life and all along the degradation pathway. The proteolytic complexes participating in the first steps of proteolysis associate linearly with hexameric ATPase complexes; their task is the recognition, unfolding, and subsequent translocation of target proteins into the respective proteolytic compartments. In general, such assemblies are labile, and thus their crystallization is challenging. Consequently, the functional assembly of most large, multicomponent proteolytic complexes has so far been visualized only by electron microscopy. Substrates en route through ATPase–protease complexes have hitherto only been observed with ClpAP/ClpXP. While the sequence of events is likely to be exemplary for the whole group of ATP-dependent protease complexes, the final objective must be the mapping of substrate-interaction sites, e.g., within the 19S subunit of the 26S proteasome, as well as the study of the mechanisms involved in substrate translocation. Also, the role of giant oligomeric superstructures formed by tricorn and TPP II is still rather enigmatic. While the proteolytic activity of tricorn capsids appears not to be different from the activity of hexamers, TPP II is fully active only when assembled into strands. For tricorn, a possible pathway for the substrates has been proposed, whereas for TPP II, no

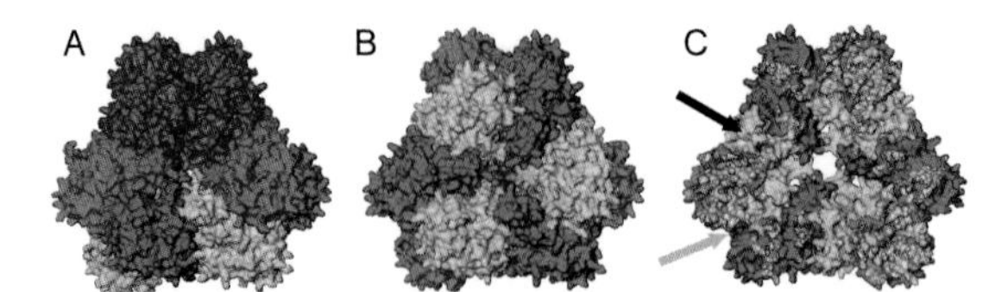

Fig. 9.7. Tetrahedral aminopeptidase. (A) Surface representation of the TET homologue FrvX from *Pyrococcus horikoshii* as a side view (PDB entry 1XFO, color-coding as in Ref. [186]). (B) Top view down the crystallographic 3-fold axis at one facet with the wide channel at the center. (C) Cut-open view down the crystallographic 3-fold axis along one edge. The black and gray arrows describe the location of the wider and the narrower channels, respectively. Around the large opening in the facet, three small openings are visible that might represent exit channels as discussed in Ref. [187].

data of that kind are available as yet. Thus, further structural studies including hybrid electron microscopy and X-ray crystallography approaches will be required to obtain insight into the modes of operation of these large proteolytic machines.

Acknowledgments

We would like to thank Dr. D. Xia (Center for Cancer Research, Bethesda, MD, USA) for sending us the PDB file of hexameric ClpA, and Dr. Peter Zwickl, Dr. Andrew Leis (MPI of Biochemistry, Martinsried, Germany), and Dr. Alasdair Steven (NIH, Bethesda, MD, USA) for critically reading the manuscript. Graphics were created with chimera [188].

References

1 A. Lupas, J. M. Flanagan, T. Tamura, W. Baumeister (1997) Self-compartmentalizing proteases. Trends Biochem. Sci. 22: 399–404.

2 R. De Mot, I. Nagy, J. Walz, W. Baumeister (1999) Proteasomes and other self-compartmentalizing proteases in prokaryotes. Trends Microbiol. 7: 88–92.

3 J. Walz, T. Tamura, N. Tamura, R. Grimm, W. Baumeister, A. J. Koster (1997) Tricorn protease exists as an icosahedral supermolecule in vivo. Mol. Cell 1: 59–65.

4 B. Rockel, J. Peters, B. Kühlmorgen, R. M. Glaeser, W. Baumeister (2002) A giant protease with a twist: the TPP II complex from *Drosophila* studied by electron microscopy. EMBO J. 21: 5979–5984.

5 A. Ruepp, C. Eckerskorn, M. Bogyo, W. Baumeister (1998) Proteasome function is dispensable under normal but not under heat shock conditions in *Thermoplasma acidophilum*. FEBS Lett. 425: 87–90.

6 T. Tamura, I. Nagy, A. Lupas, F. Lottspeich, Z. Cejka, G. Schoofs, K. Tanaka, R. Demot, W. Baumeister (1995) The First Characterization of a Eubacterial Proteasome – the 20S Complex of *Rhodococcus*. Curr. Biol. 5: 766–774.

7 J. Brannigan, G. Dodson, H.

Duggleby, P. Moody, J. Smith, D. Tomchick, A. Murzin (1995) A Protein Catalytic Framework with an N-Terminal Nucleophile Is Capable of Self-Activation. Nature 378: 416–419.

8 C. Oinonen, J. Rouvinen (2000) Structural comparison of Ntn-hydrolases. Protein Sci. 9: 2329–2337.

9 J. Löwe, D. Stock, B. Jap, P. Zwickl, W. Baumeister, R. Huber (1995) Crystal structure of the 20S proteasome from the archaeon *t-acidophilum* at 3.4 angstrom resolution. Science 268: 533–539.

10 M. Groll, L. Ditzel, J. Löwe, D. Stock, M. Bochtler, H. D. Bartunik, R. Huber (1997) Structure of 20S proteasome from yeast at 2.4 angstrom resolution. Nature 386: 463–471.

11 Y. D. Kwon, I. Nagy, P. D. Adams, W. Baumeister, B. K. Jap (2004) Crystal Structures of the *Rhodococcus* Proteasome with and without its Pro-peptides: Implications for the Role of the Pro-peptide in Proteasome Assembly. J. Mol. Biol. 335: 233–245.

12 M. Unno, T. Mizushima, Y. Morimoto, Y. Tomisugi, K. Tanaka, N. Yasuoka, T. Tsukihara (2002) The structure of the mammalian 20S proteasome at 2.75 angstrom resolution. Structure 10: 609–618.

13 M. Groll, H. Brandstetter, H. D. Bartunik, G. Bourenkov, R. Huber (2003) Investigations on the maturation and regulation of archaebacterial proteasomes. J. Mol. Biol. 327: 75–83.

14 E. Seemüller, A. Lupas, D. Stock, J. Löwe, R. Huber, W. Baumeister (1995) Proteasome from *Thermoplasma acidophilum*: A Threonine Protease. Science 268: 579–583.

15 A. F. Kisselev, T. N. Akopian, K. M. Woo, A. L. Goldberg (1999) The sizes of peptides generated from protein by mammalian 26 and 20 S proteasomes – Implications for understanding the degradative mechanism and antigen presentation. J. Biol. Chem. 274: 3363–3371.

16 I. Dolenc, E. Seemüller, W. Baumeister (1998) Decelerated degradation of short peptides by the 20S proteasome. FEBS Lett. 434: 357–361.

17 A. K. Nussbaum, T. P. Dick, W. Keilholz, M. Schirle, S. Stevanovic, K. Dietz, W. Heinemeyer, M. Groll, D. H. Wolf, R. Huber et al. (1998) Cleavage motifs of the yeast 20S proteasome beta subunits deduced from digests of enolase 1. Proc. Natl. Acad. Sci. U S A 95: 12504–12509.

18 P. Zwickl, J. Kleinz, W. Baumeister (1994) Critical elements in proteasome assembly. Nature Struct. Biol. 1: 765–770.

19 W. Heinemeyer, P. C. Ramos, R. J. Dohmen (2004) Ubiquitin-proteasome system – The ultimate nanoscale mincer: assembly, structure and active sites of the 20S proteasome core. Cell. Mol. Life Sci. 61: 1562–1578.

20 W. L. H. Gerards, J. Enzlin, M. Haner, I. Hendriks, U. Aebi, H. Bloemendal, W. Boelens (1997) The human alpha-type proteasomal subunit HsC8 forms a double ringlike structure, but does not assemble into proteasome-like particles with the beta-type subunits HsDelta or HsBPROS26. J. Biol. Chem. 272: 10080–10086.

21 Y. Yao, C. R. Toth, L. Huang, M. L. Wong, P. Dias, A. L. Burlingame, P. Coffino, C. C. Wang (1999) alpha 5 subunit in *Trypanosoma brucei* proteasome can self-assemble to form a cylinder of four stacked heptamer rings. Biochem. J. 344: 349–358.

22 T. Wenzel, W. Baumeister (1995) Conformational Constraints in Protein-Degradation by the 20S Proteasome. Nature Struct. Biol. 2: 199–204.

23 M. Groll, M. Bajorek, A. Kohler, L. Moroder, D. M. Rubin, R. Huber, M. H. Glickman, D. Finley (2000) A gated channel into the proteasome core particle. Nature Struct. Biol. 7: 1062–1067.

24 N. Benaroudj, P. Zwickl, E. Seemüller, W. Baumeister, A. L. Goldberg (2003) ATP hydrolysis by the proteasome regulatory complex

PAN serves multiple functions in protein degradation. Mol. Cell 11: 69–78.

25 C. M. Pickart, A. P. VanDemark (2000) Opening doors into the proteasome. Nature Struct. Biol. 7: 999–1001.

26 M. Rechsteiner, C. P. Hill (2005) Mobilizing the proteolytic machine: cell biological roles of proteasome activators and inhibitors. Trends Biochem. Sci. 15: 27–33.

27 M. Rechsteiner, C. Realini, V. Ustrell (2000) The proteasome activator 11 S REG (PA28) and Class I antigen presentation. Biochem. J. 345: 1–15.

28 F. G. Whitby, E. I. Masters, L. Kramer, J. R. Knowlton, Y. Yao, C. C. Wang, C. P. Hill (2000) Structural basis for the activation of 20S proteasomes by 11S regulators. Nature 408: 115–120.

29 M. Groll, M. Bochtler, H. Brandstetter, T. Clausen, R. Huber (2005) Molecular machines for protein degradation. Chembiochem 6: 222–256.

30 P. Zwickl, W. Baumeister, A. Steven (2000) Dis-assembly lines: the proteasome and related ATPase-assisted proteases. Curr. Opin. Struct. Biol. 10: 242–250.

31 P. Zwickl, W. Baumeister (1999) AAA-ATPases at the crossroads of protein life and death. Nature Cell Biol. 1: E97–E98.

32 N. Benaroudj, A. L. Goldberg (2000) PAN, the proteasome-activating nucleotidase from archaebacteria, is a protein-unfolding molecular chaperone. Nature Cell Biol. 2: 833–839.

33 A. Forster, F. G. Whitby, C. P. Hill (2003) The pore of activated 20S proteasomes has an ordered 7-fold symmetric conformation. EMBO J. 22: 4356–4364.

34 K. Tanaka, M. Kasahara (1998) The MHC class I ligand-generating system: roles of immunoproteasomes and the interferon-gamma-inducible proteasome activator PA28. Immunol. Rev. 163: 161–176.

35 W. Dubiel, G. Pratt, K. Ferrell, M. Rechsteiner (1992) Purification of an 11-S Regulator of the Multicatalytic Protease. J. Biol. Chem. 267: 22369–22377.

36 C. P. Ma, C. A. Slaughter, G. N. DeMartino (1992) Identification, Purification, and Characterization of a Protein Activator (Pa28) of the 20-S Proteasome (Macropain). J. Biol. Chem. 267: 10515–10523.

37 T. P. Dick, T. Ruppert, M. Groettrup, P. M. Kloetzel, L. Kuehn, U. H. Koszinowski, S. Stevanovic, H. Schild, H. G. Rammensee (1996) Coordinated dual cleavages induced by the proteasome regulator PA28 lead to dominant MHC ligands. Cell 86: 253–262.

38 M. Groettrup, A. Soza, M. Eggers, L. Kuehn, T. P. Dick, H. Schild, H. G. Rammensee, U. H. Koszinowski, P. M. Kloetzel (1996) A role for the proteasome regulator PA28 alpha in antigen presentation. Nature 381: 166–168.

39 K. Früh, Y. Yang (1999) Antigen presentation by MHC class I and its regulation by interferon gamma. Curr. Opin. Immun. 11: 76–81.

40 Z. G. Zhang, A. Krutchinsky, S. Endicott, C. Realini, M. Rechsteiner, K. G. Standing (1999) Proteasome activator 11S REG or PA28: Recombinant REG alpha/REG beta hetero-oligomers are heptamers. Biochemistry 38: 5651–5658.

41 T. Preckel, W. P. Fung-Leung, Z. L. Cai, A. Vitiello, L. Salter-Cid, O. Winqvist, T. C. Wolfe, M. Von Herrath, A. Angulo, P. Ghazal et al. (1999) Impaired immunoproteasome assembly and immune responses in PA28(−/−) mice. Science 286: 2162–2165.

42 C. Realini, C. C. Jensen, Z. G. Zhang, S. C. Johnston, J. R. Knowlton, C. P. Hill, M. Rechsteiner (1997) Characterization of recombinant REG alpha, REG beta, and REG gamma proteasome activators. J. Biol. Chem. 272: 25483–25492.

43 J. R. Knowlton, S. C. Johnston, F. G.

Whitby, C. Realini, Z. G. Zhang, M. Rechsteiner, C. P. Hill (1997) Structure of the proteasome activator REG alpha (PA28 alpha). Nature 390: 639–643.

44 C. W. Gray, C. A. Slaughter, G. N. DeMartino (1994) Pa28 Activator Protein Forms Regulatory Caps on Proteasome Stacked Rings. J. Mol. Biol. 236: 7–15.

45 A. J. Koster, J. Walz, A. Lupas, W. Baumeister (1995) Structural Features of Archaebacterial and Eukaryotic Proteasomes. Mol. Biol. Rep. 21: 11–20.

46 W. Y. To, C. C. Wang (1997) Identification and characterization of an activated 20S proteasome in *Trypanosoma brucei*. FEBS Lett. 404: 253–262.

47 Y. Yao, L. Huang, A. Krutchinsky, M. L. Wong, K. G. Standing, A. L. Burlingame, C. C. Wang (1999) Structural and functional characterizations of the proteasome-activating protein PA26 from *Trypanosoma brucei*. J. Biol. Chem. 274: 33921–33930.

48 J. Ortega, J. B. Heymann, A. V. Kajava, V. Ustrell, M. Rechsteiner, A. C. Steven (2005) The axial channel of the 20 S proteasome opens upon binding of the PA200 activator. J. Mol. Biol. 346: 1221–1227.

49 P. Zwickl, E. Seemüller, B. Kapelari, W. Baumeister (2001) Protein Folding in the Cell Vol. 59: pp. 187–222 (Horwich, A. L., Ed.) Academic Press, San Diego, California.

50 W. Baumeister, J. Walz, F. Zühl, E. Seemüller (1998) The proteasome: Paradigm of a self-compartmentalizing protease. Cell 92: 367–380.

51 D. Ganoth, E. Leshinsky, E. Eytan, A. Hershko (1988) A multicomponent system that degrades proteins conjugated to ubiquitin – resolution of factors and evidence for ATP-dependent complex-formation. J. Biol. Chem. 263: 12412–12419.

52 J. Driscoll, A. L. Goldberg (1990) The proteasome (multicatalytic protease) is a component of the 1500-kDa proteolytic complex which degrades ubiquitin-conjugated proteins. J. Biol. Chem. 265: 4789–4792.

53 H. Hölzl, B. Kapelari, J. Kellermann, E. Seemüller, M. Sumegi, A. Udvardy, O. Medalia, J. Sperling, S. A. Müller, A. Engel et al. (2000) The regulatory complex of *Drosophila melanogaster* 26S proteasomes: Subunit composition and localization of a deubiquitylating enzyme. J. Cell Biol. 150: 119–129.

54 M. H. Glickman, D. M. Rubin, O. Coux, I. Wefes, G. Pfeifer, Z. Cjeka, W. Baumeister, V. A. Fried, D. Finley (1998) A subcomplex of the proteasome regulatory particle required for ubiquitin-conjugate degradation and related to the COP9-signalosome and eIF3. Cell 94: 615–623.

55 J. S. Thrower, L. Hoffman, M. Rechsteiner, C. M. Pickart (2000) Recognition of the polyubiquitin proteolytic signal. EMBO J. 19: 94–102.

56 T. W. Li, N. I. Naqvi, H. Y. Yang, T. S. Teo (2000) Identification of a 26S proteasome-associated UCH in fission yeast. Biochem. Biophys. Res. Commun. 272: 270–275.

57 T. W. Li, W. Duan, H. Y. Yang, M. K. Lee, F. B. Mustafa, B. H. Lee, T. S. Teo (2001) Identification of two proteins, S14 and UIP1, that interact with UCH37. FEBS Lett. 488: 201–205.

58 Q. Deveraux, V. Ustrell, C. Pickart, M. Rechsteiner (1994) A 26-S Protease Subunit That Binds Ubiquitin Conjugates. J. Biol. Chem. 269: 7059–7061.

59 B. C. Braun, M. Glickman, R. Kraft, B. Dahlmann, P. M. Kloetzel, D. Finley, M. Schmidt (1999) The base of the proteasome regulatory particle exhibits chaperone-like activity. Nature Cell Biol. 1: 221–226.

60 A. Lupas, A. J. Koster, W. Baumeister (1993) Structural Features of 26S and 20S Proteasomes. Enzyme Protein 47: 252–273.

61 F. Confalonieri, M. Duguet (1995)

A 200-amino acid ATPase module in search of a basic function. Bioessays 17: 639–50.

62 A. Beyer (1997) Sequence analysis of the AAA protein family. Protein Sci. 6: 2043–58.

63 A. F. Neuwald, L. Aravind, J. L. Spouge, E. V. Koonin (1999) AAA(+): A class of chaperone-like ATPases associated with the assembly, operation, and disassembly of protein complexes. Genome Res. 9: 27–43.

64 Z. S. Zhang, N. Torii, A. Furusaka, N. Malayaman, Z. Y. Hu, T. J. Liang (2000) Structural and functional characterization of interaction between hepatitis B virus X protein and the proteasome complex. J. Biol. Chem. 275: 15157–15165.

65 C. Richmond, C. Gorbea, M. Rechsteiner (1997) Specific interactions between ATPase subunits of the 26 S protease. J. Biol. Chem. 272: 13403–13411.

66 C. Gordon, G. McGurk, P. Dillon, C. Rosen, N. D. Hastie (1993) Defective Mitosis Due to a Mutation in the Gene for a Fission Yeast 26S Protease Subunit. Nature 366: 355–357.

67 M. Ghislain, A. Udvardy, C. Mann (1993) *Saccharomyces-Cerevisiae* 26S protease mutants arrest cell-division in G2/metaphase. Nature 366: 358–362.

68 S. J. Russell, U. G. Sathyanarayana, S. A. Johnston (1996) Isolation and characterization of SUG2 – A novel ATPase family component of the yeast 26 S proteasome. J. Biol. Chem. 271: 32810–32817.

69 M. Seeger, C. Gordon, K. Ferrell, W. Dubiel (1996) Characteristics of 26 S proteases from fission yeast mutants, which arrest in mitosis. J. Mol. Biol. 263: 423–431.

70 D. M. Rubin, M. H. Glickman, C. N. Larsen, S. Dhruvakumar, D. Finley (1998) Active site mutants in the six regulatory particle ATPases reveal multiple roles for ATP in the proteasome. EMBO J. 17: 4909–4919.

71 H. Y. Fu, J. H. Doelling, D. M. Rubin, R. D. Vierstra (1999) Structural and functional analysis of the six regulatory particle triple-A ATPase subunits from the *Arabidopsis* 26S proteasome. Plant J. 18: 529–539.

72 J.-M. Peters, Cejka, Z., Harris, J. R., Kleinschmidt, J. A., Baumeister, W. (1993) Structural Features of the 26 S Proteasome Complex. J. Mol. Biol. 234: 932–937.

73 T. Yoshimura, K. Kameyama, T. Takagi, A. Ikai, F. Tokunaga, T. Koide, N. Tanahashi, T. Tamura, Z. Cejka, W. Baumeister et al. (1993) Molecular characterization of the 26S proteasome complex from rat-liver. J. Struct. Biol. 111: 200–211.

74 K. Fujinami, N. Tanahashi, K. Tanaka, A. Ichihara, Z. Cejka, W. Baumeister, M. Miyawaki, T. Sato, H. Nakagawa (1994) Purification and characterization of the 26-s proteasome from spinach leaves. J. Biol. Chem. 269: 25905–25910.

75 K. B. Hendil, S. Khan, K. Tanaka (1998) Simultaneous binding of PA28 and PA700 activators to 20 S proteasomes. Biochem. J. 332: 749–754.

76 F. Kopp, B. Dahlmann, L. Kuehn (2001) Reconstitution of hybrid proteasomes from purified PA700-20 s complexes and PA28$\alpha\beta$ activator: ultrastructure and peptidase activities. J. Mol. Biol. 313: 465–471.

77 P. Cascio, M. Call, B. M. Petre, T. Walz, A. L. Goldberg (2002) Properties of the hybrid form of the 26S proteasome containing both 19S and PA28 complexes. EMBO J. 21: 2636–2645.

78 J. Walz, A. Erdmann, M. Kania, D. Typke, A. J. Koster, W. Baumeister (1998) 26S proteasome structure revealed by three-dimensional electron microscopy. J. Struct. Biol. 121: 19–29.

79 F. Kopp, L. Kuehn (2003) Orientation of the 19 S Regulator Relative to the 20 S Core Proteasome: An Immunoelectron microscopy study. J. Mol. Biol. 329: 9–14.

80 C. Bult, O. White, G. J. Olsen, L. Zhou, R. D. Fleischmann, G. G. Sutton, J. A. Blake, L. M. Fitz-Gerald, R. A. Clayton, J. D. Gocayne et al. (1996) Complete Genom

Sequence of the Methanogenic Archaeon *Methanococcus jannaschii.* Science 273: 1058–1072.

81 P. Zwickl, D. Ng, K. M. Woo, H. P. Klenk, A. L. Goldberg (1999) An archaebacterial ATPase, homologous to ATPases in the eukaryotic 26 S proteasome, activates protein breakdown by 20 S proteasomes. J. Biol. Chem. 274: 26008–26014.

82 H. L. Wilson, M. S. Ou, H. C. Aldrich, J. Maupin-Furlow (2000) Biochemical and physical properties of the *Methanococcus jannaschii* 20S proteasome and PAN, a homolog of the ATPase (Rpt) subunits of the eucaryal 26S proteasome. J. Bacteriol. 182: 1680–1692.

83 A. Navon, A. L. Goldberg (2001) Proteins are unfolded on the surface of the ATPase ring before transport into the proteasome. Mol. Cell 8: 1339–1349.

84 T. Kawashima, N. Amano, H. Koike, S. Makino, S. Higuchi, Y. Kawashima-Ohya, K. Watanabe, M. Yamazaki, K. Kanehori, T. Kawamoto et al. (2000) Archaeal adaptation to higher temperatures revealed by genomic sequence of *Thermoplasma volcanium.* Proc. Natl. Acad. Sci. U S A 97: 14257–14262.

85 A. Ruepp, W. Graml, M. L. Santos-Martinez, K. K. Koretke, C. Volker, H. W. Mewes, D. Frishman, S. Stocker, A. N. Lupas, W. Baumeister (2000) The genome sequence of the thermoacidophilic scavenger *Thermoplasma acidophilum.* Nature 407: 508–513.

86 V. Pamnani, T. Tamura, A. Lupas, J. Peters, Z. Cejka, W. Ashraf, W. Baumeister (1997) Cloning, sequencing and expression of VAT, a CDC48/p97 ATPase homologue from the archaeon *Thermoplasma acidophilum.* FEBS Lett. 404: 263–268.

87 R. Golbik, A. N. Lupas, K. K. Koretke, W. Baumeister, J. Peters (1999) The janus face of the archaeal Cdc48/p97 homologue VAT: Protein folding versus unfolding. Biol. Chem. 380: 1049–1062.

88 R. M. Dai, E. Y. Chen, D. L. Longo, C. M. Gorbea, C. C. H. Li (1998) Involvement of valosin-containing protein, an ATPase co-purified with I kappa B alpha and 26 S proteasome, in ubiquitin-proteasome-mediated degradation of I kappa B alpha. J. Biol. Chem. 273: 3562–3573.

89 M. Ghislain, R. J. Dohmen, F. Levy, A. Varshavsky (1996) Cdc48p interacts with Ufd3p, a WD repeat protein required for ubiquitin-mediated proteolysis in *Saccharomyces cerevisiae.* EMBO J. 15: 4884–99.

90 M. Coles, T. Diercks, J. Liermann, A. Gröger, B. Rockel, W. Baumeister, K. K. Koretke, A. Lupas, J. Peters, H. Kessler (1999) The solution structure of VAT-N reveals a "missing link" in the evolution of complex enzymes from a simple $\beta\alpha\beta\beta$ element. Curr. Biol. 9: 1158–1168.

91 X. D. Zhang, A. Shaw, P. A. Bates, R. H. Newman, B. Gowen, E. Orlova, M. A. Gorman, H. Kondo, P. Dokurno, J. Lally et al. (2000) Structure of the AAA ATPase p97. Mol. Cell 6: 1473–1484.

92 B. Rockel, J. Walz, R. Hegerl, J. Peters, D. Typke, W. Baumeister (1999) Structure of VAT, a CDC48/p97 ATPase homologue from the archaeon *Thermoplasma acidophilum* as studied by electron tomography. FEBS Lett. 451: 27–32.

93 B. Rockel, J. Jakana, W. Chiu, W. Baumeister (2002) Electron cryo-microscopy of VAT, the archaeal p97/CDC48 homologue from *Thermoplasma acidophilum.* J. Mol. Biol. 317: 673–681.

94 I. Rouiller, B. DeLaBarre, A. P. May, W. I. Weis, A. T. Brunger, R. A. Milligan, E. M. Wilson-Kubalek (2002) Conformational changes of the multifunction p97 AAA ATPase during its ATPase cycle. Nature Struct. Biol. 9: 950–957.

95 F. Beuron, T. C. Flynn, J. P. Ma, H. Kondo, X. D. Zhang, P. S. Freemont (2003) Motions and negative cooperativity between p97 domains revealed by cryo-electron microscopy and quantised elastic deformational model. J. Mol. Biol. 327: 619–629.

96 B. DeLaBarre, A. T. Brünger (2003) Complete Structure of p97/valosin-containing protein reveals communication between nucleotide domains. Nature Struct. Biol. 10: 856–863.

97 T. Huyton, V. E. Pye, L. C. Briggs, T. C. Flynn, F. Beuron, H. Kondo, J. P. Ma, X. D. Zhang, P. S. Freemont (2003) The crystal structure of murine p97/VCP at 3.6 angstrom. J. Struct. Biol. 144: 337–348.

98 B. DeLaBarre, A. T. Brunger (2005) Nucleotide dependent motion and mechanism of action of p97/VCP. J. Mol. Biol. 347: 437–452.

99 I. Nagy, T. Tamura, J. Vanderleyden, W. Baumeister, R. De Mot (1998) The 20S proteasome of *Streptomyces coelicolor*. J. Bacteriol. 180: 5448–5453.

100 X. Zhang, K. Stoffels, S. Wurzbacher, G. Schoofs, G. Pfeifer, T. Banerjee, A. H. A. Parret, W. Baumeister, R. Demot, P. Zwickl (2004) The N-terminal coiled coil of the *Rhodococcus erythropolis* ARC AAA ATPase is neither necessary for oligomerization nor nucleotide hydrolysis. J. Struct. Biol. 146: 155–165.

101 S. Wolf, I. Nagy, A. Lupas, G. Pfeifer, Z. Cejka, S. A. Müller, A. Engel, R. Demot, W. Baumeister (1998) Characterization of ARC, a divergent member of the AAA ATPase family from *Rhodococcus erythropolis*. J. Mol. Biol. 277: 13–25.

102 K. H. Darwin, G. Lin, Z. Q. Chen, H. L. Li, C. F. Nathan (2005) Characterization of a *Mycobacterium tuberculosis* proteasomal ATPase homologue. Mol. Microbiol. 55: 561–571.

103 S. Gottesman (2003) Proteolysis in bacterial regulatory circuits. Annu. Rev. Cell Dev. Biol. 19: 565–587.

104 M. Kessel, M. R. Maurizi, B. Kim, E. Kocsis, B. L. Trus, S. K. Singh, A. C. Steven (1995) Homology in Structural Organization between *Escherichia-Coli* Clpap Protease and the Eukaryotic 26s-Proteasome. J. Mol. Biol. 250: 587–594.

105 S. Gottesman, W. P. Clark, V. Decrecylagard, M. R. Maurizi (1993) Clpx, an Alternative Subunit for the ATP-Dependent Clp Protease of *Escherichia-Coli* – Sequence and in-Vivo Activities. J. Biol. Chem. 268: 22618–22626.

106 D. Wojtkowiak, C. Georgopoulos, M. Zylicz (1993) Isolation and characterization of Clpx, a new ATP-dependent specificity component of the Clp protease of *Escherichia coli*. J. Biol. Chem. 268: 22609–22617.

107 J. M. Flynn, I. Levchenko, M. Seidel, S. H. Wickner, R. T. Sauer, T. A. Baker (2001) Overlapping recognition determinants within the ssrA degradation tag allow modulation of proteolysis. Proc. Natl. Acad. Sci. U S A 98: 10584–10589.

108 J. M. Wang, J. A. Hartling, J. M. Flanagan (1998) Crystal structure determination of *Escherichia coli* ClpP starting from an EM-derived mask. J. Struct. Biol. 124: 151–163.

109 J. M. Wang, J. A. Hartling, J. M. Flanagan (1997) The structure of clpp at 2.3 angstrom resolution suggests a model for ATP-dependent proteolysis. Cell 91: 447–456.

110 F. S. Guo, M. R. Maurizi, L. Esser, D. Xia (2002) Crystal structure of ClpA, an Hsp100 chaperone and regulator of ClpAP protease. J. Biol. Chem. 277: 46743–46752.

111 L. W. Donaldson, U. Wojtyra, W. A. Houry (2003) Solution structure of the dimeric zinc binding domain of the chaperone ClpX. J. Biol. Chem. 278: 48991–48996.

112 U. A. Wojtyra, G. Thibault, A. Tuite, W. A. Houry (2003) The N-terminal zinc binding domain of ClpX is a dimerization domain that modulates the chaperone function. J. Biol. Chem. 278: 48981–48990.

113 F. Beuron, M. R. Maurizi, D. M. Belnap, E. Kocsis, F. P. Booy, M. Kessel, A. C. Steven (1998) At sixes and sevens: Characterization of the symmetry mismatch of the ClpAP chaperone-assisted protease. J. Struct. Biol. 123: 248–259.

114 R. Grimaud, M. Kessel, F. Beuron, A. C. Steven, M. R. Maurizi (1998)

Enzymatic and structural similarities between the *Escherichia coli* ATP-dependent proteases, ClpXP and ClpAP. J. Biol. Chem. 273: 12476–12481.

115 D. Y. Kim, K. K. Kim (2003) Crystal structure of ClpX molecular chaperone from *Helicobacter pylori*. J. Biol. Chem. 278: 50664–50670.

116 S. K. Singh, F. S. Guo, M. R. Maurizi (1999) ClpA and ClpP remain associated during multiple rounds of ATP-dependent protein degradation by ClpAP protease. Biochemistry 38: 14906–14915.

117 J. Ortega, H. S. Lee, M. R. Maurizi, A. C. Steven (2004) ClpA and ClpX ATPases bind simultaneously to opposite ends of ClpP peptidase to form active hybrid complexes. J. Struct. Biol. 146: 217–226.

118 D. Xia, L. Esser, S. K. Singh, F. Guo, M. R. Maurizi (2004) Crystallographic investigation of peptide binding sites in the N-domain of the ClpA chaperone. J. Struct. Biol. 146: 166–179.

119 T. Ishikawa, M. Maurizi, R., A. C. Steven (2004) The N-terminal substrate-binding domain of ClpA unfoldase is highly mobile and extends axially from the distal surface of ClpAP protease. J. Struct. Biol. 146: 180–188.

120 J. R. Hoskins, S. K. Singh, M. R. Maurizi, S. Wickner (2000) Protein binding and unfolding by the chaperone ClpA and degradation by the protease ClpAP. Proc. Natl. Acad. Sci. U S A 97: 8892–8897.

121 I. Levchenko, L. Luo, T. A. Baker (1995) Disassembly of the Mu-Transposase Tetramer by the Clpx Chaperone. Genes Dev. 9: 2399–2408.

122 S. K. Singh, R. Grimaud, J. R. Hoskins, S. Wickner, M. R. Maurizi (2000) Unfolding and internalization of proteins by the ATP-dependent proteases ClpXP and ClpAP. Proc. Natl. Acad. Sci. U S A 97: 8898–8903.

123 E. U. Weber-Ban, B. G. Reid, A. D. Miranker, A. L. Horwich (1999) Global unfolding of a substrate protein by the Hsp100 chaperone ClpA. Nature 401: 90–93.

124 S. Wickner, S. Gottesman, D. Skowyra, J. Hoskins, K. McKenney, M. R. Maurizi (1994) A molecular chaperone, Clpa, functions like Dnak and Dnaj. Proc. Natl. Acad. Sci. U S A 91: 12218–12222.

125 T. Ishikawa, F. Beuron, M. Kessel, S. Wickner, M. R. Maurizi, A. C. Steven (2001) Translocation pathway of protein substrates in ClpAP protease. Proc. Natl. Acad. Sci. U S A 98: 4328–4333.

126 J. Ortega, S. K. Singh, T. Ishikawa, M. R. Maurizi, A. C. Steven (2000) Visualization of substrate binding and translocation by the ATP-dependent protease, ClpXP. Mol. Cell 6: 1515–1521.

127 J. Ortega, H. S. Lee, M. R. Maurizi, A. C. Steven (2002) Alternating translocation of protein substrates from both ends of ClpXP protease. EMBO J. 21: 4938–4949.

128 M. Bochtler, L. Ditzel, M. Groll, R. Huber (1997) Crystal structure of heat shock locus V (HslV) from *Escherichia coli*. Proc. Natl. Acad. Sci. U S A 94: 6070–6074.

129 M. Rohrwild, G. Pfeifer, U. Santarius, S. A. Müller, H. C. Huang, A. Engel, W. Baumeister, A. L. Goldberg (1997) The ATP-dependent HslVU protease from *Escherichia coli* is a four-ring structure resembling the proteasome. Nature Struct. Biol. 4: 133–139.

130 M. C. Sousa, C. B. Trame, H. Tsuruta, S. M. Wilbanks, V. S. Reddy, D. B. McKay (2000) Crystal and solution structures of an HslUV protease-chaperone complex. Cell 103: 633–643.

131 M. Bochtler, C. Hartmann, H. K. Song, G. P. Bourenkov, H. D. Bartunik, R. Huber (2000) The structures of HslU and ATP-dependent protease HslU-HslV. Nature 403: 800–805.

132 J. Wang, J. J. Song, M. C. Franklin, C. S. Kamtekar, Y. J. Im, S. H. Rho, I. S. Seong, C. S. Lee, C. H. Chung, S. H. Eom (2001) Crystal structures of the HslVU peptidase-ATPase complex reveal an ATP-dependent proteolysis mechanism. Structure 9: 177–184.

133 T. Ishikawa, M. R. Maurizi, D. Belnap, A. C. Steven (2000) ATP-dependent proteases – Docking of components in a bacterial complex. Nature 408: 667–668.

134 M. S. Kang, S. R. Kim, P. Kwack, B. K. Lim, S. W. Ahn, Y. M. Rho, I. S. Seong, S. C. Park, S. H. Eom, G. W. Cheong et al. (2003) Molecular architecture of the ATP-dependent CodWX protease having an N-terminal serine active site. EMBO J. 22: 2893–2902.

135 M. S. Kang, B. K. Lim, I. S. Seong, J. H. Seol, N. Tanahashi, K. Tanaka, C. H. Chung (2001) The ATP-dependent CodWX (HslVU) protease in *Bacillus subtilis* is an N-terminal serine protease. EMBO J. 20: 734–742.

136 D. Chandu, D. Nandi (2002) From proteins to peptides to amino acids: comparative genomics of enzymes involved in downstream events during cytosolic protein degradation. Applied Genomics and Proteomics 1: 235–252.

137 R. M. Balow, U. Ragnarsson, O. Zetterqvist (1983) Tripeptidyl aminopeptidase in the extralysosomal fraction of rat liver. J. Biol. Chem. 258: 11622–11628.

138 B. Tomkinson, C. Wernstedt, U. Hellman, O. Zetterqvist (1987) Active-site of tripeptidyl peptidase-II from human erythrocytes is of the subtilisin type. Proc. Natl. Acad. Sci. U S A 84: 7508–7512.

139 E. Geier, G. Pfeifer, M. Wilm, M. Lucchiari-Hartz, W. Baumeister, K. Eichmann, G. Niedermann (1999) A giant protease with potential to substitute for some functions of the proteasome. Science 283: 978–981.

140 C. Wilson, A. M. Gibson, J. R. McDermott (1993) Purification and characterization of tripeptidylpeptidase II from postmortem human brain. Neurochem. Res. 18: 743–749.

141 C. Rose, F. Vargas, P. Facchinetti, P. Bourgeat, R. B. Bambal, P. B. Bishop, S. M. Chan, A. N. Moore, C. R. Ganellin, J. C. Schwartz (1996) Characterization and inhibition of a cholecystokinin-inactivating serine peptidase. Nature 380: 403–9.

142 S. C. Renn, B. Tomkinson, P. H. Taghert (1998) Characterization and cloning of tripeptidyl peptidase II from the fruit fly, *Drosophila melanogaster.* J. Biol. Chem. 273: 19173–82.

143 B. Tomkinson (1999) Tripeptidyl peptidases: enzymes that count. Trends Biochem. Sci. 24: 355–359.

144 B. Tomkinson, A. K. Jonsson (1991) Characterization of Cdna for Human Tripeptidyl Peptidase-II – the N-Terminal Part of the Enzyme Is Similar to Subtilisin. Biochemistry 30: 168–174.

145 B. Tomkinson (1994) Characterization of cDNA for murine tripeptidyl-peptidase II reveals alternative splicing. Biochem. J. 304: 517–523.

146 W. G. B. Voorhorst, R. I. L. Eggen, A. C. M. Geerling, C. Platteeuw, R. J. Siezen, W. M. deVos (1996) Isolation and characterization of the hyperthermostable serine protease, pyrolysin, and its gene from the hyperthermophilic archaeon *Pyrococcus furiosus.* J. Biol. Chem. 271: 20426–20431.

147 E. Macpherson, R. M. Balow, S. Hoglund, B. Tomkinson, O. Zetterqvist (1986) Structure of the Exopeptidase Tripeptidyl Peptidase-II from Human-Erythrocyte and Rat-Liver. J. Ultrastruct. Mol. Struct. Res. 94: 277–277.

148 R. M. Balow, I. Eriksson (1987) Tripeptidyl peptidase II in haemolysates and liver homogenates of various species. Biochem. J. 241: 75–80.

149 A. L. Goldberg, P. Cascio, T. Saric, K. L. Rock (2002) The importance of the proteasome and subsequent proteolytic steps in the generation of antigenic peptides. Mol. Immunol. 39: 147–164.

150 I. A. York, S. C. Chang, T. Saric, J. A. Keys, J. M. Favreau, A. L. Goldberg, K. L. Rock (2002) The ER aminopeptidase ERAP1 enhances or limits antigen presentation by

trimming epitopes to 8–9 residues. Nature Immunol. 3: 1177–1184.

151 T. Saric, S. C. Chang, A. Hattori, I. A. York, S. Markant, K. L. Rock, M. Tsujimoto, A. L. Goldberg (2002) An IFN-gamma-induced aminopeptidase in the ER, ERAP1, trims precursors to MHC class I-presented peptides. Nature Immunol. 3: 1169–1176.

152 K. Falk, O. Rotzschke (2002) The final cut: how ERAPI trims MHC ligands to size. Nature Immunol. 3: 1121–1122.

153 T. Serwold, F. Gonzalez, J. Kim, R. Jacob, N. Shastri (2002) ERAAP customizes peptides for MHC class I molecules in the endoplasmic reticulum. Nature 419: 480–483.

154 F. Levy, L. Burri, S. Morel, A. L. Peitrequin, N. Levy, A. Bachi, U. Hellman, B. J. Van den Eynde, C. Servis (2002) The final N-terminal trimming of a subaminoterminal proline-containing HLA class I-restricted antigenic peptide in the cytosol is mediated by two peptidases. J. Immunol. 169: 4161–4171.

155 L. Burri, C. Servis, L. Chapatte, F. Levy (2002) A recyclable assay to analyze the NH_2-terminal trimming of antigenic peptide precursors. Protein Express. Purif. 26: 19–27.

156 R. Glas, M. Bogyo, J. S. McMaster, M. Gaczynska, H. L. Ploegh (1998) A proteolytic system that compensates for loss of proteasome function. Nature 392: 618–622.

157 E. W. Wang, B. M. Kessler, A. Borodovsky, B. F. Cravatt, M. Bogyo, H. L. Ploegh, R. Glas (2000) Integration of the ubiquitin-proteasome pathway with a cytosolic oligopeptidase activity. Proc. Natl. Acad. Sci. U S A 97: 9990–9995.

158 X. Hong, L. Lei, R. Glas (2003) Tumors acquire inhibitor of apoptosis protein (IAP)-mediated apoptosis resistance through altered specificity of cytosolic proteolysis. J. Exp. Med. 197: 1731–1743.

159 R. Gavioli, T. Frisan, S. Vertuani, G. W. Bornkamm, M. G. Masucci (2001) c-myc overexpression activates alternative pathways for intracellular proteolysis in lymphoma cells. Nature Cell Biol. 3: 283–288.

160 M. F. Princiotta, U. Schubert, W. S. Chen, J. R. Bennink, J. Myung, C. M. Crews, J. W. Yewdell (2001) Cells adapted to the proteasome inhibitor 4-hydroxy5-iodo-3-nitrophenylacetyl-Leu-Leu-leucinal-vinyl sulfone require enzymatically active proteasomes for continued survival. Proc. Natl. Acad. Sci. U S A 98: 513–518.

161 E. Macpherson, B. Tomkinson, R. M. Balow, S. Hoglund, O. Zetterqvist (1987) Supramolecular structure of tripeptidyl peptidase II from human erythrocytes as studied by electron microscopy, and its correlation to enzyme activity. Biochem. J. 248: 259–563.

162 J. R. Harris, B. Tomkinson (1990) Electron-microscopic and biochemical studies on the oligomeric states of human erythrocyte tripeptidyl peptidase 2. Micron Microsc. Acta 21: 77–89.

163 B. Rockel, J. Peters, S. A. Müller, G. Seyit, P. Ringler, R. Hegerl, R. M. Glaeser, W. Baumeister (2005) Molecular architecture and assembly mechanism of *Drosophila* Tripeptidyl peptidase II. Proc. Natl. Acad. Sci. U S A 102: 10135–10140.

164 B. Tomkinson (2000) Association and dissociation of the tripeptidyl-peptidase II complex as a way of regulating the enzyme activity. Arch. Biochem. Biophys. 376: 275–280.

165 H. Hilbi, E. Jozsa, B. Tomkinson (2002) Identification of the catalytic triad in tripeptidyl-peptidase II through site-directed mutagenesis. BBA-Proteins Proteomics 1601: 149–154.

166 B. Tomkinson, B. N. Laoi, K. Wellington (2002) The insert within the catalytic domain of tripeptidyl-peptidase II is important for the formation of the active complex. Eur. J. Biochem. 269: 1438–1443.

167 C. J. Wray, B. Tomkinson, B. W. Robb, P. O. Hasselgren (2002) Tripeptidyl-peptidase II expression and activity are increased in skeletal

muscle during sepsis. Biochem. Biophys. Res. Commun. 296: 41–47.

168 T. Tamura, N. Tamura, Z. Cejka, R. Hegerl, F. Lottspeich, W. Baumeister (1996) Tricorn protease – the core of a modular proteolytic system. Science 274: 1385–1389.

169 J. Walz, A. J. Koster, T. Tamura, W. Baumeister (1999) Capsids of tricorn protease studied by electron cryomicroscopy. J. Struct. Biol. 128: 65–68.

170 T. Tamura, N. Tamura, F. Lottspeich, W. Baumeister (1996) Tricorn protease (TRI) interacting factor 1 from *Thermoplasma acidophilum* is a proline iminopeptidase. FEBS Lett. 398: 101–105.

171 N. Tamura, F. Lottspeich, W. Baumeister, T. Tamura (1998) The role of tricorn protease and its aminopeptidase-interacting factors in cellular protein degradation. Cell 95: 637–648.

172 V. Fülöp, Z. Bocskei, L. Polgar (1998) Prolyl oligopeptidase – an unusual beta-propeller domain regulates proteolysis. Cell 94: 161–170.

173 D. Chandu, A. Kumar, D. Nandi (2003) PepN, the major Suc-LLVY-AMC-hydrolyzing enzyme in *Escherichia coli*, displays functional similarity with downstream processing enzymes in archaea and eukarya – Implications in cytosolic protein degradation. J. Biol. Chem. 278: 5548–5556.

174 N. Tamura, G. Pfeifer, W. Baumeister, T. Tamura (2001) Tricorn protease in bacteria: Characterization of the enzyme from *Streptomyces coelicolor.* Biol. Chem. 382: 449–458.

175 J. Bosch, T. Tamura, G. Bourenkov, W. Baumeister, L. O. Essen (2001) Purification, crystallization, and preliminary X-ray diffraction analysis of the tricorn protease hexamer from *Thermoplasma acidophilum.* J. Struct. Biol. 134: 83–87.

176 H. Brandstetter, J. S. Kim, M. Groll, R. Huber (2001) Crystal structure of the tricorn protease reveals a protein disassembly line. Nature 414: 466–470.

177 M. Paoli (2001) Protein folds propelled by diversity. Prog. Biophys. Mol. Biol. 76: 103–130.

178 T. Pons, R. Gomez, G. Chinea, A. Valencia (2003) Beta-propellers: Associated functions and their role in human diseases. Curr. Med. Chem. 10: 505–524.

179 H. B. Rasmussen, S. Branner, F. C. Wiberg, N. Wagtmann (2003) Crystal structure of human dipeptidyl peptidase IV/CD26 in complex with a substrate analog. Nature Struct. Biol. 10: 19–25.

180 D. Nurizzo, J. P. Turkenburg, S. J. Charnock, S. M. Roberts, E. J. Dodson, V. A. McKie, E. J. Taylor, H. J. Gilbert, G. J. Davies (2002) *Cellvibrio japonicus* alpha-L-arabinanase 43A has a novel five-blade beta-propeller fold. Nature Struct. Biol. 9: 665–668.

181 J. S. Kim, M. Groll, H. A. Musiol, R. Behrendt, M. Kaiser, L. Moroder, R. Huber, H. Brandstetter (2002) Navigation inside a protease: Substrate selection and product exit in the tricorn protease from *Thermoplasma acidophilum.* J. Mol. Biol. 324: 1041–1050.

182 H. Brandstetter, J. S. Kim, M. Groll, P. Gottig, R. Huber (2002) Structural basis for the processive protein degradation by tricorn protease. Biol. Chem. 383: 1157–1165.

183 P. Goettig, M. Groll, J. S. Kim, R. Huber, H. Brandstetter (2002) Structures of the tricorn-interacting aminopeptidase F1 with different ligands explain its catalytic mechanism. EMBO J. 21: 5343–5352.

184 L. Stoltze, M. Schirle, G. Schwarz, C. Schroter, M. W. Thompson, L. B. Hersh, H. Kalbacher, S. Stevanovic, H. G. Rammensee, H. Schild (2000) Two new proteases in the MHC class I processing pathway. Nature Immunol. 1: 413–418.

185 B. Franzetti, G. Schoehn, J. F. Hernandez, M. Jaquinod, R. W. H. Ruigrok, G. Zaccai (2002) Tetrahedral aminopeptidase: a novel

large protease complex from archaea. EMBO J. 21: 2132–2138.

186 S. Russo, U. Baumann (2004) Crystal structure of a dodecameric tetrahedral-shaped aminopeptidase. J. Biol. Chem. 279: 51275–51281.

187 L. Borissenko, M. Groll (2005) Crystal structure of TET protease reveals complementary protein degradation pathways in prokaryotes. J. Mol. Biol. 346: 1207–1219.

188 E. F. Pettersen, T. D. Goddard, C. C. Huang, G. S. Couch, D. M. Greenblatt, E. C. Meng, T. E. Ferrin (2004) UCSF chimera – A visualization system for exploratory research and analysis. J. Comput. Chem. 25: 1605–1612.

10
What the Archaeal PAN–Proteasome Complex and Bacterial ATP-dependent Proteases Can Teach Us About the 26S Proteasome

Nadia Benaroudj, David Smith, and Alfred L. Goldberg

10.1
Introduction

Much of what we have learned about biochemical pathways, gene transcription, and protein synthesis emerged initially from studies in bacteria that provided the basis for the subsequent elucidation of these processes in eukaryotic cells. Studies in prokaryotes have also provided fundamental insights into the physiological significance and mechanisms of protein degradation, although these major contributions have often been overlooked in discussions of the ubiquitin–proteasome pathway. Surprisingly, the importance of intracellular protein breakdown was not appreciated by microbiologists for a long time. In fact, until the mid-1970s, it was generally taught that in bacteria, in contrast to mammalian cells, proteins were stable after synthesis (Goldberg and Dice 1974). This conclusion was based upon classic, but over-interpreted, studies by Monod and coworkers, who showed that rates of protein breakdown (compared to rates of synthesis) are very low in *Escherichia coli* during exponential growth (Hogness et al. 1955). However, in the 1970s, our understanding of the importance of protein degradation in *E. coli* changed dramatically with the discovery that these cells rapidly degrade misfolded or incomplete proteins (Goldberg 2003; Goldberg and Dice 1974); that the overall degradation of normal proteins is regulated and increases rapidly in cells lacking amino acids or a carbon source; and that in bacteria, as in the mammalian cytosol, proteins are degraded by a process requiring ATP (Goldberg and St John 1976).

Because bacteria lack lysosomes, this discovery implied that the energy requirement for intracellular proteolysis was a universal feature of protein breakdown and was not related to the functioning of lysosomes, which were then believed to be the exclusive site of protein breakdown (Ciechanover 2005; Goldberg 2005). Our subsequent discovery of the existence of the soluble (non-lysosomal) ATP-dependent proteolytic system in reticulocytes (Etlinger and Goldberg 1977) was followed by establishment of similar cell-free systems in bacteria (Murakami et al. 1979) and subsequently in mitochondria, in which turnover of proteins uses enzyme systems quite similar to those in eubacteria (Desautels and Goldberg 1982a, 1982b). In analyzing this process in bacteria, we discovered that it depended on a new kind

Protein Degradation, Vol. 2: The Ubiquitin-Proteasome System.
Edited by R. J. Mayer, A. Ciechanover, M. Rechsteiner

ISBN: 3-527-31130-0

of enzyme, large ATP-dependent proteolytic complexes that degrade proteins and ATP in linked processes (Chung and Goldberg 1981; Gottesman 1996). As discussed below, bacteria and archaea were later found to contain several such proteolytic complexes, which function in the degradation of different types of proteins (Gottesman 1996). We had initially chosen to work in *E. coli* because of the opportunity to use genetic approaches, and, in fact, mutants lacking these ATP-dependent proteases (e.g., *lon*$^-$ strains) are defective in breakdown of misfolded and certain regulatory proteins (Gottesman 2003). Unexpectedly, with the advent of recombinant DNA, these protease-deficient strains have also proven particularly useful for expression of cloned proteins, many of which are rapidly degraded in bacteria (Baneyx and Mujacic 2004; Goldberg 2003).

Ironically, the discovery of the first ATP-dependent protease (lon/La) came at the same time as the classic discovery of the role of ubiquitin in protein breakdown in the reticulocyte system by Hershko, Ciechanover, and Rose (Ciechanover 2005; Glickman and Ciechanover 2002). This modification was proposed to explain the ATP requirement for intracellular proteolysis. Thus, two very different explanations for this requirement emerged in eukaryotes and prokaryotes, and they were initially assumed to constitute a fundamental distinction between these organisms. However, with time, it became clear that after proteins are ubiquitinated, ATP is still necessary for their breakdown (Tanaka et al. 1983), and by the late 1980s, the 26S proteasome, an ATP-dependent proteolytic complex that degrades ubiquitinated proteins, was identified. As discussed here, many of its special properties are similar to those of bacterial and archaeal ATP-dependent proteases. In fact, the isolation of the 26S proteasome by Rechsteiner's and our lab (Hough et al. 1987; Waxman et al. 1987) utilized stabilizing conditions (e.g., glycerol) and biochemical assays originally developed in studies of the bacterial ATP-dependent proteases.

As discussed below, studies of these ATP-dependent protease complexes from archaea and bacteria have proven very valuable in illuminating the structure and enzymatic mechanisms of the 26S proteasome (Voges et al. 1999). Certainly, the 20S proteasome from archaea and its regulatory ATPase complex, PAN, have been most informative in this regard. Apparently, this ancestral system evolved before protein breakdown became linked in eukaryotes to ubiquitination to enhance the selectivity and regulation of this process. Because the key properties of the archaeal complex have been conserved, its study offers many unique advantages for elucidation of the 26S proteasome function.

Eubacteria can utilize any of a number of ATP-dependent proteases (e.g., Lon, ClpAP, ClpXP, HslUV, FtsH) to eliminate short-lived regulatory or unwanted abnormal proteins. In contrast, the cytosol and nucleus of eukaryotic cells contain only one ATP-dependent proteolytic complex, the 26S proteasome, which is much larger and has a much more complex structure than these prokaryotic enzymes. In addition, although several prokaryotic ATP-dependent proteases have an architectural organization similar to that of the eukaryotic 26S proteasomes, for all of them, except HslUV, the peptidase component does not share homology with the eukaryotic 20S proteasome.

Archaea and eukarya have a common ancestor that is not shared by eubacteria. Therefore, although archaea resemble eubacteria in most cytological features, many archaeal proteins and biochemical pathways are more closely related to those of eukarya (Doolittle and Brown 1994). An excellent illustration is the presence in archaea of 20S proteasomes, which are not found in eubacteria with the exception of actinomycetes, such as *Rhodococcus erythropolis* and *Mycobacteria*. Although archaea lack ubiquitin and the lid components of the 19S regulatory complex, their 20S proteasomes function in ATP-dependent degradation. Indeed, in place of the large 19S regulatory particle and its many distinct subunits, archaeal proteasomes depend on hexameric ATPase complexes of the AAA^{+} family. Some archaeal species, including *Methanococcus jannaschii* and *Archaeoglobus fulgidus*, contain one AAA ATPase, the proteasome-activating nucleotidase (PAN), that exhibits high sequence similarity to all six ATPases of the 19S. In the presence of ATP, PAN was shown to stimulate protein degradation by 20S proteasomes from *Thermoplasma acidophilum*, the best-characterized prokaryotic proteasomes (Maupin-Furlow et al. 2004; Zwickl et al. 2000). PAN appears to be the ancestor of the 19S base, before protein degradation became linked to ubiquitin conjugation and the evolution of the 19S lid from the signalosome particle. Because of the simplicity of archaeal PAN and 20S proteasomes, their homologous subunit organization, and their ease of expression in *E. coli*, their structural and biochemical properties have been extensively studied. The discovery of the PAN complex has provided a powerful experimental system for investigating the role of ATP and the biochemical mechanisms involved in the process of substrate translocation into 20S particles. In this chapter, we describe the archaeal PAN–20S complex and review how this complex and related eubacterial ATP-dependent proteases have helped us in understanding the biochemistry of energy requirement in protein breakdown.

10.2
Archaeal 20S Proteasomes

The 20S proteasome was first identified in an archaebacterium (*T. acidophilum*) by Dahlmann et al. (1989). Since then, 20S proteasomes have been found in many other archaeal species (for reviews, see Zwickl 2002 and Zwickl et al. 2000). Dahlmann et al. (1989) pointed out that archaeal 20S particles have a cylindrical shape similar to that of eukaryotic particles, but a much simpler subunit composition and a more limited spectrum of proteolytic activities. Indeed, most archaeal 20S particles contain only two different subunits, the α- and the β-subunits, although the genome of some species such as *Haloferax volcanii* and *Pyrococcus furiosus* contain two types of α- and β-subunits. Whether these different types of subunits exist within the same 20S particle, as has been shown for *R. erythropolis* particles, or in different class of proteasomes remains to be established. As in eukaryotic 20S particles, archaeal proteasomes are composed of four stacked rings, two inner rings composed of seven β-subunits and two outer rings composed of seven α-subunits.

The outer α-rings mediate interaction with proteasome activators (in eukaryotes, 19S ATPases or 11S) and control substrate entry and/or exit.

The α-subunit from *T. acidophilum* is homologous to the seven α-subunits from *Saccharomyces cerevisiae* (27–39% similarity as shown by multiple sequence alignment in Figure 10.1). A cluster of four conserved residues located at the amino-terminal extremities (Tyr8, Asp9, Pro17, and Tyr26, based on the numbering of *T. acidophilum* sequence) stabilizes a conformation with an open entry pore (Forster et al. 2003, 2005).

The β-subunits are responsible for the proteolytic activity. The β-subunit from *T. acidophilum* and three of the β-subunits from eukaryotic 20S (β1, β2, and β5) are produced as precursors, and are processed to an active form by removal of a prosequence (Voges et al. 1999). This processing leads to a primary sequence starting with a threonine residue. The primary sequence of the processed β-subunit from *T. acidophilum* has 23–25% similarity to those of *S. cerevisiae*, as shown by multiple sequence alignment with two absolutely conserved motifs (GXXXD and GSG) (Figure 10.2). Most conserved residues are located in the N-terminal region and certain ones are of particular importance for catalysis as Thr1, Glu17, and Lys33 (based on the numbering of *T. acidophilum* sequence).

Proteasomes from *T. acidophilum* were initially reported to exhibit only chymotrypsin-like activity (cleave after hydrophobic residues using standard model fluorogenic peptide substrates) and no substantial trypsin-like cleavages (after basic residues) or caspase-like activity (after acidic residues) as found in eukaryotic 20S particles (Dahlmann et al. 1989). Similar observations were made on proteasomes from other archaea such as *H. volcanii* (Wilson et al. 1999). However, subsequent studies indicated that *T. acidophilum* particles have a clear capacity to hydrolyze standard basic and acidic peptide substrates, although much more slowly than the standard substrate of the chymotrypsin-like activity (Akopian et al. 1997). Furthermore, when the peptides produced during polypeptide degradation by *T. acidophilum* 20S were examined, chymotryptic cleavages were neither the exclusive nor the predominant type, suggesting that proteasome active sites have a broader specificity than can be assayed by studies with several fluorogenic or chromogenic peptides (Akopian et al. 1997; Wenzel et al. 1994). In fact, these particles were recently shown to rapidly cleave after glutamine residues, which the three specialized active sites of eukaryotic proteasomes cannot do (Venkatraman et al. 2004). Also, the 20S proteasomes from *Methanosarcina thermophila* and *M. jannaschii* exhibited both chymotryptic- and caspase-like activity against model substrates (Maupin-Furlow et al. 1998; Maupin-Furlow and Ferry 1995).

Certainly the clearest difference between archaeal and eukaryotic proteasomes lies in the mechanism for recognition of the substrate. In eukaryotes, the main pathway to select proteins for degradation by 26S proteasomes is by covalent attachment of multiple ubiquitin moieties through a complex enzymatic cascade involving at least three types of enzymes (E1, ubiquitin activating; E2, ubiquitin conjugating; E3, ubiquitin ligase) and ATP hydrolysis (Pickart and Eddins 2004). A ubiquitin conjugation machinery has never been found in any bacterium, nor has any other general mechanism to target proteins for degradation been identified in

```
T20S α     ----MQQGQMAYDRAITVFSPDGRLFQVEYAREAVKKG-STALGMKFANGVLLISDKKVR 55
Y20S α1    MSGAAAASAAGYDRHITIFSPEGRLYQVEYAFKATNQTNINSLAVRGKDCTVVISQKKVP 60
Y20S α2    -------MTDRYSFSLTTFSPSGKLGQIDYALTAVKQG-VTSLGIKATNGVVIATEKKSS 52
Y20S α3    ------MGSRRYDSRTTIFSPEGRLYQVEYALESISHA-GTAIGIMASDGIVLAAERKVT 53
Y20S α4    --------MSGYDRALSIFSPDGHIFQVEYALEAVKRG-TCAVGVKGKNCVVLGCERRST 51
Y20S α5    ----MFLTRSEYDRGVSTFSPEGRLFQVEYSLEAIKLG-STAIGIATKEGVVLGVEKRAT 55
Y20S α6    ------MFRNNYDGDTVTFSPTGRLFQVEYALEAIKQG-SVTVGLRSNTHAVLVALKRNA 53
Y20S α7    ----MTSIGTGYDLSNSVFSPDGRNFQVEYAVKAVENG-TTSIGIKCNDGVVFAVEKLIT 55

T20S α     SRLIE-QNSIEKIQLIDDYVAAVTSGLVADARVLVDFAR-ISAQQEKVTYGSLVNIENLV 113
Y20S α1    DKLLD-PTTVSYIFCISRTIGMVVNGPIPDARNAALRAK-AEAAEFRYKYGYDMPCDVLA 118
Y20S α2    SPLAM-SETLSKVSLLTPDIGAVYSGMGPDYRVLVDKSRKVAHTSYKRIYGEYPPTKLLV 111
Y20S α3    STLLEQDTSTEKLYKLNDKIAVAVAGLTADAEILINTAR-IHAQNYLKTYNEDIPVEILV 112
Y20S α4    LKLQDTRITPSKVSKIDSHVVLSFSGLNADSRILIEKAR-VEAQSHRLTLEDPVTVEYLT 110
Y20S α5    SPLLE-SDSIEKIVEIDRHIGCAMSGLTADARSMIEHAR-TAAVTHNLYYDEDINVESLT 113
Y20S α6    DELS---SYQKKIIKCDEHMGLSLAGLAPDARVLSNYLR-QQCNYSSLVFNRKLAVERAG 109
Y20S α7    SKLLV-PQKNVKIQVVDRHIGCVYSGLIPDGRHLVNRGR-EEAASFKKLYKTPIPIPAFA 113

T20S α     KRVADQMQQYTQ-YGG-----VRPYGVSLIFAGIDQIG--PRLFDCDPAGTINEYKATAI 165
Y20S α1    KRMANLSQIYTQRAY------MRPLGVILTFVSVDEEL-GPSIYKTDPAGYYVGYKATAT 171
Y20S α2    SEVAKIMQEATQSGG------VRPFGVSLLIAGHDEFN-GFSLYQVDPSGSYFPWKATAI 164
Y20S α3    RRLSDIKQGYTQHGG------LRPFGVSFIYAGYDDRY-GYQLYTSNPSGNYTGWKAISV 165
Y20S α4    RYVAGVQQRYTQ-SGG-----VRPFGVSTLIAGFDPRDDEPKLYQTEPSGIYSSWSAQTI 164
Y20S α5    QSVCDLALRFGEGASGEERLMSRPFGVALLIAGHDADD-GYQLFHAEPSGTFYRYNAKAI 172
Y20S α6    HLLCDKAQKNTQSYGG------RPYGVGLLIIGYDKS--GAHLLEFQPSGNVTELYGTAI 161
Y20S α7    DRLGQYVQAHTLYNS------VRPFGVSTIFGGVDKNG--AHLYMLEPSGSYWGYKGAAT 165

T20S α     GSGKDAVVSFLEREYKENLP---EKEAVTLGIKALKSSLEEGEE---------------- 206
Y20S α1    GPKQQEITTNLENHFKKSKIDHINEESWEKVVEFAITHMIDALGTEFSK----------- 220
Y20S α2    GKGSVAAKTFLEKRWNDELE---LEDAIHIALLTLKESVEGEFNGDTIELAIIGDE---- 217
Y20S α3    GANTSAAQTLLQMDYKDDMK---VDDAIELALKTLSKTTDSSALTYDRLEFATIRKGAND 222
Y20S α4    GRNSKTVREFLEKNYDRKEPPATVEECVKLTVRSLLEVVQTGAKNIEITVVKPDSD---- 220
Y20S α5    GSGSEGAQAELLNEWHSSLS---LKEAELLVLKILKQVMEEKLDE--------------- 214
Y20S α6    GARSQGAKTYLERTLDTFIK---IDGNPDELIKAGVEAISQSLRDESLT----------- 207
Y20S α7    GKGRQSAKAELEKLVDHHPEGLSAREAVKQAAKIIYLAHEDNKEKDFELEISWCSLSETN 225

T20S α     ---------LKAPEIASITVGNKYRIYDQEEVKKFL------------------------ 233
Y20S α1    ----------NDLEVGVATKDKFFTLSAENIEERLVAIAEQD------------------ 252
Y20S α2    ------NPDLLGYTGIPTDKGPRFRKLTSQEINDRLEAL--------------------- 250
Y20S α3    --GEVYQKIFKPQEIKDILVKTGITKKDEDEEADEDMK---------------------- 258
Y20S α4    ------IVALSSEEINQYVTQIEQEKQEQQEQDKKKKSNH-------------------- 254
Y20S α5    ----------NNAQLTCITKQDGFKIYDNEKTAELIKEL--------------------- 243
Y20S α6    ---------VDNLSIAIVGKDTPFTIYDGEAVAKYI------------------------ 234
Y20S α7    GLHKFVKGDLLQEAIDFAQKEINGDDDEDEDDSDNVMSSDDENAPVATNANATTDQEGDI 285

T20S α     ---
Y20S α1    ---
Y20S α2    ---
Y20S α3    ---
Y20S α4    ---
Y20S α5    ---
Y20S α6    ---
Y20S α7    HLE 288
```

Fig. 10.1. Sequence alignment of 20S proteasome α-subunits. The sequence of the *T. acidophilum* 20S proteasome α-subunit was aligned with those of the seven α-subunits of the *Saccharomyces cerevisiae* 20S proteasome with ClustalW. Identical residues in all sequences are shown in gray.

```
              *             *             *
T20S β    TTTVGITLKDAVIMATERRVTMENFIMHKNGKKLFQIDTYTGMTIAGLVGDAQVLVRYMK 60
Y20S β1   TSIMAVTFKDGVILGADSRTTTGAYIANRVTDKLTRVHDKIWCCRSGSAADTQAIADIVQ 60
Y20S β2   TTIVGVKFNNGVVIAADTRSTQGPIVADKNCAKLHRISPKIWCAGAGTAADTEAVTQLIG 60
Y20S β5   TTTLAFRFQGGIIVAVDSRATAGNWVASQTVKKVIEINPFLLGTMAGGAADCQFWETWLG 60

T20S β    AELELYRLQRRVNMPIEAVATLLSNMLNQVKYMPYMVQLLVGGIDTAP--HVFSIDAAGG 118
Y20S β1   YHLELYTSQYG-TPSTETAASVFKELCYENKDN-LTAGIIVAGYDDKNKGEVYTIPLGGS 118
Y20S β2   SNIELHSLYTSREPRVVSALQMLKQHLFKYQGH-IGAYLIVAGVDPTG-SHLFSIHAHGS 118
Y20S β5   SQCRLHELREKERISVAAASKILSNLVYQYKGAGLSMGTMICGYTRKEGPTIYYVDSDGT 120

T20S β    SVEDIYASTGSGSPFVYGVLESQYSEKMTVDEGVDLVIRAISAAKQRDSASGG----MID 174
Y20S β1   VHKLPYAIAGSGSTFIYGYCDKNFRENMSKEETVDFIKHSLSQAIKWDGSSGG---VIRM 175
Y20S β2   TDVGYYLSLGSGSLAAMAVLESHWKQDLTKEEAIKLASDAIQAGIWNDLGSGSNVDVCVM 178
Y20S β5   RLKGDIFCVGSGQTFAYGVLDSNYKWDLSVEDALYLGKRSILAAAHRDAYSGG-----SV 175

T20S β    VAVITRKDGYVQLPTDQIESRIRKLGLIL------------------------- 203
Y20S β1   VVLTAAGVERLIFYPDEYEQL--------------------------------- 196
Y20S β2   EIGKDAEYLRNYLTPNVREEKQKSYKFPRGTTAVLKESIVNICDIQEEQVDITA 232
Y20S β5   NLYHVTEDGWIYHGNHDVGELFWKVKEEEGSFNNVIG----------------- 212
```

Fig. 10.2. Sequence alignment of 20S proteasome β-subunits. The sequence of the *T. acidophilum* 20S proteasome β-subunit was aligned with those of the processed $\beta 1$, $\beta 2$, and $\beta 5$ subunits of the *S. cerevisiae* 20S proteasome with ClustalW. Identical residues in all sequences are shown in gray. The asterisks at the top of the *T. acidophilum* 20S sequence indicate the residues (Thr1, Glu17, and Lys33) that are of particular importance in catalyzing peptide bond cleavage.

these organisms. Wenzel and Baumeister (1993, 1995) showed that purified archaeal 20S particles by themselves in the absence of any ATPase component can degrade certain unfolded proteins such as phenylhydrazine-treated hemoglobin, oxidant-damaged α-lactalbumin, and reduced α-lactalbumin. Likewise, the loosely folded casein, oxidized alkaline phosphatase, and reduced insulin-like growth factor (IGF-1) are rapidly degraded by these particles (Akopian et al. 1997; Kisselev et al. 1998), although low levels of SDS can activate these archaeal particles further by facilitating entry of protein substrates, as they do in eukaryotic 20S.

The simpler organization of archaeal proteasomes has made possible major advances in our knowledge of the biochemical and structural properties of proteasomes. A key step was the cloning and efficient co-production of its α- and β-subunits in *Escherichia coli* by Zwickl et al. (1992). The first crystal structure of a 20S particle was solved by Huber and Baumeister's laboratories in 1995 using the *T. acidophilum* 20S (Lowe et al. 1995). The X-ray analysis revealed that α- and β-subunits both had a novel fold that was later defined as a characteristic feature of the N-terminal nucleophile (Ntn) hydrolase protein family. Each subunit is made of two central antiparallel β-sheets flanked by two α-helixes on one side and three α-helixes on the other side. The four stacked rings have an elongated cylindrical shape and appear tightly packed, so that peptide and protein substrates can enter the particle only through a central channel in the α-ring (Figure 10.3). The central channel has three large cavities separated by narrow constrictions. The two outer

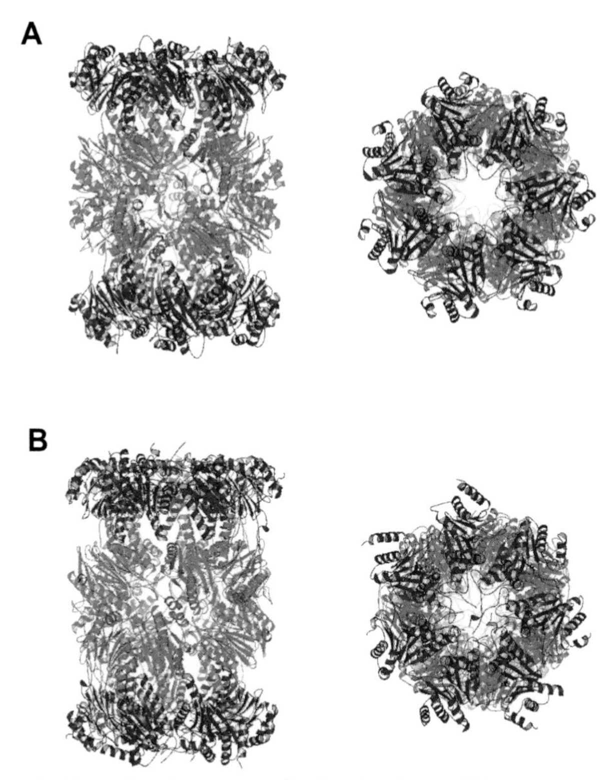

Fig. 10.3. Crystal structures of archaeal and yeast 20S proteasomes. Side (left panels) and top (right panels) views of *T. acidophilum* (panel A) and *S. cerevisiae* (panel B) 20S proteasomes. α- and β-subunits are represented in blue and red, respectively.

antechambers are located at the interface between the α- and β-rings, while the central cavity is formed by the β-rings that contain the active sites.

The catalytic mechanism of 20S proteasomes had long been unclear, and solving the tridimensional structure of archaeal 20S proteasomes, especially in the presence of a competitive inhibitor, elucidated its novel mechanism. These studies and site-directed mutagenesis (Seemuller et al. 1995) identified the amino-terminal threonine of the β-subunits as responsible for the nucleophilic attack and as the primary proton acceptor in peptide bond cleavage. These findings led to the classification of 20S proteasomes in the new superfamily of Ntn hydrolases. The ease of production of α- and β-subunits in *E. coli* has also made possible studies that clarified the processing event that generates an N-terminal threonine on the β-subunits. Seemuller et al. (1996) showed that the processing is autocatalytic and mediated by the amino-terminal threonine, another common feature of Ntn hydrolases.

Studies of the products generated during the degradation of protein substrates showed that *T. acidophilum* 20S, like eukaryotic particles, digests proteins in a processive manner to oligopeptides whose sizes range between 3 and 24 residues (with a mean size of 6–10 residues) without the release of degradation intermediates (Akopian et al. 1997; Kisselev et al. 1998; Wenzel et al. 1994). Because product size is very similar to that in eukaryotic proteasomes, this distribution of peptide size cannot be due to the number, specificity, or tridimensional location of the active sites. Therefore, the size of the products generated is probably determined by a kinetic competition between further cleavages and the ability of a product to diffuse out of the catalytic chamber (Kisselev et al. 1998; Kohler et al. 2001). These fundamental findings about the particle's structure, catalytic mechanism, and processivity were all rapidly extended to eukaryotic proteasomes (Fenteany et al. 1995; Groll et al. 1997; Kisselev et al. 1999; Nussbaum et al. 1998), where many of these questions had proven harder to resolve.

10.3 PAN the Archaeal Homologue of the 19S Complex

Although studies in bacterial and animal cells had clearly established that ATP was required for intracellular proteolysis (for reviews, see Ciechanover 2005, Goldberg 2005, and Goldberg and St John 1976), it was initially assumed that archaeal proteasomes degrade proteins independently of ATP hydrolysis. Also, several groups had failed to demonstrate proteasome-regulatory ATPases in archaea. The X-ray diffraction of the 20S had clearly demonstrated that a narrow opening controlled access to the proteolytic chamber (Lowe et al. 1995). Using nanogold-labeled insulin, Wenzel and Baumeister (1995) nicely demonstrated that protein substrates reached the central chamber by passing the narrow entry pore in the α-rings, through which only unfolded proteins could enter. In the crystalline structure, this entry pore at each end of *T. acidophilum* 20S particles appeared initially to be open (Lowe et al. 1995), whereas the equivalent entry channels in the eukaryotic particles are sealed (Groll et al. 1997; Unno et al. 2002) and necessitate an ATP-dependent mechanism for gate opening and translocation (Groll et al. 2000; Kohler et al. 2001). Because *T. acidophilum* 20S particles were able to degrade certain unfolded proteins (Wenzel and Baumeister 1993, 1995) and because no clear gate was evident in the X-ray structure (Lowe et al. 1995), it was generally assumed that protein unfolding was the only prerequisite for degradation by archaeal 20S. In eukaryotes, it has long been clear that ATP was necessary for the hydrolysis of ubiquitinated proteins and for the degradation of certain proteins that could not be conjugated to ubiquitin, because of a lack of free amino groups (Tanaka et al. 1983). These observations eventually led to the isolation by Rechteiner's (Hough et al. 1986, 1987) and Goldberg's (Waxman et al. 1987) laboratories of the large ATP-dependent complex, now known as the 26S proteasomes (for review, see Ciechanover 2005 and Goldberg 2005). At that time, this structure was believed to be distinct from the 600-kDa multicatalytic particle we later named the proteasome

(Arrigo et al. 1988). Subsequent studies using antibodies demonstrated an essential role for the 20S particle in degradation of ubiquitin conjugates (Matthews et al. 1989) and its association with other components to form the ATP-dependent 26S proteasome (Eytan et al. 1989). Further characterization of the 19S regulatory complex, especially by Rechsteiner, De Martino, Tanaka, and coworkers, led to the cloning and identification of the key ATPase subunits in eukaryotes.

The 19S ATPases, PAN, and the ATP-hydrolyzing protease complexes in bacteria and mitochondria (ClpA, ClpX, HslU, FtsH, and Lon) are all members of the AAA^+ (ATPases associated with various cellular activities) ATPase superfamily (for review, see Ogura and Wilkinson 2001). These ATPases are found in all living organisms and in all cell compartments, where they participate in a variety of essential cellular processes such as mitosis, protein folding and translocation, DNA replication and repair, membrane fusion, and proteolysis. AAA^+ ATPases are characterized by the presence of one or two conserved ATP-binding domains (200–250 residues), called AAA motifs, consisting of a Walker A and a Walker B motif (Confalonieri and Duguet 1995). The 19S-associated ATPases and PAN belong to a subfamily of AAA^+ ATPases that contains an additional motif called the second region of homology (SRH) (Lupas and Martin 2002). Despite the large variety of cellular processes in which AAA^+ ATPases participate, they have some common features. A recurrent structural feature of most AAA^+ ATPases is their assembly in oligomeric (generally hexameric), ring-shaped structures with a central pore. In addition, most appear to be involved in protein folding or unfolding, assembly or disassembly of protein complexes through nucleotide-dependent conformational changes.

Several groups had unsuccessfully attempted to demonstrate the ATP dependence of proteolysis by archaeal proteasomes. In 1996, the complete sequencing of the methanogenic archaeon *M. jannaschii* revealed the presence of two genes (named *S4* and *S8*) that are highly homologous to the genes encoding for the 19S ATPases (Bult et al. 1996). To test whether it regulates proteolysis, the *S4* gene was expressed in *E. coli*, and the 50-kDa product (PAN) was purified and characterized by Zwickl et al. (1999). The primary sequence of PAN contains only one AAA domain (residues 200–342) that includes hallmarks of this ATPase family: one P-loop motif (which includes the Walker A and B motifs) and an SRH motif at its C-terminus (Figure 10.4). PAN shares 41–45% similarity with human and yeast 19S ATPases (Zwickl et al. 1999). As seen by multiple sequence alignment with yeast Rpt1–6 19S subunits, a number of conserved residues in the PAN and 19S ATPases are found in the P-loop and SRH motifs (Figure 10.5). Both PAN and 19S ATPases possess a predicted coiled-coil motif at their N-termini (Zwickl et al. 1999).

When purified to homogeneity, PAN was shown to exist in solution as a homo-oligomeric complex of 650 kDa that has a hexameric ring structure (P. Zwickl, unpublished data; Wilson et al. 2000) and exhibits Mg^{2+}-dependent ATPase activity (Zwickl et al. 1999). In its initial characterization, PAN with ATP present was shown to promote selectively the breakdown of proteins lacking tertiary structure, including casein and oxidized RNAse A (Zwickl et al. 1999). Subsequently, PAN

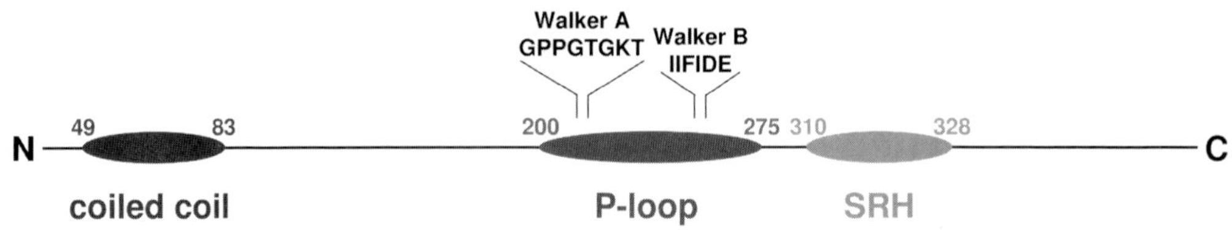

Fig. 10.4. Schematic representation of the primary sequence of PAN. PAN exhibits typical features of an AAA ATPase, i.e., a P-loop domain with Walker A and Walker B motifs and a second region of homology (SRH) at its C-terminal. The N-terminal part of PAN was predicted to contain a coiled-coil region (Zwickl et al. 1999).

was also shown to promote degradation of globular proteins such as green fluorescent protein (GFP) in a reaction requiring ATP hydrolysis (Benaroudj and Goldberg 2000). In fact, PAN by itself could catalyze ATP-dependent unfolding of stable globular proteins. As yet, no similar ATP-dependent unfolding process has been demonstrated with pure 19S particles or 26S complexes.

Although the association of an ATPase chaperone–like complex with a proteolytic particle appears to be a common feature of several systems for intracellular protein degradation (the 26S proteasome and the bacterial ClpAP, ClpXP, and HslUV complexes), an association between PAN and the 20S particle was difficult to observe by standard biochemical approaches. In fact, this failure to demonstrate such a complex led some investigators to suggest alternative mechanisms to explain the stimulation of protein degradation by PAN acting on the substrate, without formation of a PAN–20S complex (Forster et al. 2003). Alternatively, the complex between PAN and 20S proteasomes could be hard to detect because it is short-lived (e.g., compared to the complex between the 19S and eukaryotic 20S). In recent studies, the association of PAN with archaeal 20S proteasomes was demonstrated by Smith et al. (submitted for publication). Using immunoprecipitation, surface plasmon resonance, and electron microscopy, complex formation between PAN and the 20S was observed in the presence of ATP and non-hydrolyzable analogue (AMPPNP or ATPγS) but not in the presence of ADP or in the absence of any nucleotide. Thus, the association of PAN with the 20S is favored upon ATP binding and is reduced by ATP hydrolysis.

Electron microscopy of the PAN–20S complex demonstrated that PAN could associate with 20S at either one or both ends (Figure 10.6B,C). PAN appears as a two-ring structure, with a large inner ring and a smaller outer ring, and resembles a "top hat" capping the 20S cylinder. The position of the outer ring resembles the coiled coil containing the intermediate domain of HslU that protrudes outward from the HslUV complex (Bochtler et al. 2000; Sousa et al. 2000). Interestingly, the PAN–20S complex structure is remarkably similar to a 26S proteasome lacking its lid components and the non-ATPase subunits of the base (Figure 10.6D). In fact, the PAN–20S complex images can be exactly superimposed on the densities in the 26S complex and strongly suggest that an unidentified density in the 19S corresponds to the coiled-coil domain of the ATPase. These studies demonstrate

```
PAN    MVFEEFISTELKKEKK------------AFTEEFKEEKEINDNSN---LKNDL--LKEELQEKAR--------IAE 59
Rpt1   MPPKEDWEKYKAPLEDDDKKPDDDKIVPLTEGDIQVLKSYGAAPYAAKLKQTENDLKDIEARIKEKAGVKESDTGL 76
Rpt2   -MGQGVSSGQDKKKKK------------GSNQKPKYEPPVQSKFGRKKRKGGP--ATAEKLPNIYPSTRCKLKLLR 61
Rpt3   MEELGIVTPVEKAVEE------------KPAVKS-YASLLAQLNG--TVNNNS--ALSNVNSDIY------FKLKK 59
Rpt4   MS--EEQDPLLAGLG-------------ETSGDNHTQQSHEQQP--EQPQETEEHHEEEPSRVDPEQ---EAHNKA 59
Rpt5   MATLEELDAQTLPGDD------------ELDQEILNLSTQELQTRAKLLDNEIRIFRSELQRLSHEN---NVMLEK 64
Rpt6   ---------MTAAVT----------------SSNIVLETHESGI----KPYFEQKIQETELKIRSKT--------- 47

PAN    LESRILKLELEKK---------ELERE--NLQLMKENEILRRELDRMRVPP---LIVGTVVDKVGERKVVVKSSTG 116
Rpt1   APSHLWDIMGDRQRLGEEHPLQVARCTKIIKGNGESDETTTDNNNSGNSNSNSNQQSTDADEDDEDAKYVINLKQI 152
Rpt2   MERIKDHLLLEEE---------FVSNSEILKPFEKKQEEEKKQLEEIRGNP---LSIGTLEEIIDDDHAIVTSPTM 128
Rpt3   LEKEYELLTLQED---------YIKDE--QRHLKRELKRAQEEVKRIQSVP---LVIGQFLEPIDQNTGIVSSTTG 118
Rpt4   LNQFKRKLLEHRR-----YDDQLKQRRQNIRDLEKLYDKTENDIKALQSIG---QLIGEVMKELSEEKYIVKASSG 127
Rpt5   IKDNKEKIKNNRQ---------LPYLVANVVEVMDMNEIEDKENSESTTQG---GNVNLDNTAVG-KAAVVKTSSR 128
Rpt6   --ENVRRLEAQRN--------ALNDKVRFIKDELRLLQEPGS-------------YVGEVIKIVSDKKVLVKVQPE 102

PAN    PSFLVNVSHFVNPDDLAPGKRVCLNQQTLTVVDVLPENKDYRAKAMEVDERPNVRYEDIGGLEKQMQEIREVVELP 189
Rpt1   AKFVVGLGERVSPTDIEEGMRVGVDRSKYNIELPLPPRIDPSVTMMTVEEKPDVTYSDVGGCKDQIEKLREVVELP 228
Rpt2   PDYYVSILSFVDKELLEPGCSVLLHHKTMSIVGVLQDDADPMVSVMKMDKSPTESYSDIGGLESQIQEIKESVELP 201
Rpt3   MSYVVRILSTLDRELLKPSMSVALHRHSNALVDILPPDSDSSISVMGENEKPDVTYADVGGLDMQKQEIREAVELP 191
Rpt4   PRYIVGVRNSVDRSKLKKGVRVTLDITTLTIMRILPRETDPLVYNMTSFEQGEITFDGIGGLTEQIRELREVIELP 200
Rpt5   QTVFLPMVGLVDPDKLKPNDLVGVNKDSYLILDTLPSEFDSRVKAMEVDEKPTETYSDVGGLDKQIEELVEAIVLP 200
Rpt6   GKYIVDVAKDINVKDLKASQRVCLRSDSYMLHKVLENKADPLVSLMMVEKVPDSTYDMVGGLTKQIKEIKEVIELP 167

PAN    LKHPELFEKVGIEPPKGILLYGPPGTGKTLLAKAVATETNATFIRVVGSELVKKFIGEGASLVKDIFKLAKEKAPS 265
Rpt1   LLSPERFATLGIDPPKGILLYGPPGTGKTLCARAVANRTDATFIRVIGSELVQKYVGEGARMVRELFEMARTKKAC 304
Rpt2   LTHPELYEEMGIKPPKGVILYGAPGTGKTLLAKAVANQTSATFLRIVGSELIQKYLGDGPRLCRQIFKVAGENAPS 277
Rpt3   LVQADLYEQIGIDPPRGVLLYGPPGTGKTMLVKAVANSTKAAFIRVNGSEFVHKYLGEGPRMVRDVFRLARENAPS 267
Rpt4   LKNPEIFQRVGIKPPKGVLLYGPPGTGKTLLAKAVAATIGANFIFSPASGIVDKYIGESARIIREMFAYAKEHEPC 276
Rpt5   MKRADKFKDMGIRAPKGALMYGPPGTGKTLLARACAAQTNATFLKLAAPQLVQMYIGEGAKLVRDAFALAKEKAPT 276
Rpt6   VKHPELFESLGIAQPKGVILYGPPGTGKTLLARAVAHHTDCKFIRVSGAELVQKYIGEGSRMVRELFVMAREHAPS 243
                           Walker A
PAN    IIFIDEIDAIAAKRTDALTGGDREVQRTLMQLLAEMDGFDARGDVKIIGATNRPDILDPAILRPGRFDRIIEVPAP 341
Rpt1   IIFFDEIDAVGGARFDDGAGGDNEVQRTMLELITQLDGFDPRGNIKVMFATNRPNTLDPALLRPGRIDRKVEFSLP 380
Rpt2   IVFIDEIDAIGTKRYDSNSGGEREIQRTMLELLNQLDGFDDRGDVKVIMATNKIETLDPALIRPGRIDRKILFENP 353
Rpt3   IIFIDEVDSIATKRFDAQTGSDREVQRILIELLTQMDGFDQSTNVKVIMATNRADTLDPALLRPGRLDRKIEFPSL 343
Rpt4   IIFMDEVDAIGGRRFSEGTSADREIQRTLMELLTQMDGFDNLGQTKIIMATNRPDTLDPALLRPGRLDRKVEIPLP 352
Rpt5   IIFIDELDAIGTKRFDSEKSGDREVQRTMLELLNQLDGFSSDDRVKVLAATNRVDVLDPALLRSGRLDRKIEFPLP 352
Rpt6   IIFMDEIDSIGSTRVEGSGGGDSEVQRTMLELLNQLDGFETSKNIKIIMATNRLDILDPALLRPGRIDRKIEFPPP 319
       Walker B                                            SRH
PAN    -DEKGRLEILKIHTRKMNLAEDVNLEEIAKMTEGCVGAELKAICTEAGMNAIRELRDYVTMDDFRKAVEKIMEKKK 416
Rpt1   -DLEGRANIFRIHSKSMSVERGIRWELISRLCPNSTGAELRSVCTEAGMFAIRARRKVATEKDFLKAVDKVISGYK 455
Rpt2   -DLSTKKKILGIHTSKMNLSEDVNLETLVTTKDDLSGADIQAMCTEAGLLALRERRMQVTAEDFKQAKERVMKNKV 428
Rpt3   RDRRERRLIFGTIASKMSLAPEADLDSLIIRNDSLSGAVIAAIMQEAGLRAVRKNRYVILQSDLEEAYATQVKTDN 419
Rpt4   -NEAGRLEIFKIHTAKVKKTGEFDFEAAVKMSDGFNGADIRNCATEAGFFAIRDDRDHINPDDLMKAVRKVAE-VK 427
Rpt5   -SEDSRAQILQIHSRKMTTDDDINWQELARSTDEFNGAQLKAVTVEAGMIALRNGQSSVKHEDFVEGISEVQARK- 427
Rpt6   -SVAARAEILRIHSRKMNLTRGINLRKVAEKMNGCSGADVKGVCTEAGMYALRERRIHVTQEDFELAVGKVMN--K 394

PAN    VKVKEPAHLDVLYR 430
Rpt1   KFSSTSRYMQYN-- 467
Rpt2   EENLEGLYL----- 437
Rpt3   TVDKFDFYK----- 428
Rpt4   KLEGTIEYQKL--- 437
Rpt5   -SKSVSFYA----- 434
Rpt6   NQETAISVAKLFK- 405
```

Fig. 10.5. Sequence alignment of proteasomal ATPases. The sequence of *M. jannaschii* PAN was aligned with those of the six Rpt ATPase subunits of the *S. cerevisiae* 19S proteasome with ClustalW. Identical residues in all sequences are in gray. Walker A and B motifs, as well as the SRH, are underlined.

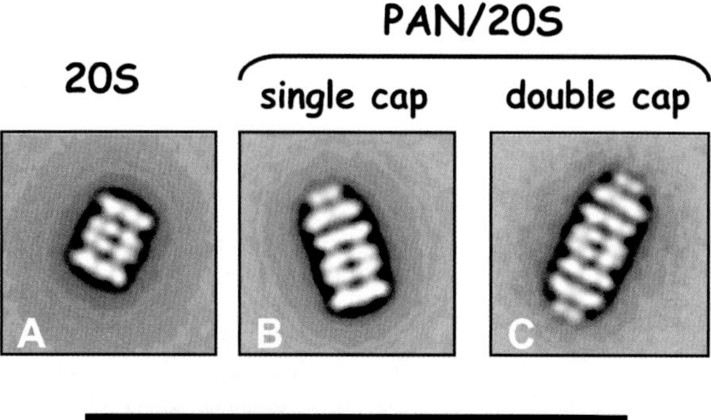

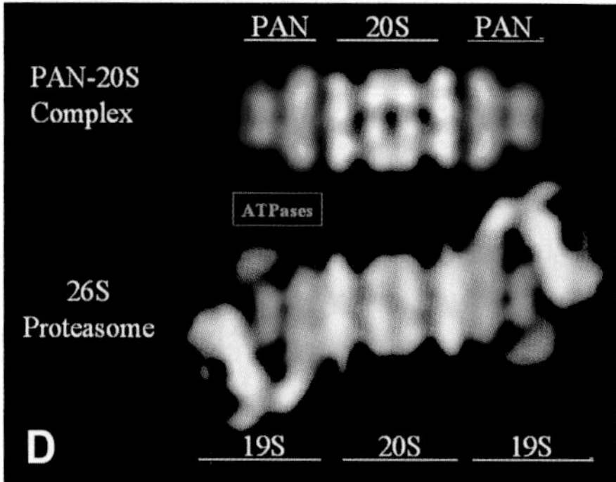

Fig. 10.6. Electron microscopy analysis of the archaeal PAN–20S and proteasomes complex. Electron micrographs of the negatively stained archaeal 20S proteasome (A), the singly capped archaeal PAN–20S complex (B), and the doubly capped archaeal PAN–20S–PAN complex (C). Negatively stained particles of the doubly capped PAN–20S complex are compared to the mammalian 26S proteasome in (D).

that PAN physically associates with 20S proteasomes for ATP-dependent protein degradation to occur.

Degradation of tri- or tetrapeptides by archaeal 20S proteasomes is not enhanced in the presence of PAN and ATP, probably because such small peptides can readily diffuse into the particle. In fact, this inability to stimulate degradation of small peptides distinguishes PAN from the 19S complex, which stimulates ATP-dependent degradation of small peptides (Kohler et al. 2001). However, Smith et al. (submitted) recently showed that PAN and ATP do stimulate degradation of peptides of seven residues or longer. Thus, a gated pore exists in the archaeal particles that allows entry of small peptides, but longer peptides (over seven residues), like proteins (Benaroudj et al. 2003), are excluded in the absence of PAN and ATP.

Thus, because PAN in the presence of ATP interacts with 20S particles and stimulates proteasomal degradation of oligopeptides and of globular and unfolded proteins, it exhibits features similar to those of the 19S complex in protein degradation. These many functional similarities between PAN and the 19S complex further confirm that PAN was the evolutionary precursor of the eukaryotic 19S complex. Moreover, because of its greater simplicity, the PAN–20S complex has proven to be tremendously useful in studies of the detailed mechanisms of protein degradation by 26S proteasomes.

10.4
VAT, a Potential Regulator of Proteasome Function

PAN is present in most archaeal species (Zwickl 2002), except *T. acidophilum* (Ruepp et al. 2000) and *T. volcanium* (Kawashima et al. 2000). In these organisms, ATP hydrolysis is probably still necessary for proteasomal degradation, which is likely supported by another AAA ATPase. One candidate is the protein VAT (VCP-like ATPase from *Thermoplasma acidophilum*), which is closely related to CDC48 of *Saccharomyces cerevisiae* and p97 (VCP) of vertebrates. These AAA family members are distantly related to the eukaryotic 19S ATPases and contain two AAA motifs. In eukaryotes, these proteins were first shown to participate in ER and Golgi membrane fusion (Latterich et al. 1995; Rabouille et al. 1995). More recent studies have demonstrated that mammalian p97/VCP and CDC48 are required for the ER-associated degradation (ERAD) of misfolded and ubiquitinated membrane proteins. Indeed, CDC48 interacts directly with the ubiquitin chains on the substrate and somehow facilitates hydrolysis of ubiquitinated protein by the 26S proteasomes (Dai and Li 2001; Rabinovich et al. 2002; Ye et al. 2001). This complex has also been proposed to be important in the degradation of subunits of cytosolic complexes, such as IκB (Dai et al. 1998; Dai and Li 2001). It is very likely that an ATP-dependent chaperone-like activity of p97/CDC48 is involved in the removal of the ubiquitinated protein from the ER and its subsequent association with 26S proteasomes (Jarosch et al. 2002; Ye et al. 2001). In archaea, where there is no ubiquitin pathway or ER or Golgi apparatus, the physiological function of VAT is still unknown. VAT may function with archaeal proteasomes in the breakdown of misfolded proteins. Purified VAT protein has an Mg^{2+}-dependent ATPase activity and assembles into a hexameric ring-shaped structure. Chaperone and potential unfoldase activities were also demonstrated for VAT (Golbik et al. 1999), but it remains to be established whether the VAT complex can activate protein degradation by archaeal 20S particles, in a fashion similar to PAN.

10.5
The Use of PAN to Understand the Energy Requirement for Proteolysis

Since the early 1970s, it has been clear that protein degradation in prokaryotes, as well as in eukaryotes, requires metabolic energy (Ciechanover 2005; Goldberg 2005; Goldberg and St John 1976). Although much has been learned about the requirement for ATP, the detailed mechanisms of the ATP-dependent proteolytic complexes are still unclear. In particular, elucidating the mechanisms whereby the 19S regulatory particle unfolds substrates and facilitates their entry into the 20S proteolytic core particle has long been a great challenge. Investigating these processes in the 26S proteasome is difficult because of the requirement for ubiquitination of substrates and the instability and complexity of the 19S particle, which contains at least 17 different subunits, including six nonidentical ATPases. To elucidate multiple roles of ATP in proteasome function in recent years, we have studied the regulatory complex PAN because many of its structural and enzymatic properties resemble those of the 19S ATPases. Moreover, because of its simple structural organization, lack of requirement for substrate ubiquitination, and ease of expression in *E. coli*, PAN offers many advantages for dissecting and clarifying the mechanism by which ATPases promote protein degradation by 20S proteasomes.

10.5.1
ATP Hydrolysis by PAN Allows Substrate Unfolding and Degradation

Because ATP hydrolysis by PAN did not enhance degradation of tri- or tetrapeptides by archaeal 20S, we initially hypothesized that the major role of PAN's ATPase activity in protein breakdown was to unfold globular substrates, a key step in facilitating their entry into the central proteolytic channel in the 20S (step 3, Figure 10.8). This unfoldase activity was verified by using as a substrate GFPssrA, a variant of GFP whose C-terminus has been fused to the 11-residue peptide ssrA. By itself, GFP is not a substrate and does not bind to PAN. In eubacteria, this unfolded C-terminal sequence is incorporated via a specific tRNA into nascent chains when ribosomes are stalled, and its presence targets the proteins for degradation by several ATP-dependant proteases (Gottesman et al. 1998; Herman et al. 1998; Keiler et al. 1996). GFP is a particularly stable protein, even at high temperatures (Tm > 65 °C) (Bokman and Ward 1981), whose unfolding can be easily monitored by a loss of its fluorescence by using the method introduced by the Horwich laboratory (Weber-Ban et al. 1999). As seen in Figure 10.7, PAN by itself catalyzed the unfolding of GFPssrA by a mechanism that required ATP hydrolysis (Benaroudj and Goldberg 2000; Benaroudj et al. 2003). This PAN-catalyzed unfolding of GFPssrA was critical in allowing its degradation by archaeal 20S proteasomes because 20S particles, in the absence of PAN, cannot degrade GFPssrA (Benaroudj and Goldberg 2000).

These findings provided the first experimental evidence that a proteasomal-associated ATPase has an unfoldase activity (Benaroudj et al. 2001). In fact, thus far no such unfoldase activity for purified eukaryotic 19S ATPases or 26S particles has been demonstrated, although such an activity seems very likely (Murakami et al. 2000), especially because GFP fusion proteins used for studying protein degradation in eukaryotes can be rapidly degraded *in vivo*. The 26S and the base of the 19S particles from yeast and mammals exhibit several activities characteristic of molecular chaperones, such as the ability to reduce protein aggregation and to promote the refolding of denatured proteins (Braun et al. 1999; Strickland et al. 2000). Also, 19S particles have been shown to be able to remodel certain substrates, e.g.,

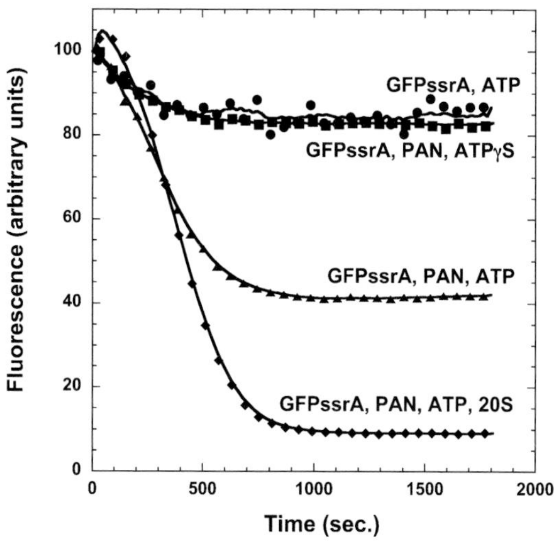

Fig. 10.7. PAN promotes ATP-dependent unfolding and proteasomal degradation of GFPssrA. The time course of fluorescence change of 500 nM of GFPssrA was followed at 45 °C (excitation at 400 nm and emission at 510 nm) in 50 mM tris (pH 7.5), 1 mM DTT, and 10 mM $MgCl_2$ in the presence of 2 mM of ATP (●); or 2 mM of ATP and 250 nM of PAN (▲); or 2 mM of ATPγS and 250 nM of PAN (■); or 2 mM of ATP, 250 nM of PAN, and 53.5 nM of archaeal 20S proteasomes (◆). The loss of GFPssrA fluorescence observed upon addition of PAN and ATP or of PAN, ATP, and 20S indicates GFPssrA unfolding and degradation, respectively.

they have the capacity to catalyze the reactivation of misfolded RNAse A and to expose otherwise buried chymotryptic sites in a folded substrate, the polyubiquitinated DHFR (Liu et al. 2002). However, ATP binding or hydrolysis by the 19S ATPases does not seem to be necessary for this remodeling activity. Moreover, a relationship between the chaperone-like activities of the 19S and protein breakdown by the 26S proteasomes remains to be established, although these activities seem very likely based upon the findings with PAN.

10.5.2
ATP Hydrolysis by PAN Serves Additional Functions in Protein Degradation

Our early findings indicated that ATP hydrolysis by PAN enhanced the degradation of various substrates that were loosely folded, such as casein (Zwickl et al. 1999). These findings suggested that ATP consumption by PAN facilitated additional steps in protein degradation aside from protein unfolding. Like the several AAA ATPases that promote protein degradation in *E. coli*, PAN's ATPase activity is stimulated two- to fivefold by protein substrates (step 1, Figure 10.8). Surprisingly, ATP hydrolysis by PAN is stimulated similarly by the globular GFPssrA, by the loosely folded casein, and even by the 11-residue ssrA recognition peptide. There-

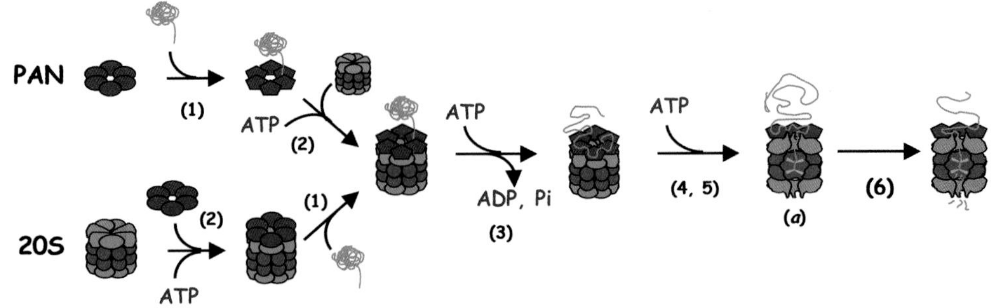

Fig. 10.8. Schematic representation of the different steps of archaeal PAN–20S-mediated protein degradation. PAN is shown as a blue hexameric ring, and the 20S proteasome is in green. Formation of the PAN–20S complex occurs upon ATP binding (step 2), and this association causes opening of the entry channel in the 20S (step 4). Gate opening by PAN requires ATP binding but not hydrolysis. Protein substrates (represented in pink) bind to PAN or to the PAN–20S complex (step 1) and trigger activation of ATP hydrolysis. The circular and pentagonal shapes of PAN subunits represent the substrate-free and substrate-bound forms of PAN, respectively. Protein substrates are unfolded in an ATP hydrolysis–dependent manner on the surface of the PAN ATPase ring (step 3). After protein unfolding, unfolded substrates are translocated into the internal chamber of 20S proteasomes (step 5). Particle *a* is a cross-section of the PAN–20S complex with an open gate that translocates an unfolded protein. 20S active sites are represented as cyan triangles inside the 20S internal chamber. Unfolded substrates studied thus far appear to be translocated by PAN into the 20S particle by a mechanism that requires only ATP binding, not ATP hydrolysis. Inside the 20S particles, proteins are processively degraded to small peptides by the multiple 20S active sites (step 6). The gate that precludes entry of protein substrates inside 20S particles also controls exit of peptide products out of the 20S particles.

fore, ATP hydrolysis is activated by substrate binding and not by the unfolding process (Benaroudj et al. 2003). Furthermore, prior denaturation of the GFPssrA did not accelerate its degradation by proteasomes, nor did it eliminate the requirement for PAN and ATP.

By measuring simultaneously the rate of ATP consumption and the rate of protein degradation, it was possible to determine the amount of ATP hydrolyzed during breakdown of different proteins. During the degradation of one molecule of the globular substrate GFPssrA, the PAN–20S complex consumed the same amount of ATP molecules (300–400 molecules) as during degradation of one molecule of denatured GFPssrA or the natively unfolded protein casein. Interestingly, this amount of ATP hydrolyzed during proteolysis corresponds to approximately one-third of the amount of ATP consumed during synthesis of these proteins. No such analysis has been carried out for the 26S proteasome, although presumably the amount of ATP utilized in degradation of proteins in eukaryotes is even higher due to the additional requirement of ATP for ubiquitination. It is also noteworthy that substrate unfolding by the ATPases does not appear to be the rate-limiting step in protein degradation.

10.5.3
PAN and ATP Regulate Gate Opening

By using a deletion variant of the archaeal 20S proteasome that lacks the N-terminal extremities of the α-subunits, we found that these residues, which correspond to those comprising the gated entry channel in the yeast proteasome (Groll et al. 2000; Kohler et al. 2001), also limit the entry of protein substrates in archaeal particles. The deletion of these residues facilitates the degradation of acid-denatured GFP and casein and eliminates the requirement for PAN and ATP for their degradation. These findings indicate that one role of PAN and ATP is to promote opening of the 20S gate (Benaroudj et al. 2003) (step 4, Figure 10.8). Recently, we found that this gateless 20S variant cleaved a variety of peptides, whose lengths range between 7 and 18 residues, at a much higher rate than did wild-type 20S (Smith et al., submitted). Therefore, the 20S particles from *T. acidophilum* possess a functional gate that excludes proteins and even peptides as small as heptamers. This discovery of a gate in the α-ring of the archaeal proteasomes was surprising because X-ray crystallography of this particle failed to indicate a specific density in this region (Lowe et al. 1995). In fact, the absence of a regulated gating mechanism for archaeal 20S proteasomes has been assumed to represent a major difference between archaeal and eukaryotic proteasomes (Groll et al. 2003; Groll and Huber 2003).

It is now clear that PAN and ATP regulate this gated entry channel into the archaeal 20S particles from *T. acidophilum*. PAN together with ATP or non-hydrolyzable ATP analogues was found to support gate opening. Thus, ATP binding by PAN, which also favors formation of the 20S–PAN complex, stimulates peptide entry through the 20S gate (Smith et al., submitted). In fact, a variety of observations strongly suggest that the association of PAN with the 20S triggers gate opening. One possible model is the non-homologous proteasome activator complex PA26 (11S) from *Trypanosoma brucei*, which (without ATP present) opens the pore into the yeast 20S proteasome by stabilizing an ordered conformation of the N-terminal extremities of α-subunits (Forster et al. 2003, 2005). Certain conserved residues in the gate of yeast 20S α-subunits appeared to be important in the stabilization of this open, ordered conformation. Because the corresponding residues in the archaeal 20S particles are also important in precluding substrate entry, it is attractive to hypothesize that PAN's association with the 20S upon ATP binding also stabilizes an open, ordered conformation of α-subunit N-terminal extremities through these residues. In support of this conclusion, we have recently shown that three conserved C-terminal residues in PAN are essential for both the ATP-dependent association with the 20S and gate opening (Smith, Chung, and Goldberg, in preparation). In fact, a peptide corresponding to the C-terminal residues by itself activates gate opening in a manner that requires the residues found in the motif that is conserved in most 19S ATPases. Most importantly, the peptide sequence that activates gate entry into the *T. acidophilum* 20S can do so in the 20S particle from rabbit muscle. These findings indicate a highly conserved mechanism for gate opening and for the role of ATP in this process. These detailed mechanisms, however, appear to differ from those controlling gate opening in the ATP-independent PA28 activators, which bear no sequence homology to PAN or the AAA family ATPases, although both activating mechanisms seem to involve C-terminal residues and binding sites on the 20S particle (Smith, Chung, and Goldberg, in preparation; Forster et al. 2005).

10.5.4
PAN and ATP Are Required for Translocation of Unfolded Substrates

A major challenge in studying the ATP requirement for protein translocation into the 20S is to dissociate this process from the process of substrate unfolding and gate opening. To study the role of ATP hydrolysis by PAN during substrate translocation, we tested whether the gateless 20S variant required ATP hydrolysis by PAN to degrade GFPssrA once it has been already unfolded by PAN. Interestingly, we found that although GFPssrA is unfolded by PAN, its degradation by 20S still requires ATP and PAN even when the 20S gate is open (Benaroudj et al. 2003). By using non-hydrolyzable analogues, we found that ATP binding is sufficient for translocation of certain unfolded substrates (Smith et al., submitted). Thus, after opening of the 20S gate, PAN in its ATP-bound form can allow translocation of unfolded proteins (casein, denatured ovalbumin, or denatured GFP) into the 20S proteolytic cavity (step 5, Figure 10.8). Thus, once a globular protein has been unfolded on the surface of PAN, it is not simply released into the medium to diffuse to nearby 20S particles. It remains possible that, with longer polypeptides or unfolded proteins with a tendency to refold, ATP hydrolysis-dependent unfolding may increase the rate of translocation and degradation. In any case, it is clear that while ATP hydrolysis is absolutely essential for the unfolding process, translocation of the bound, unfolded or loosely folded polypeptide can proceed by passive diffusion facilitated by PAN in its ATP-bound form.

10.6
Direction of Substrate Translocation

To reach the active sites within the 20S particle, substrates have to penetrate the narrow axial pore formed by the α-ring, presumably after traversing the pore in the ATPase ring of PAN or the 19S complex. These models raise the obvious questions of whether the degradation of a polypeptide chain starts from one specific end or the other and whether a substrate assumes a preferred or exclusive orientation when entering the 20S proteasome. To determine whether a polypeptide chain enters the 20S particle by its N- or C-terminus or by an internal loop, we attached bulky moieties to protein substrates that prevented their translocation through the pore in the PAN ATPase (Navon and Goldberg 2001; Navon et al., in preparation). GFPssrA was shown to be translocated exclusively in a C-to-N orientation. However, different substrates were found to be translocated in different fashions. While some proteins (maltose binding protein and GFPssrA) are transported into the proteasome by their C-terminus, others (casein) are exclusively translocated from their N-terminus, and some (calmodulin) from both directions (or by an internal loop). By contrast, the isolated 20S showed no such directional preference. Thus, the orientation of entry seems to be a property of the substrate's termini and its interaction with the ATPase. Accordingly, it has been observed that different substrates appear to enter eukaryotic 26S by different extremities (Prakash et al. 2004; Zhang and Coffino 2004; Navon et al., in preparation), and some even seem to enter by an internal loop (Liu et al. 2003). However, these observations were made with crude cell lysates or with pure 26S proteasomes and substrates that do not require ubiquitin conjugation, such as casein (Navon et al., in preparation), p21^{cip1}, and α-synuclein (Liu et al. 2003). The influence of ubiquitination on the unfolding process and the directionality of substrate translocation remain to be ascertained.

It is widely assumed that protein degradation by proteasomes is a highly processive process *in vivo*, as it is with isolated 20S particles (Akopian et al. 1997). Through studies of the degradation of a multi-domain polypeptide in cell extracts, Matouschek and colleagues (Lee et al. 2001; Prakash et al. 2004) concluded that proteasomal degradation requires not only a "degradation signal" on the polypeptide (e.g., ubiquitination) but also an unstructured region that is necessary for unfolding by the 19S complex. Once this sequence has been translocated into the proteasomes by the ATPase ring, the whole polypeptide chain is pulled into the 20S particle and degraded. Interestingly, if the polypeptide chain contains independent globular domains, it is degraded vectorially, starting from the domain that is closest to the part that first enters the 20S particle. However, if it contains a particularly stable globular domain that obstructs the pore entrance, the fragment containing the globular domain is released from the proteasomes. This finding can explain why certain substrates, such as the p105 precursor of NFκB, are degraded only partially, releasing an active p50 protein. The use of pure archaeal PAN–20S complex has allowed more rigorous investigations of ATP-dependent translocation of multi-domain proteins by proteasomal ATPases. Using protein fusions containing the GFP domain and an easily translocatable and degradable domain (e.g., the first 70 amino acids of casein or calmodulin) with the PAN–20S complex or mammalian 26S, Navon et al. (in preparation) recently found that the translocation process stops at the globular domain, which is then released from the proteasomes. Thus, simple translocation of casein or calmodulin cannot lead to the unfolding or translocation of the upstream globular GFP, unlike attachment of the tight-binding ssrA peptide, which induced efficient unfolding, translocation, and degradation of GFP by the PAN–20S complex (Benaroudj and Goldberg 2000; Navon et al., in preparation).

Remarkably, attachment of the ssrA peptide to the C-terminus of this multi-domain GFP–calmodulin fusion (generating GFP–calmodulin–ssrA) allowed translocation and degradation of the GFP by the PAN–20S complex (Navon et al., in preparation). Therefore, the nature of the terminal sequence and whether it binds tightly to the ATPase ring appears to be critical in determining whether a polypeptide is unfolded and whether a multi-domain protein is translocated processively. Surprisingly, this same ssrA sequence on the C-termini can also cause unfolding, complete translocation, and degradation of even distant domains. Other sequences such as casein and calmodulin, while readily degraded, cannot facilitate the unfolding process, and recent studies indicate that they bind to PAN much less tightly. Presumably, in eukaryotes the polyubiquitin chain functions like ssrA to promote tight binding and processive degradation.

10.7
Degradation of Polyglutamine-containing Proteins

Certain proteins in mammals (e.g., huntingtin or ataxin) contain long sequences of glutamine residues (polyQ) that can reach 20–30 residues in humans. When mutational events increase the length of these repeats to more than 35 residues, neurodegenerative disorders, such as Huntington's disease or the several spinocerebral ataxias (SCA 1–7), result (Zoghbi and Orr 2000). One of the characteristic features of these diseases is the presence of large protein inclusions in the neurons of specific regions of the brain. In addition to the aggregated polyQ proteins, these inclusions contain ubiquitin, components of 26S proteasomes, and the PA28$\alpha\beta$ activator complex. Therefore, a failure of the ubiquitin–proteasome pathway to degrade polyQ-containing proteins has been proposed to explain the accumulation of aggregated polyQ-containing proteins (Verhoef et al. 2002). In addition, several studies have suggested that the presence of these aggregated polyQ proteins in cells impairs the ability of the ubiquitin–proteasome pathway to degrade other proteins (Bence et al. 2001; Michalik and Van Broeckhoven 2004).

To test more directly the ability of pure 20S and 26S proteasomes to degrade polyQ-containing proteins, Venkatraman et al. (2004) used short peptides containing 10–30 Gln residues and found, surprisingly, that eukaryotic proteasomes could not cut in polyQ stretches in soluble peptides. Accordingly, when long glutamine repeats were fused with myoglobin, the open-gated yeast or activated 20S mammalian proteasomes hydrolyzed peptide bonds within the polypeptide chain but spared the polyQ repeat. By contrast, the less specialized active sites of the proteasome from *T. acidophilum* could rapidly degrade polyQ sequences in peptides and proteins. Therefore, even though eukaryotic and archaeal proteasomes have similar architecture, structure, and threonine-based catalytic mechanisms, they differ in their active site specificities. Presumably, the evolution of the three more specialized active sites in the eukaryotic particles provided some advantage in proteolytic rates, but because these more specialized active sites cannot bind glutamine repeats, this binding may have contributed to the occasional appearance of these relatively rare, late-onset neurodegenerative diseases.

These findings also enabled us to propose a new mechanism that may contribute to the pathogenesis of polyQ diseases and the remarkable association of these diseases with polyQ sequences longer than 30 residues. Since polyQ fragments are not degraded by mammalian proteasomes, they must be released from the proteasomes during the breakdown of polyQ-containing proteins and normally are degraded by cytosolic peptidases. However, the peptides that exit the proteasome normally range up to 25 residues, and presumably the longer the peptide, the slower the exit (Kisselev et al. 1999; Kohler et al. 2001). A failure of long polyQ fragments (>30 Gln) to diffuse out of 20S particles may lead to an inhibition of proteasome function and promote further accumulation of aggregated proteins and the inclusion formation observed in these neurodegenerative diseases.

10.8 Eubacterial ATP-dependent Proteases

Although eubacteria (aside from actinomycetes) lack 20S proteasomes and ubiquitin, they contain five different types of ATP-dependent proteases that have provided useful insights about intracellular proteolysis and the functioning of 20S proteasomes: ClpAP, ClpXP, HslUV, Lon, and the membrane-bound protease FtsH (Gottesman 2003) (see Table 10.1). Closely related proteases are also present in mitochondria and chloroplasts of eukaryotes. In ClpAP/XP and in HslUV, the ATPase and peptidase activities are located in separate subunits that form distinct sub-complexes; thus, they share with the PAN–20S and 26S complexes certain architectural and functional features. In Lon and FtsH, the ATPase and peptidase activities are located in different domains of a single polypeptide chain. Like proteasomes, the proteolytic components of these enzymes form a distinct compartment and use ATP hydrolysis to support processive protein breakdown. The ATPase components of all these ATP-dependent proteases belong to the AAA ATPase family and share identical motifs, although each ClpA subunit contains two ATP-binding domains, whereas the others (ClpX, HslU, Lon, FtsH), like the 19S and PAN ATPases, contain only one. HslUV and ClpAP/XP are the best characterized among the eubacterial proteolytic complexes, and the following section summarizes our current knowledge of their biochemical properties.

10.8.1 HslUV (ClpYQ)

In eubacteria, HslV (also called ClpQ) is a two-ring peptidase complex that, unlike ClpP, is a member of the proteasome family. HslV subunits are 18% identical to the *T. acidophilum* 20S β-subunits and share a similar fold (Bochtler et al. 1997). As in the 20S proteasome β-subunits, the N-terminal threonine of HslV acts as the nucleophile in peptide bond cleavage. HslV subunits self-associate in a dimer of two hexameric rings to form a barrel-shaped dodecamer. Thus, HslV closely resembles in structure and function the β-ring of the 20S archaeal proteasome, although it contains six rather than seven subunits and lacks the α-rings (Rohrwild et al. 1997). These complexes associate with the ATPase component HslU (also called ClpY), which is an AAA ATPase homologous to ClpA, ClpX and PAN. ATP hydrolysis by HslU is normally coupled to peptide bond cleavage in the degradation of small peptides and proteins (Rohrwild et al. 1996; Yoo et al. 1996). This feature distinguishes HslV from other eubacterial peptidases because pro-

Table 10.1. Features of ATP-dependent proteases present in eubacteria. The main ATP-dependent proteases from eubacteria are divided into two oligomeric classes, depending on whether the peptidase and ATPase activities are located on different (hetero-oligomers) or the same (homo-oligomers) polypeptide chains. Both ClpP and Lon are Ser proteases (Amerik et al. 1991; Maurizi et al. 1990), but they differ by the nature of their active-site residues. ClpP has a typical serine protease catalytic triad (Ser^{97}, His^{122}, Asp^{171}) (Maurizi et al. 1990; Wang et al. 1997), but Lon has a Ser^{679}, Lys^{722} dyad in its active site (Botos et al. 2004b). HslV, like the β-subunits of 20S proteasomes, has a single residue (Thr)-based proteolytic activity (Bochtler et al. 1997). FtsH is a metalloprotease whose active-site components include two histidine (His^{417} and His^{421}) residues and one glutamate (Glu^{479}) residue as ligands for a zinc atom (Ito and Akiyama 2005; Saikawa et al. 2002). The active-site motif of FtsH (His^{417}-Glu-Ala-Gly-His^{421}) is indicated. The oligomeric structure of ClpP has been determined by electron microscopy (Flanagan et al. 1995) and by X-ray diffraction (Wang et al. 1997). Those of HslU and HslV are based on analysis by X-ray diffraction (Bochtler et al. 1997, 2000; Sousa et al. 2000; Wang et al. 2001a). Hexameric ring structures have been observed for ClpA and ClpX by electron microscopy (Beuron et al. 1998; Grimaud et al. 1998) and modeled from the crystal structure of their monomers (Guo et al. 2002; Kim and Kim 2003). The oligomeric status of Lon and FtsH proteases is based on analogy of the crystal structures of their AAA domains and other AAA ATPases (Botos et al. 2004a; Krzywda et al. 2002).

	Family	*Peptidase*	*Active site*	*Oligomeric state of the peptidase complex*	*ATPase*	*Number of AAA domains*	*Oligomeric state of the ATPase complex*
Hetero-oligomers	ClpAP	ClpP	Ser protease: Ser^{97}, His^{122}, Asp^{171} triad	Tetradecamer (2 heptameric rings)	ClpA	2	Hexamer
	ClpXP	ClpP	Ser protease: Ser^{97}, His^{122}, Asp^{171} triad	Tetradecamer (2 heptameric rings)	ClpX	1	Hexamer
	HslUV	HslV	Threonine protease	Dodecamer (2 hexameric rings)	HslU	1	Hexamer
Homo-oligomers	Lon	C-terminal region	Ser protease: Ser^{679}, Lys^{722} dyad	Hexamer	Central region	1	Hexamer
	FtsH	C-terminal region	Zn^{2+} metalloprotease $H^{417}EAGH^{421}$ E^{479}	Hexamer	Central region	1	Hexamer

tease Lon requires ATP binding but not hydrolysis for peptidase activity (Goldberg and Waxman 1985), and ClpP does not require the presence of ATP to degrade small peptides (Thompson and Maurizi 1994; Woo et al. 1989). However, it was possible to eliminate this requirement for ATP hydrolysis for protein and peptide breakdown under certain experimental conditions such as in the presence of KCl (Huang and Goldberg 1997; Yoo et al. 1998).

X-ray diffraction studies have established that the association of HslU with HslV induces conformational changes in the peptidase active site and increases the pore size of HslV, indicating that HslU increases peptidase activity of HslV by allosteric activation and probably also by promoting peptide entry and/or products release (Huang and Goldberg 1997; Sousa et al. 2000, 2002; Wang et al. 2001a; Yoo et al. 1998). Facilitating peptide entry thus appears to be a common property among HslU, the 19S ATPases, and PAN.

Another role of the ATPase activity of HslU in protein degradation by HslV is to unfold and translocate the protein substrate. HslU must unfold globular substrates because the HslUV complex has been shown to degrade stable folded proteins (Burton et al. 2005; Kwon et al. 2004; Park et al. 2005). However, unlike for ClpX/A and the PAN complex, an unfoldase activity has not yet been directly demonstrated for HslU because of the lack of a substrate, such as GFP, whose folding status can be easily monitored and that can be recognized by HslU.

10.8.2 ClpAP and ClpXP

The most thoroughly characterized ATP-dependent proteolytic complexes from a physiological and mechanistic perspective are ClpAP and ClpXP from *E. coli*. Although often viewed as models of the proteasomes, the peptidase component of these enzymes, ClpP, is unrelated to 20S proteasome β-subunits and to HslV in both amino acid sequence and proteolytic mechanism. ClpP is a serine protease with a canonical catalytic triad instead of the N-terminal threonine active-site residue characteristic of the proteasome family (Maurizi et al. 1990). ClpP is a hollow cylindrical particle composed of a heptameric ring particle, within which are found its 14 active sites (Wang et al. 1997). Alone, ClpP is unable to degrade polypeptides longer than six residues, presumably because they cannot enter the peptidase complex. Upon binding to ClpA and ClpX, which are hexameric ring ATPase complexes of the AAA family, ClpP can degrade longer peptides and proteins in a ATP-dependent processive manner (Hwang et al. 1988; Thompson et al. 1994; Woo et al. 1989).

Much has been learned during the past 10–15 years about the ATP dependence of protein degradation by the ClpAP and ClpXP complexes. Upon nucleotide binding, the ClpA and ClpX ATPases bind polypeptide substrates through a recognition motif that can be located at the ends or middle of the polypeptide (Hoskins et al. 2002; Sauer et al. 2004). Then, ClpA and ClpX catalyze the unfolding of the substrate by a process that requires ATP hydrolysis (Hoskins et al. 2000; Kim et al. 2000; Singh et al. 2000; Weber-Ban et al. 1999). It has been suggested that unfolding is initiated from the recognition signal by sequential unraveling of the polypeptide chain (Lee et al. 2001), as is also suggested for the 19S ATPase (Prakash et al. 2004). After unfolding, substrate release from ClpA also requires hydrolysis of ATP, which suggests that the movement of the unfolded substrate from ClpA to the ClpP chamber also requires metabolic energy (Hoskins et al. 2000). Chemical inactivation of ClpP's peptidase sites has enabled investigators to capture unfolded substrates inside the catalytic chamber. Because release of the trapped unfolded substrate occurred only upon ATP hydrolysis by the ClpX, it seems likely that the ATPase allows opening of a central pore in ClpP (Kim et al. 2000). As we have found for the PAN–proteasome complex (Benaroudj et al. 2003), a large amount of ATP is consumed during degradation of a polypeptide by ClpXP (Burton et al. 2001). Once the unfolded substrate has been translocated into the ClpP catalytic chamber, it is degraded processively into small peptides (Thompson et al. 1994). If the peptides are small enough, they probably exit the ClpP chamber by passive diffusion, but longer peptides may require the ATPase to exit the chamber (Kim et al. 2000).

10.9
How AAA ATPases Use ATP to Catalyze Proteolysis

It is now clear that PAN and the bacterial AAA family of ATPases (Lon, HslU, ClpA, and ClpX) utilize ATP in multiple steps during protein degradation. As hypothesized for AAA ATPases, ATP-driven changes in the conformation of the ATPases must underlie protein unfolding, gate opening, and substrate translocation. A variety of experimental evidence indicates that PAN undergoes conformational changes upon ATP binding and/or hydrolysis. For example, since ATPγS or AMPPNP stabilize the association of PAN with 20S particles (as well as ClpA with ClpP and HslU with HslV), ATP binding must induce a conformation of PAN (or ClpA or HslU) that has a higher affinity for the 20S. This association-prone conformation is not evident in the absence of any nucleotide, or in the presence of ADP, because the PAN–20S complex is not demonstrable under those conditions (Smith et al., in preparation). Also, assays of the protease sensitivity of PAN indicate that its conformation in the ATP-bound state is different from that in its ADP-bound state (Navon et al., in preparation), as has also been demonstrated upon nucleotide binding to ClpA/X (Singh et al. 2001). Unfortunately, efforts using X-ray crystallography to resolve the structure of PAN or the 19S or to define the conformational changes upon ATP binding have thus far not been successful. However, much has been learned about the effects of ATP binding and hydrolysis on the structure of HslU (Wang et al. 2001a, 2001b). ATP binding induces a movement of its C-terminal, α-helical domain that narrows the nucleotide-binding cleft, which further narrows upon ATP hydrolysis. As a consequence, the diameter of the central pore of the HslU hexameric ring decreases, and it has been proposed that these conformational changes provide a mechanical force to thread the substrate through the pore, and perhaps to promote unfolding. Most likely, the transitions induced by ATP binding and hydrolysis are similar for all AAA ATPases and underlie their ability to unfold or remodel their substrates. Depending on the intrinsic stability of the substrate, these ATPases presumably need multiple iterative cycles of ATP-driven mechanical force for the unfolding and threading to reach completion, as indicated by the large ATP consumption during proteolysis by PAN–20S and ClpAP (Benaroudj et al. 2003; Burton et al. 2001).

Another important role of the AAA ATPases in protein degradation by certain ATP-dependent proteases is to stimulate the activity of the associated peptidase complex. In the eukaryotic 26S proteasomes, the Rpt2 ATPase controls the rate of peptide hydrolysis by regulating gate opening of 20S particles and thus limiting substrate entry and/or exit (Groll et al. 2000; Kohler et al. 2001; Rubin et al. 1998). The association of the archaeal PAN ATPase with the 20S particles causes gate opening in a similar fashion (Benaroudj et al. 2003; Smith et al., submitted).

In addition to promoting substrate entry, association of the peptidase complex with the ATPase can lead to conformational changes that enhance peptidase activity, as shown by X-ray analysis of HslUV from *Haemophilus influenza* and *E. coli* (Sousa et al. 2000; Wang et al. 2001a, 2001b). In the *H. influenza* complex, the C-terminal extremities of HslU subunits move in between two HslV protomers (Sousa et al. 2000), while two apical helixes of HslV protrude close to the nucleotide-binding cleft in HslU. As a consequence, the threonine active sites are altered, causing an allosteric activation of peptidase activity in the presence of HslU and ATP (Kwon et al. 2003; Sousa et al. 2002). In the *E. coli* complex, binding of HslU in the presence of ATP causes both HslU and HslV rings to twist round their mutual sixfold axis, thereby enlarging HslV's central pore and closing partially that of HslU (Wang et al. 2001a). In addition, upon ATP hydrolysis, a tyrosine residue in a conserved motif (GYVG) at the HslU central pore moves from inside HslU toward HslV. These findings led to the proposal that threading through the HslU ring is initiated from one end of the polypeptide chain, as has been shown for PAN (Navon and Goldberg 2001; Navon et al., in preparation), and that in the ATP-bound state, this tyrosine residue interacts with hydrophobic residues on the folded polypeptide, and upon ATP hydrolysis, the movement of the tyrosine residue toward HslV and constriction of the central pore in the HslU can promote unfolding and translocation into HslV.

It remains to be established whether this elegant model is valid and whether PAN and the 19S ATPases work in the same manner in stimulating proteasomal degradation. The many conserved motifs among AAA ATPases suggest strongly that they function through similar mechanisms.

However, the recent finding that translocation of unfolded polypeptides through PAN or the 19S ATPases into the 20S (Smith et al., in preparation) can occur in the absence of ATP hydrolysis indicates that this hypothesized "power stroke" is not essential for the degradation of most proteins. In addition, ATP-induced activation of the peptidase as shown for HslU sites in the 20S proteasome seems unlikely because PAN plus ATPγS stimulate markedly the hydrolysis of peptides excluded by the 20S gate, but they do not enhance degradation of tetrapeptides, which freely enter this particle even through the closed gate (Smith et al., in preparation)

10.10 Conclusions

A full understanding of the molecular mechanisms for protein degradation by the proteasome will require detailed structural information about the ATP- and ADP-bound forms of the PAN–20S and 19S complexes. Most likely, X-ray crystallography will first be achieved with PAN, whose many advances for study have been summarized here. Already, however, a great deal has been learned concerning the multiple steps in this process and about the multiple roles of ATP through studies of the PAN–20S complex, as well as the bacterial ATP-dependent proteases. Our present understanding of this process is illustrated by the reaction scheme in Figure 10.8:

1. Nucleotide binding to PAN promotes the association between the ATPase ring and the 20S complex (step 2).
2. Complex formation triggers gate opening in the α-ring (seen in step 4).
3. The binding of the protein substrate induces a conformational change in PAN that activates ATP hydrolysis (step 1).
4. Repeated cycles of ATP hydrolysis catalyze unfolding of globular proteins (step 3).
5. The unfolded polypeptide can diffuse through the ATPase ring (in its ATP-bound form) and the open gates in the α-ring (step 5).
6. The polypeptide in the central chamber of the 20S particle is processively degraded to small peptides (step 6).

These steps appear to be well established for the PAN–20S complex and clearly evolved before the linkage of ubiquitination to proteolysis in eukaryotes. Many detailed questions about this scheme and its general applicability to the 26S complex remain uncertain, and one outstanding issue will be resolved only through studies of the 26S proteasome, i.e., how these steps are integrated with the binding and disassembly of the polyubiquitin chain.

Acknowledgments

The authors would like to thank Mary Dethavong for her assistance in preparation of the manuscript.

References

Akopian, T. N., Kisselev, A. F., and Goldberg, A. L. (1997). Processive degradation of proteins and other catalytic properties of the proteasome from Thermoplasma acidophilum. J Biol Chem *272*, 1791–1798.

Amerik, A., Antonov, V. K., Gorbalenya, A. E., Kotova, S. A., Rotanova, T. V., and Shimbarevich, E. V. (1991). Site-directed mutagenesis of La protease. A catalytically active serine residue. FEBS Lett *287*, 211–214.

Arrigo, A. P., Tanaka, K., Goldberg, A. L., and Welch, W. J. (1988). Identity of the 19S 'prosome' particle with the large multifunctional protease complex of mammalian cells (the proteasome). Nature *331*, 192–194.

Baneyx, F., and Mujacic, M. (2004). Recombinant protein folding and misfolding in Escherichia coli. Nat Biotechnol *22*, 1399–1408.

Benaroudj, N., and Goldberg, A. L. (2000). PAN, the proteasome-activating nucleotidase from archaebacteria, is a protein-unfolding molecular chaperone. Nat Cell Biol *2*, 833–839.

Benaroudj, N., Tarcsa, E., Cascio, P., and Goldberg, A. L. (2001). The unfolding of substrates and ubiquitin-independent protein degradation by proteasomes. Biochimie *83*, 311–318.

Benaroudj, N., Zwickl, P., Seemuller, E., Baumeister, W., and Goldberg, A. L. (2003). ATP hydrolysis by the proteasome regulatory complex PAN serves multiple functions in protein degradation. Mol Cell *11*, 69–78.

Bence, N. F., Sampat, R. M., and Kopito, R. R. (2001). Impairment of the ubiquitin-proteasome system by protein aggregation. Science *292*, 1552–1555.

Beuron, F., Maurizi, M. R., Belnap, D. M., Kocsis, E., Booy, F. P., Kessel, M., and Steven, A. C. (1998). At sixes and sevens: characterization of the symmetry mismatch of the ClpAP chaperone-assisted protease. J Struct Biol *123*, 248–259.

Bochtler, M., Ditzel, L., Groll, M., and Huber, R. (1997). Crystal structure of heat shock locus V (HslV) from Escherichia coli. Proc Natl Acad Sci U S A *94*, 6070–6074.

Bochtler, M., Hartmann, C., Song, H. K., Bourenkov, G. P., Bartunik, H. D., and Huber, R. (2000). The structures of HslU and the ATP-dependent protease HslU-HslV. Nature *403*, 800–805.

Bokman, S. H., and Ward, W. W. (1981). Renaturation of Aequorea gree-fluorescent protein. Biochem Biophys Res Commun *101*, 1372–1380.

Botos, I., Melnikov, E. E., Cherry, S., Khalatova, A. G., Rasulova, F. S., Tropea, J. E., Maurizi, M. R., Rotanova, T. V., Gustchina, A., and Wlodawer, A. (2004a). Crystal structure of the AAA+ alpha domain of E. coli Lon protease at 1.9A resolution. J Struct Biol *146*, 113–122.

Botos, I., Melnikov, E. E., Cherry, S., Tropea, J. E., Khalatova, A. G., Rasulova, F., Dauter, Z., Maurizi, M. R., Rotanova, T. V., Wlodawer, A., and Gustchina, A. (2004b). The catalytic domain of Escherichia coli Lon protease has a unique fold and a Ser-Lys dyad in the active site. J Biol Chem *279*, 8140–8148.

Braun, B. C., Glickman, M., Kraft, R., Dahlmann, B., Kloetzel, P. M., Finley, D., and Schmidt, M. (1999). The base of the proteasome regulatory particle exhibits chaperone-like activity. Nat Cell Biol *1*, 221–226.

Bult, C. J., White, O., Olsen, G. J., Zhou, L., Fleischmann, R. D., Sutton, G. G., Blake, J. A., FitzGerald, L. M., Clayton, R. A., Gocayne, J. D., *et al.* (1996). Complete genome sequence of the methanogenic archaeon, Methanococcus jannaschii. Science *273*, 1058–1073.

Burton, R. E., Siddiqui, S. M., Kim, Y. I., Baker, T. A., and Sauer, R. T. (2001). Effects of protein stability and structure on substrate processing by the ClpXP unfolding and degradation machine. Embo J *20*, 3092–3100.

Burton, R. E., Baker, T. A., and Sauer, R. T. (2005). Nucleotide-dependent substrate recognition by the AAA+ HslUV protease. Nat Struct Mol Biol *12*, 245–251.

Chung, C. H., and Goldberg, A. L. (1981). The product of the lon (capR) gene in Escherichia coli is the ATP-dependent protease, protease La. Proc Natl Acad Sci U S A *78*, 4931–4935.

Ciechanover, A. (2005). Proteolysis: from the lysosome to ubiquitin and the proteasome. Nat Rev Mol Cell Biol *6*, 79–87.

Confalonieri, F., and Duguet, M. (1995). A 200-amino acid ATPase module in search of a basic function. Bioessays *17*, 639–650.

Dahlmann, B., Kopp, F., Kuehn, L., Niedel, B., Pfeifer, G., Hegerl, R., and Baumeister, W. (1989). The multicatalytic proteinase (prosome) is ubiquitous from eukaryotes to archaebacteria. FEBS Lett *251*, 125–131.

Dai, R. M., and Li, C. C. (2001). Valosin-containing protein is a multi-ubiquitin chain-targeting factor required in ubiquitin-proteasome degradation. Nat Cell Biol *3*, 740–744.

Dai, R. M., Chen, E., Longo, D. L., Gorbea, C. M., and Li, C. C. (1998). Involvement of valosin-containing protein, an ATPase Co-purified with IkappaBalpha and 26 S proteasome, in ubiquitin-proteasome-mediated degradation of IkappaBalpha. J Biol Chem *273*, 3562–3573.

Desautels, M., and Goldberg, A. L. (1982a). Demonstration of an ATP-dependent, vanadate-sensitive endoprotease in the matrix of rat liver mitochondria. J Biol Chem *257*, 11673–11679.

Desautels, M., and Goldberg, A. L. (1982b). Liver mitochondria contain an ATP-dependent, vanadate-sensitive pathway for the degradation of proteins. Proc Natl Acad Sci U S A *79*, 1869–1873.

Doolittle, W. F., and Brown, J. R. (1994). Tempo, mode, the progenote, and the universal root. Proc Natl Acad Sci U S A *91*, 6721–6728.

Etlinger, J. D., and Goldberg, A. L. (1977). A soluble ATP-dependent proteolytic system responsible for the degradation of abnormal proteins in reticulocytes. Proc Natl Acad Sci U S A *74*, 54–58.

Eytan, E., Ganoth, D., Armon, T., and Hershko, A. (1989). ATP-dependent incorporation of 20S protease into the 26S complex that degrades proteins conjugated to ubiquitin. Proc Natl Acad Sci U S A *86*, 7751–7755.

Fenteany, G., Standaert, R. F., Lane, W. S., Choi, S., Corey, E. J., and Schreiber, S. L. (1995). Inhibition of proteasome activities and subunit-specific amino-terminal threonine modification by lactacystin. Science *268*, 726–731.

Flanagan, J. M., Wall, J. S., Capel, M. S., Schneider, D. K., and Shanklin, J. (1995). Scanning transmission electron microscopy and small-angle scattering provide evidence that native Escherichia coli ClpP is a tetradecamer with an axial pore. Biochemistry *34*, 10910–10917.

Forster, A., Whitby, F. G., and Hill, C. P. (2003). The pore of activated 20S proteasomes has an ordered 7-fold symmetric conformation. Embo J *22*, 4356–4364.

Forster, A., Masters, E. I., Whitby, F. G., Robinson, H., and Hill, C. P. (2005). The 1.9 A Structure of a Proteasome-11S Activator Complex and Implications for Proteasome-PAN/PA700 Interactions. Mol Cell *18*, 589–599.

Glickman, M. H., and Ciechanover, A. (2002). The ubiquitin-proteasome proteolytic pathway: destruction for the sake of construction. Physiol Rev *82*, 373–428.

Golbik, R., Lupas, A. N., Koretke, K. K., Baumeister, W., and Peters, J. (1999). The Janus face of the archaeal Cdc48/p97 homologue VAT: protein folding versus unfolding. Biol Chem *380*, 1049–1062.

Goldberg, A. L. (2003). Protein degradation and protection against misfolded or damaged proteins. Nature *426*, 895–899.

Goldberg, A. L. (2005). Nobel committee tags ubiquitin for distinction. Neuron *45*, 339–344.

Goldberg, A. L., and Dice, J. F. (1974). Intracellular protein degradation in mammalian and bacterial cells. Annu Rev Biochem *43*, 835–869.

Goldberg, A. L., and St John, A. C. (1976). Intracellular protein degradation in mammalian and bacterial cells: Part 2. Annu Rev Biochem *45*, 747–803.

Goldberg, A. L., and Waxman, L. (1985). The role of ATP hydrolysis in the breakdown of proteins and peptides by protease La from Escherichia coli. J Biol Chem *260*, 12029–12034.

Gottesman, S. (1996). Proteases and their targets in Escherichia coli. Annu Rev Genet *30*, 465–506.

Gottesman, S. (2003). Proteolysis in bacterial regulatory circuits. Annu Rev Cell Dev Biol *19*, 565–587.

Gottesman, S., Roche, E., Zhou, Y., and Sauer, R. T. (1998). The ClpXP and ClpAP proteases degrade proteins with carboxy-terminal peptide tails added by the SsrA-tagging system. Genes Dev *12*, 1338–1347.

Grimaud, R., Kessel, M., Beuron, F., Steven, A. C., and Maurizi, M. R. (1998). Enzymatic and structural similarities between the Escherichia coli ATP-dependent proteases, ClpXP and ClpAP. J Biol Chem *273*, 12476–12481.

Groll, M., and Huber, R. (2003). Substrate access and processing by the 20S proteasome core particle. Int J Biochem Cell Biol *35*, 606–616.

Groll, M., Ditzel, L., Lowe, J., Stock, D., Bochtler, M., Bartunik, H. D., and Huber, R. (1997). Structure of 20S proteasome from yeast at 2.4 A resolution. Nature *386*, 463–471.

Groll, M., Bajorek, M., Kohler, A., Moroder, L., Rubin, D. M., Huber, R., Glickman, M. H., and Finley, D. (2000). A gated channel into the proteasome core particle. Nat Struct Biol *7*, 1062–1067.

Groll, M., Brandstetter, H., Bartunik, H.,

BOURENKOW, G., and HUBER, R. (2003). Investigations on the maturation and regulation of archaebacterial proteasomes. J Mol Biol *327*, 75–83.

GUO, F., MAURIZI, M. R., ESSER, L., and XIA, D. (2002). Crystal structure of ClpA, an Hsp100 chaperone and regulator of ClpAP protease. J Biol Chem *277*, 46743–46752.

HERMAN, C., THEVENET, D., BOULOC, P., WALKER, G. C., and D'ARI, R. (1998). Degradation of carboxy-terminal-tagged cytoplasmic proteins by the Escherichia coli protease HflB (FtsH). Genes Dev *12*, 1348–1355.

HOGNESS, D. S., COHN, M., and MONOD, J. (1955). Studies on the induced synthesis of beta-galactosidase in Escherichia coli: the kinetics and mechanism of sulfur incorporation. Biochim Biophys Acta *16*, 99–116.

HOSKINS, J. R., SINGH, S. K., MAURIZI, M. R., and WICKNER, S. (2000). Protein binding and unfolding by the chaperone ClpA and degradation by the protease ClpAP. Proc Natl Acad Sci U S A *97*, 8892–8897.

HOSKINS, J. R., SHARMA, S., SATHYANARAYANA, B. K., and WICKNER, S. (2002). Clp ATPases and their role in protein unfolding and degradation. Advances in Protein Chemistry *59*, 413–429.

HOUGH, R., PRATT, G., and RECHSTEINER, M. (1986). Ubiquitin-lysozyme conjugates. Identification and characterization of an ATP-dependent protease from rabbit reticulocyte lysates. J Biol Chem *261*, 2400–2408.

HOUGH, R., PRATT, G., and RECHSTEINER, M. (1987). Purification of two high molecular weight proteases from rabbit reticulocyte lysate. J Biol Chem *262*, 8303–8313.

HUANG, H., and GOLDBERG, A. L. (1997). Proteolytic activity of the ATP-dependent protease HslVU can be uncoupled from ATP hydrolysis. J Biol Chem *272*, 21364–21372.

HWANG, B. J., WOO, K. M., GOLDBERG, A. L., and CHUNG, C. H. (1988). Protease Ti, a new ATP-dependent protease in Escherichia coli, contains protein-activated ATPase and proteolytic functions in distinct subunits. J Biol Chem *263*, 8727–8734.

ITO, K., and AKIYAMA, Y. (2005). Cellular Functions, Mechanism of Action, and Regulation of FtsH Protease. Annu Rev Microbiol.

JAROSCH, E., TAXIS, C., VOLKWEIN, C., BORDALLO, J., FINLEY, D., WOLF, D. H., and SOMMER, T. (2002). Protein dislocation from the ER requires polyubiquitination and the AAA-ATPase Cdc48. Nat Cell Biol *4*, 134–139.

KAWASHIMA, T., AMANO, N., KOIKE, H., MAKINO, S., HIGUCHI, S., KAWASHIMA-OHYA, Y., WATANABE, K., YAMAZAKI, M., KANEHORI, K., KAWAMOTO, T., *et al.* (2000). Archaeal adaptation to higher temperatures revealed by genomic sequence of Thermoplasma volcanium. Proc Natl Acad Sci U S A *97*, 14257–14262.

KEILER, K. C., WALLER, P. R., and SAUER, R. T. (1996). Role of a peptide tagging system in degradation of proteins synthesized from damaged messenger RNA. Science *271*, 990–993.

KIM, D. Y., and KIM, K. K. (2003). Crystal structure of ClpX molecular chaperone from Helicobacter pylori. J Biol Chem *278*, 50664–50670.

KIM, Y. I., BURTON, R. E., BURTON, B. M., SAUER, R. T., and BAKER, T. A. (2000). Dynamics of substrate denaturation and translocation by the ClpXP degradation machine. Mol Cell *5*, 639–648.

KISSELEV, A. F., AKOPIAN, T. N., and GOLDBERG, A. L. (1998). Range of sizes of peptide products generated during degradation of different proteins by archaeal proteasomes. J Biol Chem *273*, 1982–1989.

KISSELEV, A. F., AKOPIAN, T. N., WOO, K. M., and GOLDBERG, A. L. (1999). The sizes of peptides generated from protein by mammalian 26 and 20 S proteasomes. Implications for understanding the degradative mechanism and antigen presentation. J Biol Chem *274*, 3363–3371.

KOHLER, A., CASCIO, P., LEGGETT, D. S., WOO, K. M., GOLDBERG, A. L., and FINLEY, D. (2001). The axial channel of the proteasome core particle is gated by the Rpt2 ATPase and controls both substrate entry and product release. Mol Cell *7*, 1143–1152.

KRZYWDA, S., BRZOZOWSKI, A. M., VERMA, C., KARATA, K., OGURA, T., and WILKINSON, A. J. (2002). The crystal structure of the AAA domain of the ATP-dependent protease FtsH of Escherichia coli at 1.5 A resolution. Structure (Camb) *10*, 1073–1083.

Kwon, A. R., Kessler, B. M., Overkleeft, H. S., and McKay, D. B. (2003). Structure and reactivity of an asymmetric complex between HslV and I-domain deleted HslU, a prokaryotic homolog of the eukaryotic proteasome. J Mol Biol *330*, 185–195.

Kwon, A. R., Trame, C. B., and McKay, D. B. (2004). Kinetics of protein substrate degradation by HslUV. J Struct Biol *146*, 141–147.

Latterich, M., Frohlich, K. U., and Schekman, R. (1995). Membrane fusion and the cell cycle: Cdc48p participates in the fusion of ER membranes. Cell *82*, 885–893.

Lee, C., Schwartz, M. P., Prakash, S., Iwakura, M., and Matouschek, A. (2001). ATP-dependent proteases degrade their substrates by processively unraveling them from the degradation signal. Mol Cell *7*, 627–637.

Liu, C. W., Millen, L., Roman, T. B., Xiong, H., Gilbert, H. F., Noiva, R., DeMartino, G. N., and Thomas, P. J. (2002). Conformational remodeling of proteasomal substrates by PA700, the 19 S regulatory complex of the 26 S proteasome. J Biol Chem *277*, 26815–26820.

Liu, C. W., Corboy, M. J., DeMartino, G. N., and Thomas, P. J. (2003). Endoproteolytic activity of the proteasome. Science *299*, 408–411.

Lowe, J., Stock, D., Jap, B., Zwickl, P., Baumeister, W., and Huber, R. (1995). Crystal structure of the 20S proteasome from the archaeon T. acidophilum at 3.4 A resolution. Science *268*, 533–539.

Lupas, A. N., and Martin, J. (2002). AAA proteins. Curr Opin Struct Biol *12*, 746–753.

Matthews, W., Driscoll, J., Tanaka, K., Ichihara, A., and Goldberg, A. L. (1989). Involvement of the proteasome in various degradative processes in mammalian cells. Proc Natl Acad Sci U S A *86*, 2597–2601.

Maupin-Furlow, J. A., and Ferry, J. G. (1995). A proteasome from the methanogenic archaeon Methanosarcina thermophila. J Biol Chem *270*, 28617–28622.

Maupin-Furlow, J. A., Aldrich, H. C., and Ferry, J. G. (1998). Biochemical characterization of the 20S proteasome from the methanoarchaeon Methanosarcina thermophila. J Bacteriol *180*, 1480–1487.

Maupin-Furlow, J. A., Gil, M. A., Karadzic, I. M., Kirkland, P. A., and Reuter, C. J. (2004). Proteasomes: perspectives from the Archaea. Front Biosci *9*, 1743–1758.

Maurizi, M. R., Clark, W. P., Kim, S. H., and Gottesman, S. (1990). Clp P represents a unique family of serine proteases. J Biol Chem *265*, 12546–12552.

Michalik, A., and Van Broeckhoven, C. (2004). Proteasome degrades soluble expanded polyglutamine completely and efficiently. Neurobiol Dis *16*, 202–211.

Murakami, K., Voellmy, R., and Goldberg, A. L. (1979). Protein degradation is stimulated by ATP in extracts of Escherichia coli. J Biol Chem *254*, 8194–8200.

Murakami, Y., Matsufuji, S., Hayashi, S., Tanahashi, N., and Tanaka, K. (2000). Degradation of ornithine decarboxylase by the 26S proteasome. Biochem Biophys Res Commun *267*, 1–6.

Navon, A., and Goldberg, A. L. (2001). Proteins are unfolded on the surface of the ATPase ring before transport into the proteasome. Mol Cell *8*, 1339–1349.

Nussbaum, A. K., Dick, T. P., Keilholz, W., Schirle, M., Stevanovic, S., Dietz, K., Heinemeyer, W., Groll, M., Wolf, D. H., Huber, R., *et al.* (1998). Cleavage motifs of the yeast 20S proteasome beta subunits deduced from digests of enolase 1. Proc Natl Acad Sci U S A *95*, 12504–12509.

Ogura, T., and Wilkinson, A. J. (2001). AAA+ superfamily ATPases: common structure–diverse function. Genes Cells *6*, 575–597.

Park, E., Rho, Y. M., Koh, O. J., Ahn, S. W., Seong, I. S., Song, J. J., Bang, O., Seol, J. H., Wang, J., Eom, S. H., and Chung, C. H. (2005). Role of the GYVG Pore Motif of HslU ATPase in Protein Unfolding and Translocation for Degradation by HslV Peptidase. J Biol Chem *280*, 22892–22898.

Pickart, C. M., and Eddins, M. J. (2004). Ubiquitin: structures, functions, mechanisms. Biochim Biophys Acta *1695*, 55–72.

Prakash, S., Tian, L., Ratliff, K. S., Lehotzky, R. E., and Matouschek, A. (2004). An unstructured initiation site is required for efficient proteasome-mediated degradation. Nat Struct Mol Biol *11*, 830–837.

RABINOVICH, E., KEREM, A., FROHLICH, K. U., DIAMANT, N., and BAR-NUN, S. (2002). AAA-ATPase p97/Cdc48p, a cytosolic chaperone required for endoplasmic reticulum-associated protein degradation. Mol Cell Biol *22*, 626–634.

RABOUILLE, C., LEVINE, T. P., PETERS, J. M., and WARREN, G. (1995). An NSF-like ATPase, p97, and NSF mediate cisternal regrowth from mitotic Golgi fragments. Cell *82*, 905–914.

ROHRWILD, M., COUX, O., HUANG, H. C., MOERSCHELL, R. P., YOO, S. J., SEOL, J. H., CHUNG, C. H., and GOLDBERG, A. L. (1996). HslV-HslU: A novel ATP-dependent protease complex in Escherichia coli related to the eukaryotic proteasome. Proc Natl Acad Sci U S A *93*, 5808–5813.

ROHRWILD, M., PFEIFER, G., SANTARIUS, U., MULLER, S. A., HUANG, H. C., ENGEL, A., BAUMEISTER, W., and GOLDBERG, A. L. (1997). The ATP-dependent HslVU protease from Escherichia coli is a four-ring structure resembling the proteasome. Nat Struct Biol *4*, 133–139.

RUBIN, D. M., GLICKMAN, M. H., LARSEN, C. N., DHRUVAKUMAR, S., and FINLEY, D. (1998). Active site mutants in the six regulatory particle ATPases reveal multiple roles for ATP in the proteasome. Embo J *17*, 4909–4919.

RUEPP, A., GRAML, W., SANTOS-MARTINEZ, M. L., KORETKE, K. K., VOLKER, C., MEWES, H. W., FRISHMAN, D., STOCKER, S., LUPAS, A. N., and BAUMEISTER, W. (2000). The genome sequence of the thermoacidophilic scavenger Thermoplasma acidophilum. Nature *407*, 508–513.

SAIKAWA, N., ITO, K., and AKIYAMA, Y. (2002). Identification of glutamic acid 479 as the gluzincin coordinator of zinc in FtsH (HflB). Biochemistry *41*, 1861–1868.

SAUER, R. T., BOLON, D. N., BURTON, B. M., BURTON, R. E., FLYNN, J. M., GRANT, R. A., HERSCH, G. L., JOSHI, S. A., KENNISTON, J. A., LEVCHENKO, I., *et al.* (2004). Sculpting the proteome with AAA(+) proteases and disassembly machines. Cell *119*, 9–18.

SEEMULLER, E., LUPAS, A., STOCK, D., LOWE, J., HUBER, R., and BAUMEISTER, W. (1995). Proteasome from Thermoplasma acidophilum: a threonine protease. Science *268*, 579–582.

SEEMULLER, E., LUPAS, A., and BAUMEISTER, W. (1996). Autocatalytic processing of the 20S proteasome. Nature *382*, 468–471.

SINGH, S. K., GRIMAUD, R., HOSKINS, J. R., WICKNER, S., and MAURIZI, M. R. (2000). Unfolding and internalization of proteins by the ATP-dependent proteases ClpXP and ClpAP. Proc Natl Acad Sci U S A *97*, 8898–8903.

SINGH, S. K., ROZYCKI, J., ORTEGA, J., ISHIKAWA, T., LO, J., STEVEN, A. C., and MAURIZI, M. R. (2001). Functional domains of the ClpA and ClpX molecular chaperones identified by limited proteolysis and deletion analysis. J Biol Chem *276*, 29420–29429.

SOUSA, M. C., TRAME, C. B., TSURUTA, H., WILBANKS, S. M., REDDY, V. S., and MCKAY, D. B. (2000). Crystal and solution structures of an HslUV protease-chaperone complex. Cell *103*, 633–643.

SOUSA, M. C., KESSLER, B. M., OVERKLEEFT, H. S., and MCKAY, D. B. (2002). Crystal structure of HslUV complexed with a vinyl sulfone inhibitor: corroboration of a proposed mechanism of allosteric activation of HslV by HslU. J Mol Biol *318*, 779–785.

STRICKLAND, E., HAKALA, K., THOMAS, P. J., and DEMARTINO, G. N. (2000). Recognition of misfolding proteins by PA700, the regulatory subcomplex of the 26 S proteasome. J Biol Chem *275*, 5565–5572.

TANAKA, K., WAXMAN, L., and GOLDBERG, A. L. (1983). ATP serves two distinct roles in protein degradation in reticulocytes, one requiring and one independent of ubiquitin. J Cell Biol *96*, 1580–1585.

THOMPSON, M. W., and MAURIZI, M. R. (1994). Activity and specificity of Escherichia coli ClpAP protease in cleaving model peptide substrates. J Biol Chem *269*, 18201–18208.

THOMPSON, M. W., SINGH, S. K., and MAURIZI, M. R. (1994). Processive degradation of proteins by the ATP-dependent Clp protease from Escherichia coli. Requirement for the multiple array of active sites in ClpP but not ATP hydrolysis. J Biol Chem *269*, 18209–18215.

UNNO, M., MIZUSHIMA, T., MORIMOTO, Y., TOMISUGI, Y., TANAKA, K., YASUOKA, N., and TSUKIHARA, T. (2002). The structure of the mammalian 20S proteasome at 2.75 A resolution. Structure (Camb) *10*, 609–618.

Venkatraman, P., Wetzel, R., Tanaka, M., Nukina, N., and Goldberg, A. L. (2004). Eukaryotic proteasomes cannot digest polyglutamine sequences and release them during degradation of polyglutamine-containing proteins. Mol Cell *14*, 95–104.

Verhoef, L. G., Lindsten, K., Masucci, M. G., and Dantuma, N. P. (2002). Aggregate formation inhibits proteasomal degradation of polyglutamine proteins. Hum Mol Genet *11*, 2689–2700.

Voges, D., Zwickl, P., and Baumeister, W. (1999). The 26S proteasome: a molecular machine designed for controlled proteolysis. Annu Rev Biochem *68*, 1015–1068.

Wang, J., Hartling, J. A., and Flanagan, J. M. (1997). The structure of ClpP at 2.3 A resolution suggests a model for ATP-dependent proteolysis. Cell *91*, 447–456.

Wang, J., Song, J. J., Franklin, M. C., Kamtekar, S., Im, Y. J., Rho, S. H., Seong, I. S., Lee, C. S., Chung, C. H., and Eom, S. H. (2001a). Crystal structures of the HslVU peptidase-ATPase complex reveal an ATP-dependent proteolysis mechanism. Structure (Camb) *9*, 177–184.

Wang, J., Song, J. J., Seong, I. S., Franklin, M. C., Kamtekar, S., Eom, S. H., and Chung, C. H. (2001b). Nucleotide-dependent conformational changes in a protease-associated ATPase HslU. Structure (Camb) *9*, 1107–1116.

Waxman, L., Fagan, J. M., and Goldberg, A. L. (1987). Demonstration of two distinct high molecular weight proteases in rabbit reticulocytes, one of which degrades ubiquitin conjugates. J Biol Chem *262*, 2451–2457.

Weber-Ban, E. U., Reid, B. G., Miranker, A. D., and Horwich, A. L. (1999). Global unfolding of a substrate protein by the Hsp100 chaperone ClpA. Nature *401*, 90–93.

Wenzel, T., and Baumeister, W. (1993). Thermoplasma acidophilum proteasomes degrade partially unfolded and ubiquitin-associated proteins. FEBS Lett *326*, 215–218.

Wenzel, T., and Baumeister, W. (1995). Conformational constraints in protein degradation by the 20S proteasome. Nat Struct Biol *2*, 199–204.

Wenzel, T., Eckerskorn, C., Lottspeich, F., and Baumeister, W. (1994). Existence of a molecular ruler in proteasomes suggested by analysis of degradation products. FEBS Lett *349*, 205–209.

Wilson, H. L., Aldrich, H. C., and Maupin-Furlow, J. (1999). Halophilic 20S proteasomes of the archaeon Haloferax volcanii: purification, characterization, and gene sequence analysis. J Bacteriol *181*, 5814–5824.

Wilson, H. L., Ou, M. S., Aldrich, H. C., and Maupin-Furlow, J. (2000). Biochemical and physical properties of the Methanococcus jannaschii 20S proteasome and PAN, a homolog of the ATPase (Rpt) subunits of the eucaryal 26S proteasome. J Bacteriol *182*, 1680–1692.

Woo, K. M., Chung, W. J., Ha, D. B., Goldberg, A. L., and Chung, C. H. (1989). Protease Ti from Escherichia coli requires ATP hydrolysis for protein breakdown but not for hydrolysis of small peptides. J Biol Chem *264*, 2088–2091.

Ye, Y., Meyer, H. H., and Rapoport, T. A. (2001). The AAA ATPase Cdc48/p97 and its partners transport proteins from the ER into the cytosol. Nature *414*, 652–656.

Yoo, S. J., Seol, J. H., Shin, D. H., Rohrwild, M., Kang, M. S., Tanaka, K., Goldberg, A. L., and Chung, C. H. (1996). Purification and characterization of the heat shock proteins HslV and HslU that form a new ATP-dependent protease in Escherichia coli. J Biol Chem *271*, 14035–14040.

Yoo, S. J., Kim, H. H., Shin, D. H., Lee, C. S., Seong, I. S., Seol, J. H., Shimbara, N., Tanaka, K., and Chung, C. H. (1998). Effects of the cys mutations on structure and function of the ATP-dependent HslVU protease in Escherichia coli. The Cys287 to Val mutation in HslU uncouples the ATP-dependent proteolysis by HslvU from ATP hydrolysis. J Biol Chem *273*, 22929–22935.

Zhang, M., and Coffino, P. (2004). Repeat sequence of Epstein-Barr virus-encoded nuclear antigen 1 protein interrupts proteasome substrate processing. J Biol Chem *279*, 8635–8641.

Zoghbi, H. Y., and Orr, H. T. (2000). Glutamine repeats and neurodegeneration. Annu Rev Neurosci *23*, 217–247.

Zwickl, P. (2002). The 20S proteasome. Curr Top Microbiol Immunol *268*, 23–41.

Zwickl, P., Lottspeich, F., and Baumeister, W. (1992). Expression of functional Thermoplasma acidophilum proteasomes in Escherichia coli. FEBS Lett *312*, 157–160.

Zwickl, P., Ng, D., Woo, K. M., Klenk, H. P., and Goldberg, A. L. (1999). An archaebacterial ATPase, homologous to ATPases in the eukaryotic 26 S proteasome, activates protein breakdown by 20 S proteasomes. J Biol Chem *274*, 26008–26014.

Zwickl, P., Goldberg, A. L., and Baumeister, W. (2000). Proteasomes in Prokaryotes, In Proteasomes: The World of Regulatory Proteolysis, W. Hilt, and D. H. Wolf, eds. (Georgetown: Eureka.com/Landes Bioscience), pp. 8–20.

11
Biochemical Functions of Ubiquitin and Ubiquitin-like Protein Conjugation

Mark Hochstrasser

Abstract

Protein modification by ubiquitin and ubiquitin-like proteins (Ubls) plays a pervasive role in eukaryotic cell regulation. One aim of this chapter is to survey the ubiquitin and Ubl conjugation systems in order to highlight key mechanistic and functional features. Another is to discuss some of the gaps in our understanding of both the evolutionary origins of these conjugation systems and the changes Ubl attachment can impart on a conjugated protein. The ubiquitin and Ubl systems use related enzymes to activate and attach ubiquitin and Ubls to proteins (and, in at least one case, to phospholipids). Most ubiquitin and Ubl attachments are dynamic, with efficient reversal of the modifications by a battery of deconjugating enzymes. The versatility of these systems is reflected in the enormous array of biological processes they control. It is likely that ubiquitin and Ubl attachments function fundamentally as a means of regulating macromolecular interactions. Best known is the ability of polyubiquitinated protein to bind with high affinity to polyubiquitin receptor sites on the proteasome, causing the rapid degradation of the tagged protein. Specific examples of physiological deployment of ubiquitin and Ubl attachment will be used to illustrate distinct mechanisms of regulation by these highly conserved protein modifiers.

11.1
Introduction

The biological functions of many proteins are altered by their covalent attachment to polypeptide modifiers [1–5]. Among these types of modification, probably the best known is ubiquitination. Ubiquitin can target proteins for degradation by the 26S proteasome, but additional effects of protein ubiquitination are now well documented. Ubiquitin is joined reversibly to proteins by amide linkage between the carboxy terminus of ubiquitin and primary amino groups of the acceptor proteins [3, 5]. The primary amine is usually a lysine ε-amino group (the bond with ubiqui-

Protein Degradation, Vol. 2: The Ubiquitin-Proteasome System.
Edited by R. J. Mayer, A. Ciechanover, M. Rechsteiner

ISBN: 3-527-31130-0

tin is then called an isopeptide bond) but can also be the N-terminal N^{α} amino group [6].

11.1.1
The Ubiquitin Conjugation Pathway

The C-terminus of ubiquitin must be activated before it can form a covalent bond with another protein [3, 5] (see Figure 11.1, which depicts a more general Ubl cycle). Initially, the ubiquitin is adenylated by the ubiquitin-activating enzyme E1. A high-energy mixed anhydride bond links the ubiquitin–AMP, which remains bound to the E1. This bond is then attacked by a sulfhydryl group of a cysteine in the E1 enzyme, yielding a high energy E1–ubiquitin thioester intermediate. The activated ubiquitin is subsequently passed to one of a large number of distinct

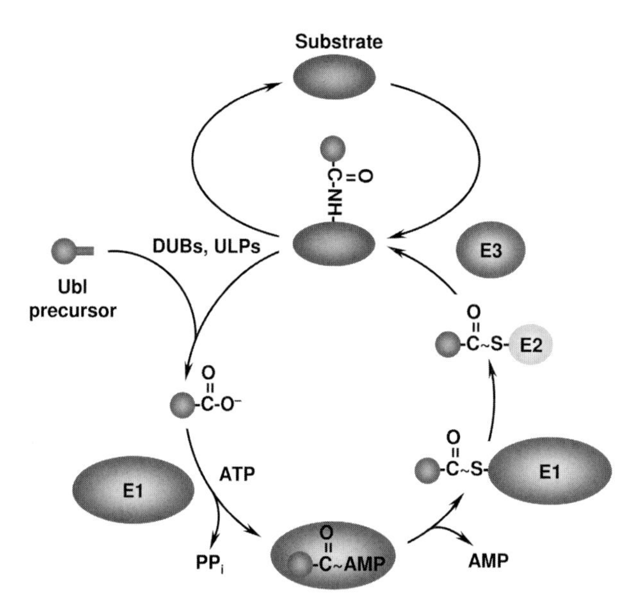

Fig. 11.1. A general ubiquitin-like protein (Ubl) conjugation cycle. After precursor processing, the C-terminal carboxyl group of the Ubl is activated by an E1 Ubl-activating enzyme, which catalyzes formation of a Ubl–AMP intermediate from ATP and the Ubl (the high-energy bond between the AMP phosphate and the C-terminal carboxyl group is indicated by "~"). This Ubl adenylate remains noncovalently bound to the E1 but is then attacked by an active-site cysteine of the E1, leading to formation of a thioester bond between the E1 and Ubl and release of AMP. An E2 Ubl-conjugating enzyme receives the Ubl from the E1, creating an E2–Ubl thioester intermediate. An E3 Ubl–protein ligase then stimulates transfer of the Ubl to a substrate amino group. Additional Ubl molecules can be added either to other lysine side chains on the substrate or to the Ubl itself, the latter leading to polymeric Ubl chains. Ubl chain formation is well documented for ubiquitin; SUMO can form chains *in vivo*, but their functional significance is uncertain. Ubl modifications are usually dynamic and can be removed by deubiquitinating enzymes (DUBs) or Ubl-specific proteases (ULPs). Most Ubls are also synthesized in precursor forms, and the C-terminal extensions are removed by DUBs or ULPs as well.

ubiquitin-conjugating (E2) enzymes by transthiolation to a conserved cysteine side chain of the E2. The E2 proteins catalyze substrate ubiquitination in conjunction with a ubiquitin–protein ligase (E3). For one structural class of E3 proteins (the "HECT domain" E3s), the ubiquitin is first transferred to a conserved cysteine of the E3 before the final transfer to a substrate amine. For most other ubiquitination reactions, a primary role for the E3 appears to be as an adaptor that positions the substrate in close proximity to the reactive E2–ubiquitin thioester bond. The majority of such E3s are characterized by a RING domain, which coordinates a pair of zinc ions and participates in E2 binding [7].

Additional roles for E3s in the catalytic cycle, such as allosteric activation of the E2, remain a distinct possibility [8, 9]. For instance, all E2s have an asparagine residue upstream of the active site cysteine. This asparagine is implicated in the formation of an oxyanion hole that stabilizes the tetrahedral intermediate formed by nucleophilic attack of a substrate amino group on the activated carbonyl of ubiquitin [9]. However, the side chain of the asparagine is fully hydrogen-bonded and oriented away from the active cysteine in the atomic structures determined for isolated E2 enzymes. E3 binding to the E2 and/or ubiquitin thioester formation on the E2 may trigger local structural changes that allow rotation of the E2 asparagine side chain to a position where it can help generate a functional oxyanion hole [10].

11.1.2 Ubiquitin Polymers

In many cases, particularly for proteolytic substrates, more than one ubiquitin is attached to the substrate protein. These ubiquitin molecules can be attached to different substrate amino groups, or they can be attached to each other to form a polyubiquitin chain that is linked to a single substrate site [11, 12]. The ubiquitin molecules in these polymers are linked through the lysine side chain of one ubiquitin with the C-terminal carboxyl of the next ubiquitin. Ubiquitin has several different lysines that contribute to such linkages. For instance, the polyubiquitinated proteins recognized by the proteasome usually have ubiquitin Lys48-linked chains, and the chain must include at least four ubiquitins for tight binding to the proteasome [13]. Ubiquitin chain formation is also essential for certain types of DNA repair and signal transduction pathways, but these chains have Lys63 linkages and do not target the proteins to the proteasome. Ubiquitin polymers of distinct topology are generally thought to have intrinsically different binding affinities for particular target proteins [14]. However, some proteins, such as the S5a proteasome subunit, can bind different types of ubiquitin chains with comparable affinity [15]. Monoubiquitination has distinct signaling functions, as will be discussed below.

11.1.3 Ubiquitin Attachment Dynamics

Ubiquitinated proteins are in a dynamic state, subject to either further rounds of ubiquitin addition or ubiquitin removal by deubiquitinating enzymes (DUBs) (analogous enzymes act on Ubls; see Figure 11.1). The DUBs comprise one of the largest classes of ubiquitin-system enzymes, but their individual functions are just now beginning to come into view [16]. Many DUBs have negative roles in ubiquitin-dependent signaling. For example, removal of a polyubiquitin chain from a proteolytic substrate prior to its binding to the proteasome will prevent degradation of the substrate. Several DUBs have been demonstrated to have substrate-specific deubiquitinating activity. An illustrative example is the herpesvirus-associated ubiquitin-specific protease (HAUSP). HAUSP can specifically deubiquitinate the p53 tumor suppressor protein; this limits p53 degradation, thereby enhancing p53 pro-apoptotic and growth inhibitory functions [17]. Other DUBs can have positive functions in ubiquitin-dependent processes. The best-known examples of this are the enzymes that recover ubiquitin from proteasome-bound polyubiquitinated substrates [16, 18]. Failure to remove the polyubiquitin from the tagged proteins severely impedes their degradation, presumably because it is difficult to unfold and degrade the highly structured ubiquitin molecules [19]. Ubiquitin (and most Ubls) is synthesized in C-terminally extended precursor forms, which are also processed by DUBs to expose a terminal Gly–Gly dipeptide that is necessary for ubiquitin activation and conjugation.

The fates of ubiquitinated proteins vary greatly. Ubiquitin-induced functional changes depend on whether a single ubiquitin or a polyubiquitin chain is attached (and, if a chain, with what topology), where and when in the cell the modification occurs, and exactly what protein receives the modification. Similar considerations apply to the Ubls and their targets, although Ubl chain formation is not widely observed. Dynamic modification of proteins by ubiquitin and Ubls allows reversible switches between different functional states, as is true for other transient covalent protein modifications such as phosphorylation. A major focus of the remainder of this chapter will be on the general properties of the resulting ubiquitin and Ubl conjugates and the functional consequences of these modifications.

11.2
Ubls: A Typical Modification Cycle by an Atypical Set of Modifiers

The existence of potential ubiquitin-related protein modifiers first became apparent in the late 1980s with the discovery that an interferon-stimulated gene product of 15 kDa, or ISG15, shares significant sequence similarity with ubiquitin and is recognized by anti-ubiquitin antibodies [20]. In 1992, ISG15 was shown to modify other proteins by what seemed likely to be a similar post-translational enzymatic mechanism [21]. Like ubiquitin, ISG15 is synthesized in precursor form, and its C-terminal tail is processed off to expose a diglycine motif that is essential for ISG15–protein conjugation [22]. The mature protein is composed of two domains, both with substantial sequence and structural similarity to ubiquitin (Table 11.1) [23]. Although ISG15 is the prototypical Ubl, it remains one of the least understood. Only very recently have the ISG15 E1 and E2 enzymes been identified [24–26]. These enzymes, like the Ubl itself, are strongly induced by type I interferons,

Table 11.1. Known or suspected Ubls.

Modifiera[a]	*Identity with Ub (%)*	*E1*[a]	*E2*[a]	*Comments*	*Reference*
Ubiquitin (Ub)	100	Uba1	Many	Viral form more diverged (75% identity)	123
ISG15	32/37[b]	Ube1L	UbcH8	First Ubl identified	21
Rub1/NEDD8	55	Uba3-Ula1	Ubc12	Substrates: cullins, p53	110
Smt3/SUMO1–4	18	Uba2-Aos1	Ubc9	Vertebrates have 4 distinct *SUMO* genes	111
Atg12	NS[c]	Atg7	Atg10	~20% identical to ATG8	112
Atg8	NS	Atg7	Atg3	3 known human isoforms; has the Ub fold	33
Urm1	NS	Uba4	–	Related to MoaD, ThiS	44
UFM1	NS	Uba5	Ufc1	Has the Ub fold	115
FUBI/MNSFβ	38	–	–	Derived from ribosomal precursor	113
FAT10	32/40[b]	–	–	Substrates unknown	116
Ubl-1	40	–	–	Nematode ribosomal precursor	114
Hub1	22	–	–	Might bind only noncovalently to targets	117–119
BUBL1, 2	variable (up to 80%)	–	–	Ciliate putative autoprocessed proteins	32
SF3a120	30	–	–	Ubl at C-terminus; no data for conjugation	120
Oligo(A) synthetase	42	–	–	Ubl at C-terminus; no data for conjugation	121

[a] Yeast names are listed for E1s and E2s except for the ISG15 and UFM1 systems, which are not found in *S. cerevisiae*; for Ubl names, the yeast names are given (listed first if a vertebrate ortholog is known and goes by a different name) if present in yeast;
[b] Two Ub-related domains;
[c] Not statistically significant.

and they are presumed to be important in antiviral responses; to date, however, the only genetic evidence supporting this idea is the finding that mice lacking a protease with ISG15-deconjugating activity, UBP43, have a more vigorous innate immune response to viral infections [27]. Current data also suggest that ISG15 functions in signal transduction, particularly the Jak–STAT pathway, but its exact role is unclear. Multiple ISG15–protein conjugates are likely, but only a small number of substrates have been reported so far [28].

Since the discovery of ISG15, at least 10 additional ubiquitin-related proteins have been identified that can covalently modify other macromolecules or are strongly suspected of having this ability (Table 11.1). The widespread occurrence of Ubls underscores the potential regulatory importance of protein attachment to other proteins (or to lipids). Ubiquitin can modify hundreds if not thousands of different proteins [29, 30]. Some Ubls, such as small ubiquitin-related modifier (SUMO), appear to rival ubiquitin in the number and diversity of substrates targeted, while others are likely to have a very limited number of substrates. Atg12 (autophagy protein 12), for example, is thought to have but a single target (Atg5), and Atg8 is specifically attached to phosphatidylethanolamine, a phospholipid.

Most Ubl modification pathways utilize highly similar enzymatic mechanisms involving E1-like, E2-like, and often E3-like enzymes as well as specific Ubl-deconjugating enzymes (Figure 11.1). Several unusual ubiquitin and Ubl conjugation mechanisms have been proposed as well. These noncanonical mechanisms, which at this point still have little supporting experimental evidence, range from ubiquitin hydrolases effectively working in reverse [31] to a set of unusual ciliate self-splicing polyproteins [32]. The ciliate polyproteins consist of a series of Ubl domains flanked by self-splicing bacterial intein-like (BIL) domains. The BIL domains are postulated to excise the flanking Ubl segments and attach them to substrates during autocatalytic *in cis* processing reactions. Such atypical mechanisms most likely account for only a very small percentage of Ubl conjugation reactions. The major pathways of protein conjugation appear to have evolved by repeated rounds of duplication and diversification of enzymes and protein modifiers derived from ancient enzyme–cofactor biosynthetic pathways (see below).

It should be noted that there are also many ubiquitin-related proteins in which a ubiquitin-like domain (UBL) is built into a larger polypeptide, but the UBL is neither excised nor attached to other proteins. Such UBLs may impart properties on a protein similar to those conferred by a transferable Ubl, but the UBL is locked into a single target. The UBL-containing proteins include several proteins that also have ubiquitin-binding domains. These multi-domain proteins are thought to help transfer polyubiquitinated proteins from E2 and E3 complexes to the proteasome and will be discussed later in the chapter.

11.2.1
Some Unusual Ubl Conjugation Features

Among the eight Ubl conjugation pathways for which E1 enzymes have been identified (the first eight entries in Table 11.1), there are several with unusual properties that deserve additional comment. First, the Atg8 and Atg12 Ubls share a single E1-like enzyme but have different E2-like proteins [33]. This remains the only known example in which a single E1, Atg7, can activate two different Ubls. The ability of Atg7 to transfer Atg8 and Atg12 to distinct E2s suggests that these E2s bind to the E1–Ubl complex in ways that are productive for transfer only of the E2's cognate Ubl. The structural basis of this discrimination remains to be determined. Second, there is now also an example of different E1 enzymes specific for different Ubls sharing the same E2 enzyme to effect cognate substrate modification. UBEL1, the E1 for ISG15, and UBA1, the E1 for ubiquitin, can both use the same E2, UbcH8 [25, 26]. How Ubl substrate targeting specificity is maintained is an important question, although it might be explained largely by mass action. ISG15 and UBEL1 are only strongly expressed when cells are induced with interferon. Potentially, activated ISG15 would be present at sufficiently high levels under these conditions to ensure that UbcH8 picks up a sizeable amount of it and, together with ISG15-specific E3s, which might also be interferon-inducible, will be able to transfer it to the appropriate targets. No such E3 has yet been described, but it would be predicted to be able to recognize ISG15-charged, but not ubiquitin-charged, UbcH8. For the SUMO pathway, recent studies reveal direct contacts between an E3 and the SUMO protein [10, 34]. Similarly, ISG15-specific E3s might be able to use direct ISG15 contacts to identify the cognate ISG15–E2 thioester even if the same E2 is sometimes linked to ubiquitin.

In the remainder of the review, several areas will be emphasized. First, I will expand on earlier speculations about the possible evolutionary origins of the ubiquitin and Ubl modification systems [1]. Such an evolutionary perspective is useful when potential mechanistic variations in Ubl activation and conjugation are considered. It also suggests an explanation for the otherwise mysterious existence of the widespread E1–E2 couple in Ubl conjugation. Second, I will take examples from both the ubiquitin and Ubl literature to highlight common and divergent themes regarding the biochemical functions of ubiquitin and Ubl ligation. In other words, the emphasis will be on what happens after the modifier is attached to its target rather than focusing on what determines how a target is chosen for modification in the first place. The latter topic is extensively reviewed elsewhere in these volumes. Finally, I will attempt to generalize these examples to give a broader account of the biochemical and physiological consequences of ubiquitin and Ubl conjugation.

11.3
Origins of the Ubiquitin System

For many years, the evolutionary antecedents to the ubiquitin system were completely mysterious. Ubiquitin itself was regarded as perhaps the most highly conserved of all eukaryotic proteins, yet no eubacterial or archaeal proteins shared any obvious primary sequence similarity to it [35]. An early hint to ubiquitin's origins came from the cloning and sequencing of the gene encoding the E1 ubiquitin-activating enzyme; the protein displays weak but significant similarity to ChlN/MoeB, an *E. coli* protein required for the biosynthesis of the molybdenum cofactor (Moco) [36]. At the time, the biochemical function of MoeB was unknown, and thus this similarity was not informative by itself. During the late 1990s, however, the protein sequences and catalytic mechanisms of the enzymes used to synthesize Moco (and thiamin [vitamin B1]) began to be deciphered, and intriguing similarities to ubiquitin activation were noted [37–39].

11.3.1
Sulfurtransferases and Ubl Activation Enzymes

Biosynthesis of both Moco and thiamin requires insertion of sulfur atoms into their precursor forms. In each case, the sulfur is donated by a small sulfur carrier protein termed MoaD and ThisS, respectively. The donor sulfur derives from a thiocarboxylate group generated at the C-termini of these proteins (see Figure 11.2). MoaD and ThiS are related and, like ubiquitin, end with a pair of glycines. Most interestingly, conversion of the C-terminal glycine carboxylate to a thiocarboxylate

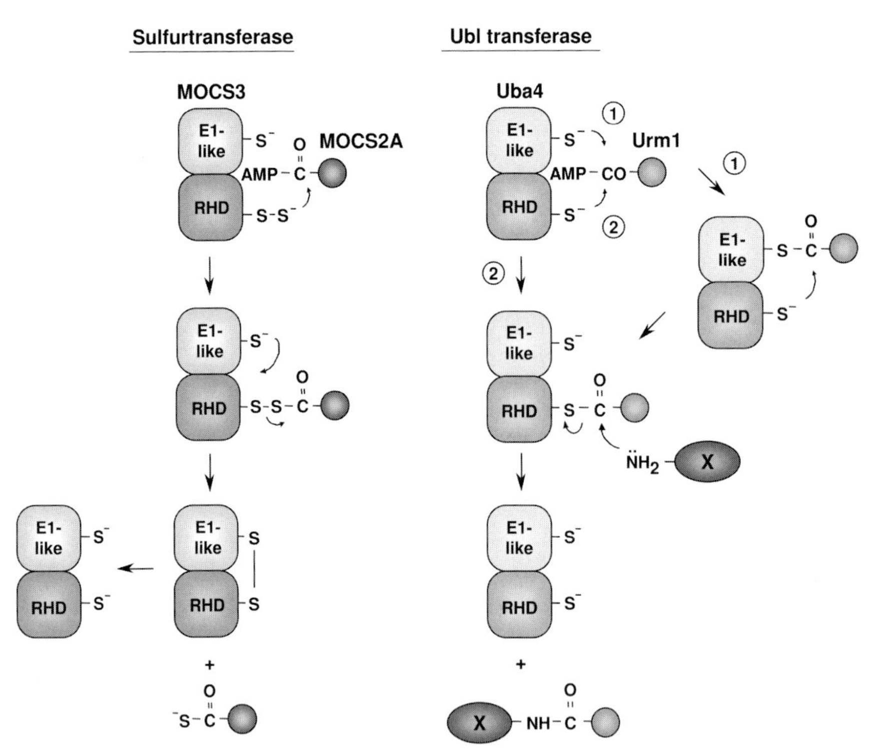

Fig. 11.2. Potential parallels between the MOCS2A–MOCS3 (MoaD–MoeB) and Urm1–Uba4 pathways. In MOCS2A activation, the C-terminally adenylated protein is thought to be attacked by a persulfide on the C-terminal RHD in MOCS3 (for *E. coli* MoeB, a persulfide on a separate, unidentified protein can be postulated). Release of the MOCS2A thiocarboxylate from its persulfide linkage to the MOCS3 RHD requires a second thiol group, which is proposed to be the E1-like domain cysteine, but other reducing factors may serve as well, depending on the species. For Uba4, it is not yet known whether a thioester between Urm1 and the conserved cysteine in the E1-like domain is formed (as depicted in pathway 1). Pathway 2 supposes that the adenylated Urm1 is directly attacked by a thiol from the RHD (a persulfide is also conceivable). No experimental evidence for the postulated RHD-linked Urm1 intermediate is currently available. Nucleophilic attack by substrate (X) transfers Urm1 from Uba4 to the substrate.

in these proteins is preceded by C-terminal adenylation by an E1-related enzyme: MoeB for MoaD and ThiF for ThiS [38–40]. Both MoaD and ThiS were later shown to share the ubiquitin fold despite the lack of obvious overall sequence similarity to ubiquitin [41, 42]. Therefore, ubiquitin, MoaD, and ThiS are all structurally related proteins whose C-termini are activated through adenylation by homologous E1-related enzymes [5, 43].

Further evidence for an evolutionary link between these sulfur transfer systems and ubiquitin activation came from a bioinformatics analysis of proteins that might be related to MoaD or ThiS in the predicted *S. cerevisiae* proteome [44]. This eukaryote lacks any Moco-containing enzymes and uses a different mechanism for thiamin synthesis; therefore, the aim of the sequence searches was to identify potential new Ubls that might have been missed in scans with ubiquitin. One previously uncharacterized protein was uncovered and named ubiquitin-related modifier-1 (Urm1), although no sequence similarity to ubiquitin could be detected. By yeast two-hybrid interaction screening with Urm1, Furukawa et al. [44] then identified a novel E1-related protein, which they named Uba4. Uba4 is more closely related to ThiF and MoeB than it is to the E1 ubiquitin-activating enzyme. Nevertheless, Uba4 appears to form a thioester intermediate with Urm1 and to stimulate covalent addition of Urm1 to cellular proteins. (Subsequent analysis revealed that a major target of Urm1 conjugation is Ahp1, a thiol-specific antioxidant protein [45].) The conclusion from these findings is that Urm1 and Uba4 function as a Ubl-protein conjugation system despite bearing much closer sequence relatedness to biosynthetic sulfur transfer factors than to ubiquitin and ubiquitin-activating enzyme. As such, the Urm1–Uba4 system might represent a kind of "missing link" in the evolution of the ubiquitin system from these sulfur transfer pathways [1, 44].

11.3.2 The E1–E2 Couple

Of course, the ubiquitin system involves a number of additional enzymes beyond the E1-activating enzyme. E2, E3, and ubiquitin- and Ubl-deconjugating enzymes are also central components, although E3s and Ubl-cleaving enzymes might not be part of all Ubl conjugation systems. On the other hand, an E1–E2 couple may be obligatory in all conjugation pathways that utilize an E1-like enzyme (Figure 11.2). For the eight Ubl conjugation systems in which an E1 has been identified (Table 11.1), all but one – the Urm1 system – is known to require a separate E2 protein. The transfer of Ubls from a cysteine side chain of an E1 to a cysteine on an E2 is not chemically necessary insofar as the Ubl C-terminus is already activated when it is bound to the E1, and the transfer to E2 also yields an enzyme–Ubl thioester bond. Enhanced regulatory flexibility or substrate specificity might help to explain the existence of E2s. Multiple E2 isozymes characterize the ubiquitin pathway, and in conjunction with different E3s, this enzyme diversity might increase the range or specificity of substrate modification. An E2 might also be needed to generate polymerized forms of ubiquitin or of certain Ubls [46, 47]. However, these explanations do not account for the evolutionary appearance of E2s in the first place, and no protein obviously related in sequence to any E2s has been identified in the biosynthetic sulfur donor pathways.

A possible exception to the E2 requirement for E1-catalyzed Ubl-protein conjugation is Urm1. Because Urm1 ligation appears to be poised, evolutionarily speaking, between the MoaD/ThiS activation and Ubl-protein ligation mechanisms, this exception is potentially instructive. A unique and conserved feature of Uba4 (the Urm1 E1 enzyme) compared to other E1-like proteins is the presence of a rhodanese homology domain (RHD) in the protein (Figure 11.2). Rhodanese and a number of RHD proteins are sulfurtransferases that form a persulfide (–S–S–H) on their active-site cysteine. Many MoeB family proteins, such as human MOCS3, have a similar domain organization, with an E1-like domain followed by an RHD. Based on these and other similarities, we previously proposed that thiocarboxylate formation in MoaD (MOCS2A in humans) catalyzed by MoeB/MOCS3 is closely related mechanistically to Uba4-catalyzed Urm1 activation and transfer [1] (see Figure 11.2). The Uba4 RHD in this scenario functions as a kind of built-in E2.

In the part of this speculative model pertaining to the sulfurtransferase, it was suggested that a persulfide is generated at the RHD active site of MoeB and that this attacks the activated MoaD to form an acyl disulfide intermediate (Figure 11.2, left side). In a second step, reductive cleavage of the MoaD acyl disulfide by the conserved E1-domain cysteine releases the MoaD thiocarboxylate. Recent experiments on human MOCS2A thiocarboxylate formation support many features of this model [48, 49]. We and others [1, 37, 39] had also initially suggested that the cysteine in the E1-like domain forms a thioester with MoaD. This appears not to occur, and the importance of this cysteine for Moco synthesis varies between species. The model in Figure 11.2 incorporates this revision. Interestingly, an acyl disulfide intermediate linking *E. coli* ThiF and ThiS has also been isolated, but this occurs on the conserved E1 domain cysteine [50]; such a covalent complex is not universally observed [51].

What are the implications for the Uba4 mechanism of Urm1 activation? As noted, the RHD of Uba4 was proposed to function as a built-in E2, using its active-site cysteine to attack a thioester intermediate on the E1-like domain [1] (Figure 11.2). We have found that the conserved RHD cysteine of Uba4 is required for its physiological function and for Urm1-protein conjugation *in vivo* (I. Velichutina, M. Hochstrasser, unpublished data). However, it remains to be shown that Urm1 is transferred onto the RHD or even that it forms a thioester at the E1 site, as has been assumed (Figure 11.2, right side). More generally, the E1-to-E2 transthiolation of Ubls may reflect the derivation of these protein-conjugation systems from sulfurtransferases that mobilize sulfur from a protein-linked persulfide through reductive cleavage by a second enzyme thiol group. Urm1 conjugation may retain features of the more ancient sulfurtransferases, while a process of "molecular takeover" might have occurred for other Ubl ligation pathways such that the RHD was replaced by a distinct E2 species. It is noteworthy that the E2s for the Atg8 and Atg12 pathways, while very weakly related to each other, are not detectably similar in sequence to the identified E2s for the remaining Ubl pathways, suggesting the possibility of convergent evolution of E2s from at least two separate lineages.

11.4 Ubiquitin-binding Domains and Ubiquitin Receptors in the Proteasome Pathway

Most early work on ubiquitin focused on its role in proteolysis [52–54]. Once it became clear that the 26S proteasome was responsible for the degradation of polyubiquitinated proteins, an obvious question was how the polyubiquitin chain facilitated degradation of the substrate protein. Several functions for such chains were proposed, but the most patent was that they enhanced the association between substrate and proteasome by directly binding to the proteasome. Elegant biochemical experiments eventually demonstrated that direct binding between the Lys48-linked polyubiquitin chain and the proteasome could fully account for the observed affinity of model polyubiquitinated proteins for the protease complex [13].

11.4.1 A Proteasome "Ubiquitin Receptor"

Beginning with the assumption that a polyubiquitinated protein could directly bind to a single subunit of the proteasome, Rechsteiner and colleagues used far-Western analysis to search for a ubiquitin receptor within the proteasome [55]. The subunits of purified human 26S proteasomes, which consist of one or two 19S regulatory complexes bound to a 20S proteasome core, were resolved by denaturing gel electrophoresis, blotted onto a membrane, and incubated with a radiolabeled polyubiquitinated protein. Remarkably, a single subunit of the 19S regulatory complex called S5a (Rpn10 in yeast) bound tightly to the radiolabeled substrate despite having been denatured initially in SDS and separated from the other subunits of the complex. Binding in solution was later shown as well. Two related ~30-residue hydrophobic segments in S5a are responsible for the binding to polyubiquitin [56]. A subsequent bioinformatics analysis recognized a more general ~20-residue core related to the S5a ubiquitin-binding element, and this element was christened the ubiquitin interaction motif (UIM) [57].

11.4.2 A Plethora of Ubiquitin-binding Domains

Since the description of the UIM, a substantial number of distinct ubiquitin-binding modules have been discovered by a combination of bioinformatic, biochemical, and structural studies [58]. Ubiquitin-binding domains include the UBA, CUE, UEV, NZF, DAUP/ZnF-UBP/PAZ, and GAT domains [59–61]. These domains range between ~35 and ~145 amino acids in length and vary considerably in structure. Nevertheless, several generalizations unite them. First, all of the structurally characterized ubiquitin-binding proteins contact the same general region on the ubiquitin molecule. The interface centers on a hydrophobic surface on the β-sheet of ubiquitin, which includes Ile44 and is sometimes called the Ile44 face. Hydrophobic interactions dominate the binding. Second, the binding to a single ubiquitin is generally very weak, with apparent dissociation constants ranging from ~20 μM to nearly 1 mM. Despite this unimpressive affinity, mutational studies have made clear the physiological relevance of these weak binding interactions in many cases [58, 59].

Amplification of the ubiquitin signal by linking multiple ubiquitin moieties into a chain can have dramatic effects on the affinity or avidity of interaction with target proteins. Again, the first and probably clearest example comes from studies on the proteasome and S5a/Rpn10. In the original far-Western analyses that identified S5a as a ubiquitin-binding subunit, binding of S5a to ubiquitin chains that had at least four ubiquitin units was clearly far tighter than to shorter chains [55]. Later, quantitative studies of Lys48–ubiquitin chain binding to full 26S proteasomes revealed a similar discontinuity in binding affinity [13]. Using competitor ubiquitin chains of various lengths, Thrower et al. [13] measured the inhibition of degradation of a model substrate. They found that tetrameric chains displayed very strong inhibition ($K_i \sim 170$ nM), i.e., high affinity, whereas inhibition was extremely weak with trimeric chains ($K_i \sim 1.9$ μM) and undetectable with dimeric chains ($K_i > 15$ μM). Such nonlinear effects imply that a unique binding signal is created by formation of a tetrameric chain rather than independent binding of multiple monoubiquitin moieties [62].

11.4.3
Ubiquitin-Conjugate Adaptor Proteins

Although S5a/Rpn10 was the first identified ubiquitin-binding subunit in the proteasome, it is not the only one, and for the degradation of many proteasome substrates, S5a/Rpn10 is completely dispensable. In *S. cerevisiae* for instance, loss of Rpn10 leads to only very minor phenotypic abnormalities [63]. Examination of individual substrates *in vivo* also suggests that only a subset are affected by loss of Rpn10 [63, 64]. Another major way by which polyubiquitinated proteins bind to the proteasome is through "adaptor proteins" that are thought to shuttle on and off the proteasome, ferrying their cargo from ubiquitin–ligase complexes or other intermediaries to sites on the 19S proteasome regulatory complex [64–67].

Although the details of the apparent substrate handoffs to and from these adaptors are still unclear, some common features of the adaptor proteins are emerging. One commonality is that the adaptors bear separate modules for proteasome and polyubiquitin binding. Three structurally related adaptors, Rad23, Dsk2, and Ddi1, were first characterized in yeast. They have an N-terminal ubiquitin-like domain (UBL) and one or two UBA elements, which, as noted earlier, are ubiquitin-binding domains. The UBL of these adaptors binds directly to either of two specific subunits in the proteasome 19S regulatory complex [68, 69]. These subunits, Rpn1 and Rnp2, share a series of leucine-rich repeats (LRRs). The LRRs, at least for

Rpn1, directly bind to the UBL of Rad23, and presumably the same holds for the other adaptors [68].

Only a subset of polyubiquitinated proteasome substrates requires any of these UBL–UBA adaptor proteins [64, 70]. Genetic studies suggest that Rpn10 and Rad23 have overlapping functions in substrate targeting, but degradation of some ubiquitinated proteins requires neither [64, 71]. Such substrates might be targeted by unidentified adaptors, or they might be recognized by an integral proteasome subunit [122]. One question raised by these data is how adaptor proteins with apparently generic proteasome- and polyubiquitin-binding modules can discriminate between different polyubiquitinated substrates (or even whether there is any functional difference between targeting such substrates directly to the proteasome or to a proteasome-binding adaptor protein). Moreover, while the bipartite nature of these adaptors suggests how they can bind simultaneously to both the polyubiquitinated substrate and the proteasome, it does not indicate how substrate transfer, e.g., between the adaptor and the proteasome, takes place.

A provocative recent study suggests some unexpected features of polyubiquitinated substrate transfers and also addresses the question of substrate specificity. This work indicates that the UBL of Rad23 binds not only to the proteasomal Rpn1 subunit but also to a second protein, Ufd2, which participates in polyubiquitination of a limited set of proteins [66]. Rpn1 and Ufd2 compete for Rad23 binding. One interpretation of these findings is that binding of Rad23 to a Ufd2-containing ubiquitin–ligase complex engaged in substrate polyubiquitination displaces the substrate-linked polyubiquitin chain from Ufd2. This in turn could facilitate binding of the polyubiquitin chain to the UBA domains of Rad23. The net effect will be the transfer of the polyubiquitinated substrate from Ufd2 to Rad23. Subsequently, the Rad23 UBL must somehow release Ufd2 and bind the proteasome, initiating the final transfer of substrate. Whether additional factors are required for these transfer reactions remains to be determined.

11.5
Ubiquitin-binding Domains and Membrane Protein Trafficking

Targeting proteins to the proteasome is not the only function of ubiquitin. An intricate array of dynamic ubiquitinated protein–protein target interactions has been described in a completely different arena, namely, membrane protein trafficking [58]. As with polyubiquitinated protein trafficking to the proteasome, sequential interactions of ubiquitinated substrate proteins with multiple ubiquitin-binding factors is a central feature of membrane protein sorting. However, in membrane protein trafficking, monoubiquitin is the predominant signal. As noted earlier, the binding of monoubiquitin to a ubiquitin-binding domain is generally weak; therefore, association is usually very transient unless additional substrate-binding sites are combined with the ubiquitin-binding motif. In principle, this allows considerable flexibility and sensitivity in the regulation of endocytosis and other membrane protein trafficking events.

Initial evidence for the importance of membrane protein ubiquitination for trafficking came from yeast [72–74]. These early studies demonstrated that cell surface receptors are ubiquitinated at the plasma membrane and that this ubiquitination correlates with their endocytosis and eventual degradation in the vacuole (the yeast equivalent of the lysosome). Subsequent research from many laboratories working in a variety of organisms has revealed that monoubiquitin attachment to receptors at the cell surface is a commonly employed endocytic signal [75]. Moreover, monoubiquitination of transmembrane proteins also helps to sort them once they have entered the endosomal membrane system.

An intensively investigated ubiquitin-dependent trafficking pathway in higher eukaryotes is signal transduction by receptor tyrosine kinases, the best studied of which is the epidermal growth factor (EGF) receptor [58, 76]. Beginning at the cell surface, ubiquitin modification functions at several stages in the endocytosis and intracellular sorting of EGF receptors to the lysosome, where the receptors are ultimately degraded. Thus, as with yeast plasma membrane proteins, ubiquitination is a means of downregulating or attenuating the surface expression of EGF receptors.

Once in endosomal vesicles, EGF receptors either can be sorted to a recycling compartment and thence back to the plasma membrane, or they can continue on toward late endosomes [76]. Late endosomes mature by a processing of invagination and vesiculation to form multivesicular bodies (MVBs). Receptors either stay in the limiting membrane of the MVB, which allows them to make their way back to the plasma membrane, or sort into the internal vesicles. MVBs eventually fuse with lysosomes, and the internalized vesicles and their cargo receptors are destroyed by lysosomal lipases and proteases.

Surprisingly, the E3 that ubiquitinates the EGF receptor at the cell surface continues to colocalize with the receptor along the endocytic pathway all the way to the internal vesicles of MVBs. This sustained colocalization may be crucial for maintaining the EGF receptor in its ubiquitin-modified state and for its sorting to the lysosome [77]. In yeast, there is no requirement for continued E3 association with the ubiquitinated receptor during trafficking [78]. It could be that in mammalian cells, the endocytosed receptors are more susceptible to deubiquitination by DUBs and therefore require repeated rounds of ubiquitin re-addition. This sustained requirement for ubiquitin on receptor proteins reflects ubiquitin-dependent endosomal sorting steps that occur within both yeast and mammalian cells. When either endocytosed proteins or biosynthetic membrane cargo proteins moving from the Golgi to the vacuole or lysosome are ubiquitinated, they are sorted into the invaginating regions of the late endosome. This sorting requires the sequential action of at least four highly conserved protein complexes that act at the endosome surface [79].

The first of these complexes is a heteromultimer formed by Hrs and STAM. Both the Hrs and STAM subunits contain ubiquitin-binding UIMs, which are necessary for sorting of cargo into internal MVB vesicles in yeast and might directly bind ubiquitinated membrane cargo proteins [80]. In mammalian cells, the Hrs–STAM complex is required for EGF receptor sorting in the MVB and degradation in the lysosome. Based on these and other data, this ubiquitin-binding complex has been proposed to be the sorting receptor for ubiquitinated membrane proteins at the endosome [80, 81].

Following initial recognition of ubiquitinated cargo by Hrs–STAM, the cargo is passed on to the ESCRT-I complex, which also includes a ubiquitin-binding subunit, Tsg101. ESCRT-I is recruited to the endosome through direct interactions between Tsg101 and both a tetrapeptide motif in Hrs and, apparently, the monoubiquitinated membrane protein cargo [82, 83]. Therefore, as was true for trafficking of certain polyubiquitinated proteins to the proteasome, ubiquitinated membrane proteins destined for the lysosome also need to be exchanged between a series of ubiquitin-binding factors. In both cases, these escort or adaptor proteins might shield the substrate from DUBs that could otherwise prematurely remove the ubiquitin signal. By interrogating the ubiquitin–substrate conjugate multiple times along the pathway to degradation, these factors might also enhance substrate selectivity.

Finally, two other complexes important for protein sorting at the late endosome membrane, ESCRT-II and ESCRT-III, help to drive the ubiquitinated cargo into invaginating membrane domains. Prior to scission of an invaginating region, a specific DUB is recruited to the ESCRT-III complex to recover the ubiquitin from the targeted receptors [84, 85]. This prevents degradation of ubiquitin and the depletion of cellular ubiquitin pools.

For simplicity, only ubiquitination of the endocytic or biosynthetic membrane protein cargo itself was noted in the preceding discussion. However, this is not the only point at which ubiquitination is important in membrane protein trafficking. Multiple endocytotic factors and membrane protein-sorting factors are also monoubiquitinated, sometimes in response to the same ligands that trigger receptor ubiquitination, e.g., EGF binding to EGF receptor [58, 75]. Many of these factors also contain ubiquitin-binding domains. These observations have led to the proposal that these soluble factors form dynamic protein networks in which ubiquitination of one factor allows intramolecular or intermolecular binding to other trafficking factors. This might contribute to or regulate the assembly of the ubiquitin-binding factors at plasma membrane sites for endocytosis or at endosomal sites for MVB sorting.

11.5.1 The MVB Pathway and RNA Virus Budding

A source of considerable excitement in the area of ubiquitin-dependent trafficking has been the discovery that enveloped RNA viruses such as HIV-1 and Ebola commandeer the MVB machinery to bud from the surface of the cell [86]. Budding is the way by which such viruses detach from the membrane of infected host cells. Outward budding of the plasma membrane and inward vesiculation of the late endosome membrane are topologically equivalent, and thus it is not entirely surprising that the same vesiculation machinery might be used in both cases. A key issue is how these viruses recruit MVB components to sites of virus particle assembly on the plasma membrane. Expression of the HIV-1 Gag protein is sufficient for membrane budding and release, and a specific region in Gag called the late or L domain is necessary for these events. Interestingly, the L domain in HIV-1 (and other retroviruses) includes a tetrapeptide similar to the sequence in Hrs mentioned earlier; as with Hrs, this sequence provides a docking site for the Tsg101 subunit of the ESCRT-1 complex. In this case, however, recruitment of ESCRT-I serves to hijack the MVB machinery to viral budding sites. In this way, the L domain of Gag binds to the MVB machinery, bringing the ESCRT complexes to Gag-associated plasma membrane sites and thereby stimulating the budding and release of virus particles.

What is unclear in all this is the exact role of ubiquitin [86]. Ubiquitin is somehow necessary for enveloped RNA virus budding based on ubiquitin depletion and mutagenesis experiments. Many retroviruses incorporate ubiquitin into their virions, and their Gag proteins are ubiquitinated in infected cells. However, Gag ubiquitination does not appear to be necessary for efficient viral budding and release, at least for the HIV-1 and MuLV retroviruses [87]. Components of the MVB machinery such as Hrs are also modified by ubiquitin, so it could be that this is where ubiquitin ligation is most important for viral egress from the cell.

11.6 Sumoylation and SUMO-binding Motifs

Besides ubiquitin, SUMO has been the most extensively studied ubiquitin-related protein modifier, and it is already clear that it can modify many proteins, possibly hundreds, *in vivo*. Recent results from studies on SUMO ligation and sumoylated protein interactions underscore and extend many of the ideas about ubiquitin–protein interactions, which were summarized in the preceding two sections.

RanGAP1 was the first SUMO-modified protein to be described, and it has been a useful prototype for understanding how conjugation to SUMO can modify a protein's function [88]. RanGAP1, the Ran GTPase-activating protein, is modified at a specific lysine that conforms to what has turned out to be a widely, but not universally, utilized sumoylation consensus sequence (ΨKxD/E, where Ψ is a large aliphatic residue and x is any residue). Protein sumoylation is also different from ubiquitination in that site-specific modification can occur with just the cognate E1 and E2 but no E3. This can be explained by the observation that the consensus sumoylation site provides a set of specific side-chain interactions with the E2 active-site pocket [89].

Attachment of SUMO to RanGAP1 dramatically changes the subcellular localization of the enzyme, resulting in its concentration on the cytoplasmic fibrils of the nuclear pore complex (NPC) where it binds to RanBP2/Nup358. Binding of SUMO–RanGAP1 to RanBP2 is not competed off by an excess of either free RanGAP1 or SUMO, suggesting either that essential RanBP2-binding determinants are present in both the RanGAP1 and SUMO portions of the SUMO–RanGAP1 conjugate or that formation of the conjugate somehow alters the conformation of

RanGAP1 or SUMO to generate a RanBP2-binding site [90]. Recent work on noncovalent SUMO interactions with RanBP2 and other proteins points toward the first of these two models and has helped to identify the first SUMO-binding motif (SBM).

11.6.1 A SUMO-binding Motif

Unlike the description of ubiquitin-binding motifs, which has become something of a cottage industry, there are relatively few studies thus far that address noncovalent binding of SUMO to other proteins. The first published foray into this area came from a two-hybrid screen for proteins that interacted with the p53-related protein p73α [91]. Both SUMO and a series of known SUMO-interacting proteins were found, and SUMO was shown to be conjugated to p73α. Comparison of the SUMO-interacting proteins suggested a short common motif that was potentially important for SUMO binding, and this was supported by subsequent mutagenesis experiments. The consensus derived from this small group of proteins was hhXSXS/Taaa, where *h* is a hydrophobic amino acid, *a* is an acidic amino acid, and *X* is any residue.

Building on these findings, Song et al. [34] carried out quantitative binding and NMR chemical shift perturbation studies to map the interface between SUMO and peptides bearing an SBM. With as few as nine residues, these peptides are substantially smaller than the domains that typically bind ubiquitin (20–145 residues); despite this small size and the fact that the peptides and SUMO form 1:1 complexes, the dissociation constants are between 5 μM and 10 μM, which is much tighter than the binding seen with all the ubiquitin-binding motifs discussed earlier. The refined SBM consensus sequence derived by Song et al. emphasized a central group of 3–4 hydrophobic residues. A region of RanBP2 known to bind sumoylated RanGAP1 has a sequence that conforms to this consensus. Mutagenesis experiments supported the significance of these hydrophobic residues but also indicated a contribution from flanking acidic residues to SUMO binding. Viewed together, the data from Minty et al. [91] and Song et al. [34] suggest that the SBM is composed primarily of a hydrophobic amino acid cluster flanked on one or both sides by acidic residues.

Interestingly, a two-hybrid screen for proteins that interacted with yeast SUMO, Smt3, pointed to the presence of a potential Smt3-binding sequence related to the above SBM sequences, namely, a cluster of 3–4 hydrophobic residues flanked by acidic amino acids [92]. This finding suggests that the mechanism of noncovalent SUMO interaction with target proteins is conserved from yeast to mammals. There will almost certainly be additional SUMO-binding domains distinct from the SBM, but it is notable that unrelated screens as different as the yeast and mammalian two-hybrid studies could yield such similar consensus sequences.

A very recent report on the crystal structure for a complex containing a SUMO1-RanGAP1 conjugate, a segment of RanBP2, and the SUMO E2 revealed the structure of the RanBP2 SBM and its mode of binding to SUMO [10]. The SBM peptide from RanBP2 forms a β strand that sits in a hydrophobic surface depression on SUMO and extends the SUMO β sheet; the ends of the SBM segment are acidic residues that interact with basic residues on the SUMO surface. The SBM–SUMO interface determined from this structure is therefore consistent with the mutagenesis and binding data mentioned above.

Importantly, both the crystallographic analysis of SUMO–RanGAP1 and an earlier NMR study of the same conjugate [93] revealed minimal direct interaction between SUMO and RanGAP1, except around the isopeptide linkage, and no obvious structural changes from the unligated proteins. From the NMR analysis, the loop of RanGAP1 linked to SUMO and the SUMO tail were both highly dynamic. Therefore, the mechanism of SUMO–RanGAP1 binding to RanBP2 does not appear to be through a conformational switch in the conjugate but rather by cooperative and simultaneous interaction of RanBP2 to a bipartite binding site created by the physical linkage of SUMO to RanGAP1. This type of binding is likely to typify many other SUMO–protein conjugate interactions.

11.6.2
A SUMO-induced Conformational Change

In contrast to the structural independence of the SUMO and substrate moieties in the SUMO–RanGAP1 conjugate, an elegant series of mechanistic studies on the function of SUMO conjugation to the DNA repair enzyme thymine-DNA glycosylase (TDG) suggests that in this case, SUMO conjugation to the substrate induces a substantial and functionally important conformation change in the protein [94]. TDG initiates base excision repair of a mismatched thymine or uracil nucleotide by removing the base, leaving an abasic site in the DNA. *In vitro*, TDG dissociation from the abasic site is strongly rate limiting; tight binding to the potentially harmful abasic site likely shields the site until downstream enzymes can complete the repair process. Nonetheless, a mechanism for enzyme turnover is required, and unexpectedly, site-specific SUMO ligation to TDG turns out to greatly accelerate its release from the DNA [95].

Even more surprisingly, the stimulation of TDG–DNA dissociation by sumoylation of the TDG C-terminal domain appears to operate through a sumoylation-induced conformational change in the TDG N-terminal domain, which causes the enzyme to loosen its grip on the abasic site [94]. Based on a protease sensitivity assay, the N-terminal domain of unsumoylated TDG triggers a conformational change in the enzyme when it binds to a G–U DNA mismatch substrate, enhancing its affinity but preventing catalytic turnover. If, instead, a sumoylated version of TDG is used in the reaction, TDG conformation no longer changes upon incubation with this DNA substrate, suggesting that in its sumoylated form, TDG does not assume the high-affinity DNA-binding state. N-terminally deleted TDG and sumoylated versions of both full-length and N-terminally truncated TDG show identical kinetic behavior when base excision is assayed (in all cases, turnover is still very slow: $k_{cat} \sim 0.05\ \text{min}^{-1}$). These data suggest that sumoylation allows enzymatic turnover by somehow facilitating an N-terminal domain-dependent conformational switch back to a state with low DNA-binding affinity. In the cell, one would assume, TDG sumoylation should occur only after the base excision step. This way, DNA damage recognition and base excision can occur efficiently (non-sumoylated, tight binding state), but substrate release and handoff to the downstream DNA endonuclease will then also be possible (sumoylated, weak binding state). What controls the timing of sumoylation and desumoylation relative to DNA binding and base excision is an interesting question that remains to be examined.

11.6.3 Interactions Between Different Sumoylated Proteins

While SUMO ligation to TDG is likely to be critical to its normal *in vivo* function, a noncovalent interaction between the two also seems to be important [96]. A mutation in TDG that blocks noncovalent SUMO binding also blocks its covalent attachment. In addition, TDG associates with the promyelocytic leukemia (PML) protein, and the same mutation that blocks TDG binding to SUMO blocks *in vivo* colocalization with PML; this association is not impaired by simple elimination of the TDG sumoylation site (sumoylated TDG does bind slightly better to PML *in vitro*). In other words, noncovalent association of SUMO with TDG is important for TDG interaction with PML. Interestingly, PML is itself a sumoylated protein and also has an SBM, so an obvious model is that the SUMO portion of sumoylated PML binds to TDG, and the SUMO on TDG binds to the PML SBM. Indeed, mutations that eliminate PML sumoylation inhibit TDG binding as well.

PML functions as a scaffold for the assembly of so-called PML nuclear bodies, subnuclear structures that have been implicated in transcriptional regulation and DNA repair. PML must be sumoylated to associate with these bodies and to concentrate a number of other sumoylated proteins there as well [97]. By combining sumoylation sites and SUMO-binding motifs into the same polypeptide, networks of protein interactions, typified by the TDG-PML association, can be created (also see Ref. [92]). This is reminiscent of what seems to occur with ubiquitin modification of and association with components of the endocytic machinery. As we saw earlier, a number of endocytic factors are ubiquitinated and also carry ubiquitin-binding domains. This is thought to contribute to the activation and interaction of such factors and/or to the exchange of ubiquitinated cargos between them.

Besides being relevant to the question of how SUMO–protein conjugates interact specifically and noncovalently with their protein partners, these data might also be significant when considering SUMO ligation specificity. As noted earlier, site-specific protein sumoylation is frequently observed *in vitro* with only the E1 and E2 enzymes, at least when they are at high concentration. We saw that one element of this specificity comes from E2–substrate contacts at consensus sumoylation sites. However, direct but noncovalent SUMO–substrate binding could provide an important additional binding determinant, particularly in combination with a consensus sumoylation site. This notion is supported by the finding that TDG is not sumoylated if its noncovalent SUMO-binding site is mutated [96].

11.7
General Biochemical Functions of Protein–Protein Conjugation

The examples chosen in the previous sections illustrate various ideas about the biochemical functions of protein modification by ubiquitin and Ubls. The simplest generalization to come out of all these examples is that attachment of ubiquitin or a Ubl to a protein (or other macromolecule) creates a distinct physiological state by altering the protein's interactions with other macromolecules (Figure 11.3). This is an almost trivial assertion, but more specific versions of the statement will be for-

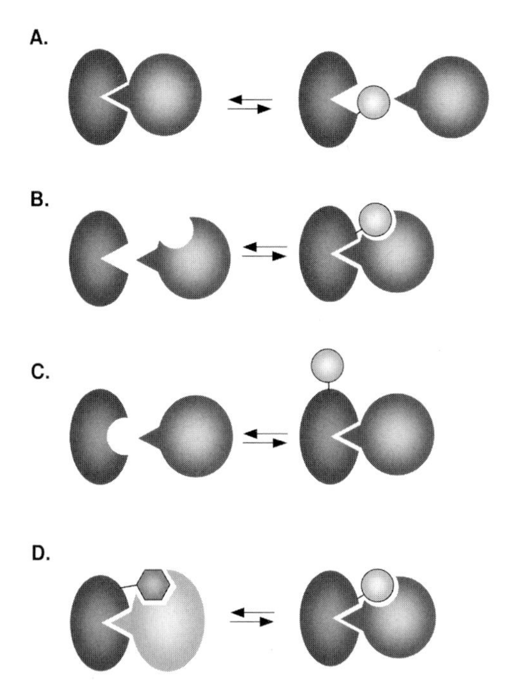

Fig. 11.3. Possible molecular functions for ubiquitin or Ubl-protein modification. (A) Ubl conjugation directly blocks an interaction between two proteins. A potential example of this is the sumoylation of the vaccinia A40R protein, which prevents association and aggregation between A40R monomers. (B) Ubl conjugation facilitates protein binding by providing a (additional) binding site. The best-documented case for this type of regulation is the sumoylation of RanGAP1, which leads to the binding of the conjugate to the nuclear pore protein RanBP2. (C) Ubl-conjugation causes a conformational change that enhances binding (or could have the reverse effect). For instance, it appears that SUMO attachment to thymine-DNA glycosylase (TDG) triggers a conformational change in TDG that lowers its affinity for DNA. (D) Modification by one Ubl helps to recruit a factor that is different from the protein that would be recruited were the substrate modified by another type of Ubl. Alternatively, one Ubl (red hexagon) might simply block conjugation of the substrate to another Ubl (or to another molecule), thereby preventing the substrate's interaction with another protein (green). The modification of IκB by SUMO on the same sites used by ubiquitin has been proposed to reflect such a mechanism. (Note: Not all possible variations on these basic mechanisms are shown.)

mulated below that may help give a sense of which mechanisms of altering macromolecular interaction by ubiquitin or Ubl attachment are more common and why this might be so.

11.7.1 Negative Regulation by Ubl Conjugation

Because ubiquitin and Ubls are bulky modifiers, one obvious way they could function is by steric occlusion: the attached Ubl simply blocks the ability of its substrate to bind to another protein (or another part of the same protein) (Figure 11.3A). There are still relatively few well-established examples of this inhibitory mode of action. One possible reason is that for such a mechanism to operate effectively in many cases, a very large fraction of the protein would need to be modified by the Ubl. However, for many proteins, only a very small fraction is observed in the conjugate form. This can sometimes be attributed to artifactual deconjugation during protein isolation, but even when such deconjugation is largely prevented, often only a few percent of a particular protein is modified, e.g., with SUMO or ISG15.

Nevertheless, if the small fraction of modified protein were localized to some functionally unique cellular site, or if a transient modification were sufficient to put the protein in a new state, such an inhibitory mechanism could still operate. An example that combines both of these mechanisms is the transient ubiquitination of histone H2B at chromosomal sites of induced transcription [98]. (The exact biochemical consequences of this histone ubiquitination are not yet known, so this might not be an example of a modification that inhibits interaction.) Local ubiquitination is brought about by the recruitment of a histone H2B-specific ubiquitin ligase complex. Such transient histone H2B ubiquitination triggers histone H3 methylation, and this new histone state, which is necessary but not sufficient for gene activation, no longer requires that ubiquitin remain on histone H2B. Indeed, deubiquitination of the histone is needed for completion of the switch to the transcriptionally active state. Another example of such molecular memory is the sumoylation of TDG discussed earlier: SUMO attachment is needed to weaken the interaction with DNA after base scission, but once TDG has released from the DNA, the SUMO is no longer necessary or desirable [94]. In this case, however, SUMO attachment seems to inhibit macromolecular interaction (DNA binding) indirectly by inducing a change in enzyme conformation.

If a large percentage of a protein were Ubl-modified, the notion of negative regulation of protein interaction and function would be more straightforward. A recent example of a protein that is nearly quantitatively modified by SUMO is the vaccinia virus A40R early protein [99]. A40R sumoylation is required for its localization to ER viral replication sites. Mutation of the A40R sumoylation site causes the protein to self-associate and aggregate into long rods. Thus, SUMO attachment to A40R appears to block its interaction with another protein (another copy of A40R). A second potential example of quantitative sumoylation is the plasma membrane K2P1 potassium leak channel [100]. The apparent SUMO modification is proposed to block channel opening and thereby leakage of K^+ ions from the cell.

11.7.2 Positive Regulation by Ubl Conjugation

When the Ubl modifier enhances an interaction with another macromolecule, it usually does so by participating directly in the formation of part or all of the binding interface with the target molecule (Figure 11.3B). In principle, such an interaction can also be modulated if an allosteric change in a target binding site were induced by the attached Ubl (Figure 11.3C). Many examples of Ubl regulation fall into the former category. Here it is easy to see that even if only a small fraction of a particular protein were modified, its new activity could suffice to effect a change in physiological state. Many of the examples in the preceding sections reflect this kind of mechanism. As discussed, noncovalent ubiquitin–protein or Ubl–protein interactions tend to be weak. Binding can be greatly enhanced either by polymerization of the ubiquitin signal (no clear example of obligatory Ubl chain formation is known) or by combining the weak binding from ubiquitin or Ubl with additional weak binding sites (possibly created by additional ubiquitin or Ubl modifications). The combination of multiple weak interactions to give highly specific protein–protein binding is a well-established idea in the signal transduction field. An example of such multivalent binding was discussed earlier for SUMO–RanGAP1 binding to the nuclear pore complex.

11.7.3 Cross-regulation by Ubls

Interestingly, alternative Ubl or ubiquitin modifications sometimes occur on the same substrate, and these can direct the protein to different targets. The clearest illustration of this is proliferating cell nuclear antigen (PCNA), which can be monoubiquitinated, polyubiquitinated, or sumoylated, and each of these forms results in the recruitment of distinct downstream effector proteins. PCNA functions as a DNA polymerase processivity factor in various modes of DNA replication and DNA repair. It forms a homotrimer that encircles the DNA double helix and associates with multiple DNA polymerases. Ubiquitin or a Lys63-linked polyubiquitin chain is attached to a single PCNA lysine, and SUMO is primarily attached to this same site, with a small amount at a second lysine [101]. When the DNA replication machinery encounters a DNA lesion and cannot replicate past it, the type of post-translational modification of PCNA determines which of several distinct mechanisms will be engaged to correct or bypass the lesion.

A model that accounts for these different outcomes was recently outlined [102]. The ubiquitin E2 Rad6 and the E3 Rad18 are responsible for monoubiquitination of PCNA, and this ubiquitin can be extended by a heterodimeric E2, Ubc13-Mms2, and the E3 Rad5 into a Lys63-linked polyubiquitin chain [101, 103]. During normal replication, the replicative DNA polymerase is associated with PCNA at the primer–template junction. Monoubiquitination of PCNA is proposed to prevent the polymerase from accessing the junction when DNA damage is encountered, allowing a translesion synthesis (TLS) polymerase to enter the complex [102, 104].

These TLS polymerases bind directly to PCNA and promote either error-free or mutagenic replication through the lesion, depending on the TLS polymerase and the type of DNA damage. If a Lys63–ubiquitin chain has formed on PCNA, it is postulated to cause complete dissociation of the replicative DNA polymerase complex, allowing a template-switching mechanism with error-free copying of the other replicated DNA strand.

The consequences of SUMO ligation to PCNA have been less clear, but PCNA sumoylation, which occurs during a normal S phase without induced DNA damage [101], prevents recombinational bypass of lesions [102]. Such recombination during normal DNA replication can be deleterious because of the risk of chromosome rearrangements. Very recent results strongly support this function for SUMO–PCNA, and argue that the sumoylated form specifically recruits the Srs2 DNA helicase to the replication fork [105, 106]. Srs2 disassembles the nucleoprotein filaments that are necessary intermediates for DNA strand invasion and homologous recombination [107, 108]. The exact protein–protein interactions affected by the different ubiquitin and SUMO modifications have not been worked out in full, but it appears that both positive and negative regulation of such interactions occurs (Figure 11.3). Because the same site of PCNA is used for both ubiquitin and SUMO attachment, there also appears to be some antagonism between these two modifications [105] (Figure 11.3D). Similar competition between ubiquitin and SUMO for the same substrate lysine has been seen with IκB, an inhibitor of the NF-κB signaling pathway [109]. For IκB, Lys48 polyubiquitination leads to IκB degradation and NF-κB activation, but sumoylation of the same IκB lysines prevents this.

11.8 Conclusions

As should be evident from the above survey, ubiquitin and Ubl modification of proteins represents a highly versatile means of regulating protein function. Nature has made widespread use of such conjugation through the elaboration of multiple variants of the same basic enzymatic mechanism. On the order of a dozen or so Ubls have been documented to date, and for eight of these, at least one enzyme in the pathway for substrate conjugation has been identified. Ubiquitin itself can attach to proteins in the form of polymers of different topology, and these topological variants impart differences in function as well. The fundamental E1–E2 couple, which probably arose very early in the evolution of the ubiquitin system from more ancient sulfur transfer pathways, has been supplemented with an array of specificity factors (E3s) in some of the pathways, especially the ubiquitin pathway. Deconjugating enzymes have turned ubiquitin and many of the Ubls into dynamic modifiers whose attachments are tightly regulated both spatially and temporally. The basic biochemical consequence of protein modification by ubiquitin or Ubls is usually a change in the target's association with other proteins. This change can occur by both direct and indirect mechanisms and can either stimulate or inhibit particular protein–protein interactions. Given the intricacy of the ubiquitin–Ubl system, research into its functions and mechanisms should continue to tax and reward investigators for years to come.

Acknowledgments

I thank Cecile Pickart, Rachael Felberbaum, Oliver Kerscher, Stefan Kreft, Alaron Lewis, and Tommer Ravid for many helpful comments on the manuscript. Work from my laboratory was supported by grants from the U.S. National Institutes of Health (GM46904 and GM53756).

References

1 Hochstrasser, M. Evolution and function of ubiquitin-like protein-conjugation systems. *Nat Cell Biol* **2**, E153–E157 (2000).
2 Jentsch, S. & Pyrowolakis, G. Ubiquitin and its kin: how close are the family ties? *Trends Cell Biol* **10**, 335–42 (2000).
3 Pickart, C. M. Mechanisms Underlying Ubiquitination. *Annu Rev Biochem* **70**, 503–533 (2001).
4 Schwartz, D. C. & Hochstrasser, M. A superfamily of protein tags: ubiquitin, SUMO and related modifiers. *Trends Biochem Sci* **28**, 321–8 (2003).
5 Huang, D. T., Walden, H., Duda, D. & Schulman, B. A. Ubiquitin-like protein activation. *Oncogene* **23**, 1958–71 (2004).
6 Ciechanover, A. & Ben-Saadon, R. N-terminal ubiquitination: more protein substrates join in. *Trends Cell Biol* **14**, 103–6 (2004).
7 Zheng, N., Wang, P., Jeffrey, P. D. & Pavletich, N. P. Structure of a c-Cbl-UbcH7 complex: RING domain function in ubiquitin-protein ligases. *Cell* **102**, 533–9 (2000).
8 Furukawa, M., Ohta, T. & Xiong, Y. Activation of UBC5 ubiquitin-conjugating enzyme by the RING finger of ROC1 and assembly of active ubiquitin ligases by all cullins. *J Biol Chem* **277**, 15758–65 (2002).
9 Wu, P. Y. et al. A conserved catalytic residue in the ubiquitin-conjugating enzyme family. *Embo J* **22**, 5241–50 (2003).
10 Reverter, D. & Lima, C. D. Insights into E3 ligase activity revealed by a SUMO-RanGAP1-Ubc9-Nup358 complex. *Nature* **435**, 687–92 (2005).
11 Pickart, C. M. & Cohen, R. E. Proteasomes and their kin: proteases in the machine age. *Nat Rev Mol Cell Biol* **5**, 177–87 (2004).
12 Hochstrasser, M. Ubiquitin signalling: what's in a chain? *Nat Cell Biol* **6**, 571–2 (2004).
13 Thrower, J. S., Hoffman, L., Rechsteiner, M. & Pickart, C. M. Recognition of the polyubiquitin proteolytic signal. *Embo J* **19**, 94–102 (2000).
14 Raasi, S., Orlov, I., Fleming, K. G. & Pickart, C. M. Binding of polyubiquitin chains to ubiquitin-associated (UBA) domains of HHR23A. *J Mol Biol* **341**, 1367–79 (2004).
15 Baboshina, O. V. & Haas, A. L. Novel multiubiquitin chain linkages catalyzed by the conjugating enzymes $E2_{EPF}$ and RAD6 are recognized by 26S proteasome subunit 5. *J. Biol. Chem.* **271**, 2823–2831 (1996).
16 Amerik, A. Y. & Hochstrasser, M. Mechanism and function of deubiquitinating enzymes. *Biochim Biophys Acta* **1695**, 189–207 (2004).
17 Li, M. et al. Deubiquitination of p53 by HAUSP is an important pathway

for p53 stabilization. *Nature* **416**, 648–53 (2002).

18 Guterman, A. & Glickman, M. H. Deubiquitinating enzymes are IN/(trinsic to proteasome function). *Curr Protein Pept Sci* **5**, 201–11 (2004).

19 Yao, T. & Cohen, R. E. A cryptic protease couples deubiquitination and degradation by the proteasome. *Nature* **419**, 403–407 (2002).

20 Haas, A. L., Ahrens, P., Bright, P. M. & Ankel, H. Interferon induces a 15-kilodalton protein exhibiting marked homology to ubiquitin. *J Biol Chem* **262**, 11315–23 (1987).

21 Loeb, K. R. & Haas, A. L. The interferon-inducible 15-kDa ubiquitin homolog conjugates to intracellular proteins. *J Biol Chem* **267**, 7806–13 (1992).

22 Potter, J. L., Narasimhan, J., Mende-Mueller, L. & Haas, A. L. Precursor processing of pro-ISG15/UCRP, an interferon-beta-induced ubiquitin-like protein. *J Biol Chem* **274**, 25061–8 (1999).

23 Narasimhan, J. et al. Crystal structure of the interferon-induced ubiquitin-like protein ISG15. *J Biol Chem* (2005).

24 Yuan, W. & Krug, R. M. Influenza B virus NS1 protein inhibits conjugation of the interferon (IFN)-induced ubiquitin-like ISG15 protein. *Embo J* **20**, 362–71 (2001).

25 Zhao, C. et al. The UbcH8 ubiquitin E2 enzyme is also the E2 enzyme for ISG15, an IFN-alpha/beta-induced ubiquitin-like protein. *Proc Natl Acad Sci U S A* **101**, 7578–82 (2004).

26 Kim, K. I., Giannakopoulos, N. V., Virgin, H. W. & Zhang, D. E. Interferon-inducible ubiquitin E2, Ubc8, is a conjugating enzyme for protein ISGylation. *Mol Cell Biol* **24**, 9592–600 (2004).

27 Ritchie, K. J. et al. Role of ISG15 protease UBP43 (USP18) in innate immunity to viral infection. *Nat Med* **10**, 1374–8 (2004).

28 Malakhov, M. P. et al. High-throughput immunoblotting. Ubiquitiin-like protein ISG15 modifies key regulators of signal transduction. *J Biol Chem* **278**, 16608–13 (2003).

29 Peng, J. et al. A proteomics approach to understanding protein ubiquitination. *Nat Biotechnol* **21**, 921–6 (2003).

30 Hitchcock, A. L., Auld, K., Gygi, S. P. & Silver, P. A. A subset of membrane-associated proteins is ubiquitinated in response to mutations in the endoplasmic reticulum degradation machinery. *Proceedings of the National Academy of Sciences of the United States of America* **100**, 12735–40 (2003).

31 Liu, Y., Fallon, L., Lashuel, H. A., Liu, Z. & Lansbury, P. T., Jr. The UCH-L1 gene encodes two opposing enzymatic activities that affect alpha-synuclein degradation and Parkinson's disease susceptibility. *Cell* **111**, 209–18 (2002).

32 Dassa, B., Yanai, I. & Pietrokovski, S. New type of polyubiquitin-like genes with intein-like autoprocessing domains. *Trends Genet* **20**, 538–42 (2004).

33 Ichimura, Y. et al. A ubiquitin-like system mediates protein lipidation. *Nature* **408**, 488–92 (2000).

34 Song, J., Durrin, L. K., Wilkinson, T. A., Krontiris, T. G. & Chen, Y. Identification of a SUMO-binding motif that recognizes SUMO-modified proteins. *Proc Natl Acad Sci U S A* **101**, 14373–8 (2004).

35 Sharp, P. M. & Li, W.-H. Molecular evolution of ubiquitin genes. **2**, 328–332 (1987).

36 McGrath, J. P., Jentsch, S. & Varshavsky, A. *UBA1*: an essential yeast gene encoding ubiquitin-activating enzyme. *Embo J* **10**, 227–36 (1991).

37 Rajagopalan, K. V. Biosynthesis and processing of the molybdenum cofactors. *Biochem Soc Trans* **25**, 757–61 (1997).

38 Taylor, S. V. et al. Thiamin biosynthesis in Escherichia coli. Identification of this thiocarboxylate as the immediate sulfur donor in the thiazole formation. *J Biol Chem* **273**, 16555–60 (1998).

39 Appleyard, M. V. et al. The Aspergillus nidulans cnxF gene and its involvement in molybdopterin

biosynthesis. Molecular characterization and analysis of in vivo generated mutants. *J Biol Chem* **273**, 14869–76 (1998).
40 Leimkuhler, S., Wuebbens, M. M. & Rajagopalan, K. V. Characterization of Escherichia coli MoeB and its involvement in the activation of molybdopterin synthase for the biosynthesis of the molybdenum cofactor. *J Biol Chem* **276**, 34695–701 (2001).
41 Rudolph, M. J., Wuebbens, M. M., Rajagopalan, K. V. & Schindelin, H. Crystal structure of molybdopterin synthase and its evolutionary relationship to ubiquitin activation. *Nat Struct Biol* **8**, 42–6 (2001).
42 Wang, C., Xi, J., Begley, T. P. & Nicholson, L. K. Solution structure of ThiS and implications for the evolutionary roots of ubiquitin. *Nat Struct Biol* **8**, 47–51 (2001).
43 Duda, D. M., Walden, H., Sfondouris, J. & Schulman, B. A. Structural Analysis of Escherichia Coli ThiF. *J Mol Biol* **349**, 774–86 (2005).
44 Furukawa, K., Mizushima, N., Noda, T. & Ohsumi, Y. A protein conjugation system in yeast with homology to biosynthetic enzyme reaction of prokaryotes. *J Biol Chem* **275**, 7462–5 (2000).
45 Goehring, A. S., Rivers, D. M. & Sprague, G. F., Jr. Attachment of the ubiquitin-related protein Urm1p to the antioxidant protein Ahp1p. *Eukaryot Cell* **2**, 930–6 (2003).
46 Silver, E. T., Gwozd, T. J., Ptak, C., Goebl, M. & Ellison, M. J. A chimeric ubiquitin conjugating enzyme that combines the cell cycle properties of CDC34 (UBC3) and the DNA repair properties of RAD6 (UBC2): implications for the structure, function and evolution of the E2s. *Embo J* **11**, 3091–8 (1992).
47 Chen, P., Johnson, P., Sommer, T., Jentsch, S. & Hochstrasser, M. Multiple ubiquitin-conjugating enzymes participate in the in vivo degradation of the yeast MATα2 repressor. *Cell* **74**, 357–369 (1993).
48 Matthies, A., Rajagopalan, K. V., Mendel, R. R. & Leimkuhler, S. Evidence for the physiological role of a rhodanese-like protein for the biosynthesis of the molybdenum cofactor in humans. *Proc Natl Acad Sci U S A* **101**, 5946–51 (2004).
49 Matthies, A., Nimtz, M. & Leimkuhler, S. Molybdenum Cofactor Biosynthesis in Humans: Identification of a Persulfide Group in the Rhodanese-like Domain of MOCS3 by Mass Spectrometry. *Biochemistry* **44**, 7912–7920 (2005).
50 Xi, J., Ge, Y., Kinsland, C., McLafferty, F. W. & Begley, T. P. Biosynthesis of the thiazole moiety of thiamin in Escherichia coli: identification of an acyldisulfide-linked protein–protein conjugate that is functionally analogous to the ubiquitin/E1 complex. *Proc Natl Acad Sci U S A* **98**, 8513–8 (2001).
51 Park, J. H. et al. Biosynthesis of the thiazole moiety of thiamin pyrophosphate (vitamin B1). *Biochemistry* **42**, 12430–8 (2003).
52 Hershko, A. & Ciechanover, A. The ubiquitin system for protein degradation. *Annu Rev Biochem* **61**, 761–807 (1992).
53 Goldberg, A. L. & Rock, K. L. Proteolysis, Proteasomes and Antigen Presentation. *Nature* **357**, 375–379 (1992).
54 Hochstrasser, M. Ubiquitin-dependent protein degradation. *Ann. Rev. Genet.* **30**, 405–439 (1996).
55 Deveraux, Q., Ustrell, V., Pickart, C. & Rechsteiner, M. A 26 S Protease Subunit That Binds Ubiquitin Conjugates. *J. Biol. Chem.* **269**, 7059–7061 (1994).
56 Young, P., Deveraux, Q., Beal, R. E., Pickart, C. M. & Rechsteiner, M. Characterization of two polyubiquitin binding sites in the 26 S protease subunit 5a. *J Biol Chem* **273**, 5461–7 (1998).
57 Hofmann, K. & Falquet, L. A ubiquitin-interacting motif conserved in components of the proteasomal and lysosomal protein degradation systems. *Trends Biochem Sci* **26**, 347–50 (2001).

58 Di Fiore, P. P., Polo, S. & Hofmann, K. When ubiquitin meets ubiquitin receptors: a signalling connection. *Nat Rev Mol Cell Biol* **4**, 491–7 (2003).

59 Schnell, J. D. & Hicke, L. Non-traditional functions of ubiquitin and ubiquitin-binding proteins. *J Biol Chem* **278**, 35857–60 (2003).

60 Fisher, R. D. et al. Structure and ubiquitin binding of the ubiquitin-interacting motif. *J Biol Chem* **278**, 28976–84 (2003).

61 Prag, G. et al. Structural mechanism for ubiquitinated-cargo recognition by the Golgi-localized, gamma-ear-containing, ADP-ribosylation-factor-binding proteins. *Proc Natl Acad Sci U S A* **102**, 2334–9 (2005).

62 Varadan, R. et al. Solution conformation of Lys63-linked di-ubiquitin chain provides clues to functional diversity of polyubiquitin signaling. *J Biol Chem* **279**, 7055–63 (2004).

63 van Nocker, S. et al. The multiubiquitin-chain-binding protein Mcb1 is a component of the 26S proteasome in Saccharomyces cerevisiae and plays a nonessential, substrate-specific role in protein turnover. *Mol. Cell. Biol.* **16**, 6020–6028 (1996).

64 Verma, R., Oania, R., Graumann, J. & Deshaies, R. J. Multiubiquitin chain receptors define a layer of substrate selectivity in the ubiquitin-proteasome system. *Cell* **118**, 99–110 (2004).

65 Chen, L. & Madura, K. Rad23 promotes the targeting of proteolytic substrates to the proteasome. *Mol Cell Biol* **22**, 4902–13 (2002).

66 Kim, I., Mi, K. & Rao, H. Multiple interactions of rad23 suggest a mechanism for ubiquitylated substrate delivery important in proteolysis. *Mol Biol Cell* **15**, 3357–65 (2004).

67 Elsasser, S., Chandler-Militello, D., Muller, B., Hanna, J. & Finley, D. Rad23 and Rpn10 serve as alternative ubiquitin receptors for the proteasome. *J Biol Chem* **279**, 26817–22 (2004).

68 Elsasser, S. et al. Proteasome subunit Rpn1 binds ubiquitin-like protein domains. *Nature Cell Biology* **4**, 725–30 (2002).

69 Saeki, Y., Sone, T., Toh-e, A. & Yokosawa, H. Identification of ubiquitin-like protein-binding subunits of the 26S proteasome. *Biochemical & Biophysical Research Communications* **296**, 813–9 (2002).

70 Medicherla, B., Kostova, Z., Schaefer, A. & Wolf, D. H. A genomic screen identifies Dsk2p and Rad23p as essential components of ER-associated degradation. *EMBO Rep* **5**, 692–7 (2004).

71 Lambertson, D., Chen, L. & Madura, K. Pleiotropic defects caused by loss of the proteasome-interacting factors Rad23 and Rpn10 of Saccharomyces cerevisiae. *Genetics* **153**, 69–79 (1999).

72 Kölling, R. & Hollenberg, C. P. The ABC-transporter Ste6 accumulates in the plasma membrane in a ubiquitinated form in endocytosis mutants. *EMBO J.* **13**, 3261–3271 (1994).

73 Hicke, L. & Riezman, H. Ubiquitination of a yeast plasma membrane receptor signals its ligand-stimulated endocytosis. *Cell* **84**, 277–287 (1996).

74 Roth, A. F. & Davis, N. G. Ubiquitination of the yeast a-factor receptor. *J Cell Biol* **134**, 661–74 (1996).

75 Hicke, L. & Dunn, R. Regulation of membrane protein transport by ubiquitin and ubiquitin-binding proteins. *Annu Rev Cell Dev Biol* **19**, 141–72 (2003).

76 Marmor, M. D. & Yarden, Y. Role of protein ubiquitylation in regulating endocytosis of receptor tyrosine kinases. *Oncogene* **23**, 2057–70 (2004).

77 Longva, K. E. et al. Ubiquitination and proteasomal activity is required for transport of the EGF receptor to inner membranes of multivesicular bodies. *J Cell Biol* **156**, 843–54 (2002).

78 Hicke, L. Protein regulation by monoubiquitin. *Nat Rev Mol Cell Biol* **2**, 195–201 (2001).

79 Katzmann, D. J., Odorizzi, G. & Emr, S. D. Receptor downregulation

and multivesicular-body sorting. *Nat Rev Mol Cell Biol* **3**, 893–905 (2002).

80 Bilodeau, P. S., Urbanowski, J. L., Winistorfer, S. C. & Piper, R. C. The Vps27p Hse1p complex binds ubiquitin and mediates endosomal protein sorting. *Nat Cell Biol* **4**, 534–9 (2002).

81 Komada, M. & Kitamura, N. The Hrs/STAM complex in the down-regulation of receptor tyrosine kinases. *J Biochem (Tokyo)* **137**, 1–8 (2005).

82 Pornillos, O. et al. HIV Gag mimics the Tsg101-recruiting activity of the human Hrs protein. *J Cell Biol* **162**, 425–34 (2003).

83 Sundquist, W. I. et al. Ubiquitin recognition by the human TSG101 protein. *Mol Cell* **13**, 783–9 (2004).

84 Amerik, A. Y., Nowak, J., Swaminathan, S. & Hochstrasser, M. The Doa4 deubiquitinating enzyme is functionally linked to the vacuolar protein-sorting and endocytic pathways. *Mol Biol Cell* **11**, 3365–80 (2000).

85 Luhtala, N. & Odorizzi, G. Bro1 coordinates deubiquitination in the multivesicular body pathway by recruiting Doa4 to endosomes. *J Cell Biol* **166**, 717–29 (2004).

86 Morita, E. & Sundquist, W. I. Retrovirus budding. *Annu Rev Cell Dev Biol* **20**, 395–425 (2004).

87 Ott, D. E., Coren, L. V., Chertova, E. N., Gagliardi, T. D. & Schubert, U. Ubiquitination of HIV-1 and MuLV Gag. *Virology* **278**, 111–21 (2000).

88 Melchior, F. SUMO – nonclassical ubiquitin. *Annu Rev Cell Dev Biol* **16**, 591–626 (2000).

89 Bernier-Villamor, V., Sampson, D. A., Matunis, M. J. & Lima, C. D. Structural basis for E2-mediated SUMO conjugation revealed by a complex between ubiquitin-conjugating enzyme Ubc9 and RanGAP1. *Cell* **108**, 345–56 (2002).

90 Mahajan, R., Delphin, C., Guan, T., Gerace, L. & Melchior, F. A small ubiquitin-related polypeptide involved in targeting RanGAP1 to nuclear pore complex protein RanBP2. *Cell* **88**, 97–107 (1997).

91 Minty, A., Dumont, X., Kaghad, M. & Caput, D. Covalent modification of p73alpha by SUMO-1. Two-hybrid screening with p73 identifies novel SUMO-1-interacting proteins and a SUMO-1 interaction motif. *J Biol Chem* **275**, 36316–23 (2000).

92 Hannich, J. T. et al. Defining the SUMO-modified proteome by multiple approaches in Saccharomyces cerevisiae. *J Biol Chem* **280**, 4102–10 (2005).

93 Macauley, M. S. et al. Structural and dynamic independence of isopeptide-linked RanGAP1 and SUMO-1. *J Biol Chem* **279**, 49131–7 (2004).

94 Steinacher, R. & Schar, P. Functionality of human thymine DNA glycosylase requires SUMO-regulated changes in protein conformation. *Curr Biol* **15**, 616–23 (2005).

95 Hardeland, U., Steinacher, R., Jiricny, J. & Schar, P. Modification of the human thymine-DNA glycosylase by ubiquitin-like proteins facilitates enzymatic turnover. *Embo J* **21**, 1456–64 (2002).

96 Takahashi, H., Hatakeyama, S., Saitoh, H. & Nakayama, K. I. Noncovalent SUMO-1 binding activity of thymine DNA glycosylase (TDG) is required for its SUMO-1 modification and colocalization with the promyelocytic leukemia protein. *J Biol Chem* **280**, 5611–21 (2005).

97 Duprez, E. et al. SUMO-1 modification of the acute promyelocytic leukaemia protein PML: implications for nuclear localisation. *J Cell Sci* **112**, 381–93 (1999).

98 Henry, K. W. et al. Transcriptional activation via sequential histone H2B ubiquitylation and deubiquitylation, mediated by SAGA-associated Ubp8. *Genes Dev* **17**, 2648–63 (2003).

99 Palacios, S. et al. Quantitative SUMO-1 Modification of a Vaccinia Virus Protein Is Required for Its Specific Localization and Prevents Its Self-Association. *Mol Biol Cell* **16**, 2822–35 (2005).

100 Rajan, S., Plant, L. D., Rabin, M. L., Butler, M. H. & Goldstein, S. A. Sumoylation silences the plasma membrane leak K+ channel K2P1. *Cell* **121**, 37–47 (2005).

101 Hoege, C., Pfander, B., Moldovan, G. L., Pyrowolakis, G. & Jentsch, S. RAD6-dependent DNA repair is linked to modification of PCNA by ubiquitin and SUMO. *Nature* **419**, 135–41 (2002).

102 Haracska, L., Torres-Ramos, C. A., Johnson, R. E., Prakash, S. & Prakash, L. Opposing effects of ubiquitin conjugation and SUMO modification of PCNA on replicational bypass of DNA lesions in Saccharomyces cerevisiae. *Mol Cell Biol* **24**, 4267–74 (2004).

103 Stelter, P. & Ulrich, H. D. Control of spontaneous and damage-induced mutagenesis by SUMO and ubiquitin conjugation. *Nature* **425**, 188–91 (2003).

104 Kannouche, P. L., Wing, J. & Lehmann, A. R. Interaction of human DNA polymerase eta with monoubiquitinated PCNA: a possible mechanism for the polymerase switch in response to DNA damage. *Mol Cell* **14**, 491–500 (2004).

105 Pfander, B., Moldovan, G. L., Sacher, M., Hoege, C. & Jentsch, S. SUMO-modified PCNA recruits Srs2 to prevent recombination during S phase. *Nature* (2005).

106 Papouli, E. et al. Crosstalk between SUMO and ubiquitin on PCNA is mediated by recruitment of the helicase Srs2p. *Mol Cell*, Published online Jun. 9, 2005 10.1016/S1097276505013766 (2005).

107 Krejci, L. et al. DNA helicase Srs2 disrupts the Rad51 presynaptic filament. *Nature* **423**, 305–9 (2003).

108 Veaute, X. et al. The Srs2 helicase prevents recombination by disrupting Rad51 nucleoprotein filaments. *Nature* **423**, 309–12 (2003).

109 Desterro, J. M. P., Rodriguez, M. S. & Hay, R. T. SUMO-1 modification of IκBα inhibits NF-κB activation. *Molec. Cell* **2**, 233–239 (1998).

110 Hori, T. et al. Covalent modification of all members of human cullin family proteins by NEDD8. *Oncogene* **18**, 6829–34 (1999).

111 Johnson, E. S. Protein modification by SUMO. *Annu Rev Biochem* **73**, 355–82 (2004).

112 Mizushima, N. et al. A protein conjugation system essential for autophagy. *Nature* **395**, 395–398 (1998).

113 Nakamura, M. & Tanigawa, Y. Characterization of ubiquitin-like polypeptide acceptor protein, a novel pro-apoptotic member of the Bcl2 family. *Eur J Biochem* **270**, 4052–8 (2003).

114 Jones, D. & Candido, E. P. Novel ubiquitin-like ribosomal protein fusion genes from the nematodes Caenorhabditis elegans and Caenorhabditis briggsae. *J Biol Chem* **268**, 19545–51 (1993).

115 Komatsu, M. et al. A novel protein-conjugating system for Ufm1, a ubiquitin-fold modifier. *Embo J* **23**, 1977–86 (2004).

116 Raasi, S., Schmidtke, G. & Groettrup, M. The ubiquitin-like protein FAT10 forms covalent conjugates and induces apoptosis. *J Biol Chem* **276**, 35334–43 (2001).

117 Dittmar, G. A., Wilkinson, C. R., Jedrzejewski, P. T. & Finley, D. Role of a ubiquitin-like modification in polarized morphogenesis. *Science* **295**, 2442–6 (2002).

118 Luders, J., Pyrowolakis, G. & Jentsch, S. The ubiquitin-like protein HUB1 forms SDS-resistant complexes with cellular proteins in the absence of ATP. *EMBO Rep* **4**, 1169–74 (2003).

119 Yashiroda, H. & Tanaka, K. Hub1 is an essential ubiquitin-like protein without functioning as a typical modifier in fission yeast. *Genes Cells* **9**, 1189–97 (2004).

120 Kramer, A., Mulhauser, F., Wersig, C., Groning, K. & Bilbe, G. Mammalian splicing factor SF3a120 represents a new member of the SURP family of proteins and is homologous to the essential splicing factor PRP21p of Saccharomyces cerevisiae. *Rna* **1**, 260–72 (1995).

121 Yamamoto, A. et al. Two types of chicken 2′,5′-oligoadenylate synthetase mRNA derived from alleles at a single locus. *Biochim Biophys Acta* **1395**, 181–91 (1998).

122 Lam, Y. A., Lawson, T. G., Velayutham, M., Zweier, J. L. & Pickart, C. M. A proteasomal ATPase subunit recognizes the polyubiquitin-degration signal. *Nature* **416**, 763–67 (2002).

123 Haas, A. L., Katzung, D. J., Reback, P. M., Guarino, L. A. Functional characterization of the ubiquitin variant encoded by the baculovirus Autographa californica. *Biochemistry* **35**, 5385–94 (1996).

Index

Protein Degradation, Vol. 2: The Ubiquitin-Proteasome System.
Edited by R. J. Mayer, A. Ciechanover, M. Rechsteiner

ISBN: 3-527-31130-0